The Moon, Lamentations, Pancakes, and Much More

Research on Subjects of Jewish Law

by

Rabbi Dr. Chaim Simons

Grosvenor House
Publishing Limited

This book is published by
Grosvenor House Publishing Ltd
Link House
140 The Broadway, Tolworth, Surrey, KT6 7HT.
www.grosvenorhousepublishing.co.uk

A CIP record for this book
is available from the British Library

ISBN 978-1-80381-310-3

INDEX

PREFACE

Over the course of forty years, I have researched and written a large number of papers, some of which are interdisciplinary. A good proportion of these papers are on halachic (Jewish Law) analysis. They were written in Hebrew and almost all were published in the journal "Sinai" of the Rabbi Kook Institute in Jerusalem. The precise location of these papers in the "Sinai" journals was given in my book "A Miraculous Escape" on pages 253-254.

I have, during the year 2022, translated these papers into English and most of them appear in this book. The order in which these papers appear in this book is approximately the order in which these subjects appear in the Orach Chaim section of the Shulchan Aruch.

My other papers on halachic analysis were published in my previous books:

"An Unusual Case History of Adoption and Conversion with Particular Reference to Jewish Law," "Analysis according to Jewish Law of the Murders in Agatha Christie's Book 'And Then There Were None'." These appear in my book "From Keyboard to Computer" published by Grosvenor House Publishing, 2021, ISBN 9781839757860.

"Wormy Cheese, Cloned Pig Meat and Much More for a Kosher Table," "Unpleasant Odours as discussed in Rabbinic Literature with accompanying Scientific Background," and "The Chinese Etrog," appear in my book "Riddles, Donkeys. Wormy Cheese and Much More" published by Grosvenor House Publishing, 2022, ISBN 978183975048.

"The Chanukah Miracle of the Cruse of Oil – Is it to be interpreted literally or is it just an allegory?," "Lighting Chanukah Candles with Pig Fat" and "The Date of Purim in Kiryat Arba," appear in my book "A Miraculous Escape and Selected Writings on Jewish Themes" published by Grosvenor House Publishing 2022, ISBN 9781803810300.

"Jewish Religious Observance by the Jews of Kaifeng China" (book of over 250 pages), was published by the Sino-Judaic Institute, Seattle WA USA, 2010, ISBN 9781304179821.

I give grateful acknowledgements to the many organisations and people who supplied me with the information needed to write these papers. They include the Israel National Library in Jerusalem, including their microfilm department, the Hebrew University's National Medical Library in Jerusalem, the Library of the Rabbi Kook Institute in Jerusalem, the Kiryat Arba Municipal Library, the Yeshivat Nir Library in Kiryat Arba, the HebrewBooks website on the Internet and other material from the Internet,

This book contains thirteen of these papers. In every case, the source material has been meticulously documented. Readers may put in a request to my e-mail chaimsimons@gmail.com for original photocopied source materials quoted in these papers to be sent to their e-mails without charge.

I have included in the name of the title of the book the words "the Moon, Lamentations, Pancakes" since these are just a few of the subjects contained in this volume.

The various papers in this volume were written in Hebrew at different periods of time, and as a result there are some small differences in the style and set out of the contents. These differences also appear in the English translation of the papers.

Numerous Hebrew words had to be transliterated into English, and in different books the same word which is transliterated in several different ways. I was not consistent in the transliteration, and the same word may well have been transliterated in different ways in the course of this book. (I should mention that the publisher of this book does not have Hebrew letter fonts, and thus requiring transliteration.)

Numerous Hebrew words are given (transliterated) in the course of this book. In most cases when using this word for the first time in a particular paper, I have put an English translation in brackets after the Hebrew word.

The intention of the book is not to give halachic rulings, but just to enable the reader to study the subject. Needless to say, halachic rulings can only be given by Poskim (Rabbinical arbiters).

BLACKENING THE RETZUOT OF TEFILLIN INCLUDING THE INNER SIDE

(retzuot = straps of tefillin)

(Throughout this paper the term "clean" living creature is used for one which is permitted to be eaten, and the term "unclean" living creature is used for one which is forbidden to be eaten.)

In his Mishneh Torah, the Rambam writes: "There are eight requirements in the making of tefillin. All of them are Halacha l'Moshe miSinai (laws transmitted to Moshe on Mount Sinai), and therefore it is necessary to fulfil all of them, and if one deviates with any of them, the tefillin are possul (not kosher). They are ... the retzuot of the tefillin must be black,"[1] namely one has to paint a black colour on the retzuot.

What things are prohibited in the paint composition?

It is written in the Gemara: "For sacred work, only the skin of a clean animal was considered as fit. For what halacha was this said? It was said for tefillin where it is written 'so that the Torah of G-d may be in your mouth." It is written in the Gemara that this halacha also applies to the parts of the tefillin, such as the (sinews used to sew up the batim) and the retzuot.[3] [In other religious artifacts there are different opinions as to whether this halacha applies, and also, if there is a difference between holy artifacts and mitzvah artifacts.[4]]

It is written in the Gemara: "You may write tefillin on the skin of a nevailah (animal which died) and a tereifah (an animal with certain physical defects),"[5] provided the animal itself is a clean species. A question which therefore arises, is what is the ruling regarding the chailev (forbidden fat), the gid hanasheh (sinew – sciatic nerve), and the blood all from clean animals, when

1

used in the ingredients of the colour used to paint the retzuot of the tefillin?

Rabbi Shimon Greenfeld (19th-20th centuries) in his book entitled "Shu't Maharshag" refers to chailev, and he wrote: "Chailev that comes from a clean animal, even though it is forbidden to eat the chailev, since it comes from a clean animal it is considered as something that is permitted to put in one's mouth."[6] Rabbi Aharon Hoffman brings without disputing it the Maharshag on this subject.[7] The same answer could also be given in connection with blood and the gid hanasheh.

The question about the use of the gid hanasheh is also found in connection with the sewing up the tefillin batim, and this question was already asked about two hundred years ago by Rabbi Yonah Landsofer (17th-18th centuries) in his book "Bnei Yonah." On this he wrote: "One needs to investigate the sewing the "yeriot" (sections) of a Sefer Torah using gid hanasheh since it is not permitted to be put in one's mouth."[8] Rabbi Shlomo Ganzfried (the author of Kitzur Shulchan Aruch) in his book "Kesat haSofer" brings the "Bnei Yonah" but adds "but it seems to me that one should be strict."[9] On the other hand, Rabbi Uri Feivel Halevi Schreier in his commentary "Gedolei haKodesh" on "Da'at Kedoshim" sided with those who were lenient in this matter.[10]

Rabbi Avraham Genichovsky gave an explanation for these differences of opinion: "The side which ruled that it was permitted, held that it was like nevailot and tereifot, since even though it was forbidden to eat them, they were to be regarded as "species" which one was permitted to put in one's mouth. However, on the other hand, one can also rule that they are forbidden since a gid hanasheh is always forbidden to be eaten, and at no stage in the life of the animal was it permitted to eat it. Thus it cannot be regarded as a species that one could put in one's mouth, and this is unlike nevailot and tereifot which only due to an occurrence during the animal's life become forbidden to be eaten.[11] It would appear that one could bring the same reasoning to forbid the chailev and blood of a clean animal.

There is also another aspect which can be discussed on this subject. Rabbi Uri Schreier, in his commentary "Gedolei haKodesh," states that when the animal was still a fetus in the mother's womb, the gid hanasheh was permitted, and therefore it can be used in the sewing of Torah scrolls and tefillin.[12]

Also, in a "ben pekuah" (an animal just about to be born, but is still in its mother's womb) both the chailev and the gid hanasheh are permitted to be eaten. On the other hand, this does not apply to blood, because the blood of a ben pekuah is forbidden.[13]

One should note that there is a difference between the blackening of retzuot, and that of the blackening of the tefillin batim, and this issue is discussed by Rabbi Yechezkel Landau who was known as the "Noda biYehudah," (18th century). With regards to the retzuot, the blackening is Halacha l'Moshe miSinai, so there can be no compromises.[14] One must use a black colour whose origin does not come from an unclean animal. However, regarding the blackening of the tefillin batim, there are differences of opinion as to whether the source is Halacha l'Moshe miSinai, and even whether the tefillin batim indeed have to be black in colour. Therefore, regarding the blackening material for the tefillin batim, there is room to be lenient and use material that comes from unclean animals; however the "Noda biYehudah" prefers to be strict on the matter.[15]

The question arises as to which substances are included in the prohibition of blackening of the retzuot, namely is the prohibition exclusive to animals, or is a flora which if forbidden to be eaten, for example, orlah (fruit in the first three years after planting), included in the prohibition. The answer is: "All the prohibitions of eating and certainly the prohibitions of getting benefit are invalid for this blackening, and it is thus not only limited to that which comes from an animal source." [16]

On this subject, Rabbi Shlomo Zalman Auerbach (20th century, Eretz Yisrael) wrote: "It must be discussed whether it is permissible to blacken tefillin using peels of shemitta products, however, b'diavad (in retrospect) if they had been used for the blackening it is kosher." This is discussed further in the "Dvar Halacha" of "Halichot Shlomo" and it concludes that after using these peels in the blackening of the tefillin, the wearer will be happy that he is then able to put on these tefillin every day, and this would be like benefitting from this shemitta produce, and it would thus be permitted.[17]

The staff of the Institute in Jerusalem for training Dayanim (Religious Judges) in practical halacha summarised this subject: "The rule in this matter is that any species which is permitted to be eaten, but lacks something which thus forbids it to be eaten, such as the absence of shechitah, may be utilised, but something which from the outset is not permitted is prohibited."[18] However, there are still questions that remain, for example, the gid hanasheh that is forbidden to be eaten, (but as stated above), there are some who permit using the gid hanasheh in order to sew tefillin, and furthermore there are some who permit the use of chailev in the composition of the material for dyeing. It is also possible to ask what is the ruling on mixing animal meat with animal milk? Each one is permitted by itself, but the mixture is prohibited. One requires further discussion on this matter.

Another question that can be asked is about "stam yeinam" (wine prepared or handled by non-Jews). Is it permitted to use it as one of the ingredients for the ink for blackening the retzuot? This question was discussed in the year 5434 (close to the end of the seventeenth century), by Rabbi Shabti Ber, who published a book of responsa that includes the use of stam yeinam in preparing the ink to write the tefillin parashiot. Rabbi Ber held that since one can use nevailot and tereifot in making the leather for the tefillin, and this process requires the specific intention that one is making it for the sake of the mitzvah, he thus argues that how much more so one can use stam yeinam for the ink which does not require making it for the sake of the mitzvah.[19] Thirty years later, Rabbi Shmuel Avuhav (17th century) wrote on this subject in his book "Dvar Shmuel": "Why would the law of wine, which does not contain the prohibition of enjoyment, when according to the halacha it is part of a mixture be worse than the prohibition of nevailah and tereifah which are permitted according to all opinions for the writing of Sifrei Torah, tefillin and mezuzot."[20] We can thus see that these two Poskim (Rabbinical arbiters) permit the use of stam yeinam in the ink, (and probably also for the blackening for the retzuot). However, about a century later, Rabbi Elazar

Fleckeles (18th-19th centuries) in his book "Teshuva m'Ahava" disagreed with these two Rabbis and wrote "According to my opinion, one requires further study on this subject." [21]

However, they failed to mention that there are two ways that wine can reach the state of becoming stam yeinam. The first is that the wine was prepared by a non-Jew, and therefore this wine from the outset was stam yeinam, and thus was not from the outset permitted by the halacha, and it is therefore not permitted to use it to make blackening for the retzuot of the tefillin. The second way is kosher (non-boiled) wine which a non-Jew has touched, and his very act of touching it turns it into stam yeinam. The question is that since the wine was initially kosher and was therefore allowed to be used in a blackening for retzuot, but due to the fact that it was touched by a non-Jew it became forbidden to drink; it seems that this is similar to the situation of nevailah and tereifah, and therefore one should be allowed to use such wine as an ingredient in the preparation of the blackening material for retzuot? The Rabbis of the above quoted Institute in Jerusalem answered that the distinction between these two types of wine is "interesting" and at the same time said that no answer had been found on this matter, adding: "For both methods it is similar to the nevailah and tereifa from a clean animal."[22]

In addition to the actual blackening material for the retzuot, it is customary to mix this colouring matter with the fat of an unclean fish in order "to improve the appearance and soften the leather of the retzuot:" The question to be asked is how is it possible to utilise material from an unclean fish? On this the "Noda biYehudah" rules that it does not disqualify the tefillin because this unclean material is only there for the sake of appearance and therefore has no significance.[23] Rabbi Yitzchak Dov Halevi Bamberger (19th century, Germany) brings the "Noda biYehudah but adds" and in any case it is good to be meticulous and use kosher fat."[24]

At a later period, a question was asked about adding glycerin (a substance made from animal oils) to the colouring matter that blackens the retzuot. The function of the glycerin is "to stick the colour on the skin of the retzuot, because without it the colour will not be absorbed." It was ruled that using glycerin is more prohibitive than using the fat of an unclean fish, and therefore could not be used. Even though the amount of glycerin was less than one sixtieth (which is generally regarded as a quantity insufficient to disqualify something), it is still forbidden because it is a "davar hamamid" (an indispensable ingredient).[25]

Is it necessary to prepare the blackening material for the specific purpose of the mitzvah of tefillin?

The material for the painting is not like the leather to make the tefillin batim, or the gidim for sewing the tefillin, or the parchment for the parashiot, all of which must be prepared specifically for the purpose of the mitzvah of tefillin. However, this is not so for the material to be used for the blackening. Any black colour can be used after checking that it does not contain substances that disqualify the material for blackening the retzuot.[26] For example, the company that manufactures shoe polish called "Kiwi," publishes the composition of their shoe polish, and it can be seen that it does not contain prohibited substances.[27] However, there is a blackening colour that is under Rabbinical supervision that was specially prepared for this purpose,[28] and in an answer that was asked on this question to the above mentioned Rabbinical organisation in Jerusalem, they answered: "Lechatchila (from the outset) it is proper to use a paint which was prepared specially for this purpose."[29] Some of these materials come in a bottle with a small brush (like in "Tipp-Ex") in order to apply the paint to the retzuot, and some are made like a marker ("tush"). In cases where it comes as a paste, it is recommended to apply it with a rag or tissue so that the colour is smooth and will look beautiful on the retzuah.[30]

There was an occasion when a question was asked by a person who used a simple black marker to blacken his tefillin. Although the constituents of the markers were kosher, the marker was not under Rabbinical supervision. Before the colouring material was finished, this person bought material under Rabbinical supervision. His question whether the halacha regarding "holy artifacts" applied to his first material, and if so, how should he deal with it? The answer he received was: "You do not need to put in a genizah (a depositary for timeworn sacred books and ritual objects) the simple marker you have, but just stop using it in tefillin, and use only the special blackening which had been prepared for the purpose of tefillin." [31]

What is the meaning of this "black" colour?

Rabbi Chaim Vital, (the foremost disciple of the Ari, 16th-17th centuries, Eretz Yisrael), in his book "Shaar haKavanot" wrote that the meaning of the word "black" in this connection is "black as a crow." [32] Likewise, Rabbi Avraham ben Moshe from Sinsheim, the author of the book "Baruch She'amar" defines the colour as a "crow." [33]

All this is lechatchila. In the "Piskei Teshuvot" it is written: "But [the colour being black as a crow] is not l'ikuva (even by not doing this, one still keeps the halacha), and that everything that people call black even if it is pale blue, or brown or pale gray is also kosher." [34]

From the discussion in the "Mishnah Berurah" one can see that there is a difference between what is called "black" for the retzuot, and what is called "black" ink used to write the parashiot. He wrote: "The retzuot which are also Halacha l'Moshe miSinai, must be sufficiently black that it is like blue, which is not so [with the ink to write the parashiot] because the halacha is that one must write in ink .... a blue appearance is certainly not classed as ink." [35]

So that the black is not easily erased, the "Baruch She'amar," suggests how to blacken the retzuot. He writes: "The blackening of the retzuot is Halacha l'Moshe miSinai, and when the ink dries a little, one should blacken it a second time and even a third time, until is sufficiently black just like a crow, and this is for the retzuot, the batim, the titora (the leather base to close up the bayit) and the ma'avarta (the hollow extension of the bayit through which the retzuot are passed)." [36]

During the period of the "Chazon Ish," there were those who blackened their retzuot with "black plastic paints." After seeing that after the retzuot get wet, the plastic colour peels off in thin layers. Rabbi Yaakov Yisrael Kanievsky (the "Steipler"), whose tefillin had been painted with this material said, "I was very upset about this and it pained me. I have been putting on these tefillin for about a year and now it will turn out that they were possul." But he himself performed this test, "and the colour did not come off and they were thus kosher and he continued to wear them." [37]

What about blackening retzuot whose length is longer than the halacha requires?

Lechatchila, the length of the retzua of the arm tefillin is to be able to encircle the arm, wind it around the arm seven times, and then wind it three times around the finger and finally tie it.[38] For the head tefillin the length of the retzua is, to be able to encircle the head, and then the retzuot that will hang in front of him should reach to the navel or a little above it.[39] The Shulchan Aruch also gives the minimum length of the retzuot. There it states that the retzua of the arm tefillin, needs to encircle the arm, then without the seven windings stretch up to the middle finger, and then wind it three times on the finger and tie it. [40] In the head tefillin, the length of the retzua is to encircle the head;[41] Rabbi Meir Leibush (19th century), known as the "Malbim" adds that the retzuot in addition to circling the head, should hang down in front of him "two tefachim" (almost twenty centimetres).[42]

In a situation where the length of the retzua of the tefillin is above the minimum length, and the excess length has not been blackened properly, but the rest of the retzua (namely the length required according to the halacha) is properly blackened, one can ask whether the tefillin are then kosher?

The "Mishnah Berurah" discusses this issue, but ends with the words "one needs further investigation." He wrote that he was doubtful if according to the Torah it was sufficient to be blackened just for the minimum length, or, since the excess part is joined to the obligatory length, the whole retzua has to be black. He continues and explains that one can learn from the Gemara that if the Halacha l'Moshe miSinai applied only to the minimal part of the retzua, it would be permissible to paint with other colours on the excess part of the retzua. Although this subject was mentioned in the Gemara, the Gemara was silent on the outcome of this subject. On the other hand, it is possible to say that according to the Torah one can paint in different colours on the excess part, but the Sages forbade it because "it would look speckled."[43]

What percentage of the retzua needs to be black?

Lechatchila, the entire minimum required length of the retzua, (or according to other opinions, the entire retzua) should be painted black.[44] But in practice, over time the black layer may begin to peel off the retzua. Rabbi Avraham Aharon Broda (17th-18th centuries) wrote about this in his book "Lishkat haSofer": "If over a period of time the retzua becomes white, one should blacken it again." [45]

In order for the colour to be fully absorbed on the retzuot, one should use glue in combination with the black paint. The question is whether when using glue there will be an apprehension of a physical separation between the retzua and one's body. Rabbi Moshe Sternbuch (20th-21st centuries) discusses the issue and decided "There is no fear of a physical separation since because the glue will have been forever nullified, it will not be called a physical separation."[46]

The part of the retzua that goes into the ma'avarta is always hidden. That's why the Mishnah Berurah wrote: "I don't know if the Halacha l'Moshe miSinai applies to this part of the retzua.[47] However, Rabbi Yaakov Meir Stern gives an answer: "One must blacken the part that goes into the ma'avarta."[48]

A question which arises is, is it necessary to locate points that are not black on the retzua? Rabbi Shlomo Zalman Auerbach wrote about this and ruled: "A measure of the blackness of the tefillin is that everything that is considered black according to the visibility of the eye, even though there are points that are not black."[49]

What is the percentage lacking the black colour in the retzua that can cause the tefillin to be possul? Some Poskim use the rule "the majority is as the whole." Among them is Rabbi Yosef Chaim Sonnenfeld (19th-20th centuries) who writes in his responsa "Salmat Chaim" for those who cannot find retzuot that are entirely black, but "only the majority is black, of course it is kosher."[50] The "Maharshag" wrote in similar words, but without Rabbi Sonnenfeld's condition: "As long as the majority of the required length of the retzuot are black they are kosher, and we do not say that what is missing from the required length that is not black, is not regarded as if it is missing."[51] However, Rabbi Avraham David from Butschatsch, in his book the "Da'at Kedoshim" is less sure about this because he writes "on the blackness of the retzuot perhaps one can benefit from the principle that the majority is as the whole.[52] According to the opinion that only the minimum area needs to be black, "one does not need to be so careful about the blackness" of the area in excess of that.[53]

In contrast to the above, Rabbi Yaakov Yisrael Kanievsky (the "Steipler") "was very careful to blacken the entire retzua, and one needs to be careful to check their blackness since it is sometimes found that one needs to blacken them."[54]

Is one allowed to stick a black plastic material on the retzuot instead of painting them?

In 5775, the Bet Din (Rabbinical Court) of Rabbi Nissim Karelitz "published an announcement warning against the phenomenon regarding tefillin retzuot which were found on the market, and it had turned out that they were made of two layers, namely, a layer of leather and a layer of thick or thin plastic which had been glued to the leather. The halacha of these retzuot is explained in the Mishnah Berurah, chapter 32, paragraph 185, that if one is able completely peel off from the tefillin a layer which is like black paper, the tefillin is possul. And how much more so in our case that the black strip is pasted on the leather and the leather itself was not blackened at all."[55]

At the time, this subject caused a great stir in the press, and especially in the newspaper "Yated Na'eman" where it publicised that "thousands of Jews are putting on possul tefillin every day."[56]

In response to this, Rabbi Menachem Goldberg, from the "Stam Institute," wrote: "Synthetic-plastic colouring material does not harm the retzuot, except only when it is attached and glued."[57]

A question that arises, is, does gluing black "tapet" in order to shine and properly decorate blackened retzuot possul the tefillin according to the halacha? There are differences of opinion on this, namely, there are those who allow it according to the principle "anything done for beautification is not regarded as a physical separation," and there are those who reject it because it is the covering which is the determinant.[58]

How can blackening of retzuot that was done without the specific intention to fulfil the mitzvah be corrected?

A question which arises is that should the blackening of the retzuot be done with the specific intention to fulfil the mitzvah? On this there is a dispute between the "Beit Yosef" who ruled, that b'diavad it does not possul the tefillin, and in contrast the Rema who ruled that even b'diavad the tefillin would be possul.[59]

One may well ask that in the case when the retzuot have already been blackened without this specific intention, would it be beneficial to go back and blacken them again on top of the existing colour, but this time with the specific intention? On this, Rabbi Avraham Gombiner (17th century), the "Magen Avraham" wrote: "It seems to me that if the Jew blackened them again with the specific intention the tefillin would then be kosher," and he cites Talmudic evidence from the writing of a get (divorce document).[60] However, Rabbi Yosef Teomim (18th century), the "Pri Megadim" and other Acharonim (great Rabbis who lived from about the sixteenth century) criticised his proof and the issue remained requiring further study.[61]

There are a number of ways to correct blackening in order that it will be with the specific intention. The "Piskei Teshuvot" suggest: "the original colour should be scraped off and painted again this time with the specific intention."[62] However, in practice this is not a simple process! Another piece of advice can be found in the Biur Halacha, namely one paints on the black retzua with different colours such as green or white and in this way one cancels out the black,[63] or in other words, go over with "tippex" (white colour) the black colour which was done without the specific intention and then paint it with the colour black, with the correct intention.

Also, the "Pri Megadim" suggests an alternative method on this subject: "And it seems that if there are no other retzuot, one should completely blacken the other side of the retzua with the correct intention,[64] namely that one does not have to blacken specifically on the on the hair side of the retzua, but can turn over the retzua so that the black will then be worn on the outside. Rabbi Sternbuch explains that according to the Pri Megadim: "The main obligation of blackness depends on the appearance."[65] (However, it is stated below that there are Poskim who rule that if one blackens the side of the retzua that was on the side of the animal's flesh, then the tefillin will be possul.)

Can a woman, a slave, a non-Jew, a child or a machine paint the retzuot?

It is written in the Torah "and you shall bind them" and after that it states that "and you shall write them"[66] and from there one learns that the one who does not "bind" namely, the one who does not have the obligation to bind (put on) tefillin, cannot write the parashiot in tefillin.[67] It is further written in the Shulchan Aruch: "Anyone that is disqualified for writing [the parashiot] is also disqualified from making corrections."[68] This means that the number of stages required to prepare the batim can only be done by those who are obligated to put on tefillin. These actions include making the "shin" for the head tefillin, sewing up the batim[69] and making the knots on the retzuot.[70] However, there are also certain actions that even those who are exempt from the mitzvah of tefillin can do. Among them are the processing of the parchment[71] and the cutting of the retzuot.[72] However, there are also actions in which there are differences of opinion as to whether those who do not have the obligation to wear tefillin can do them. Amongst them are the blackening of the retzuot.[73] In addition to this, the question arises as to whether the action should be done specifically for the sake of the mitzvah of tefillin?

It is written in the Shulchan Aruch: "Women ... are exempt from tefillin because it is a positive mitzvah which has a fixed time,"[74] and therefore it depends on the dispute (mentioned above) whether she can blacken the tefillin. According to the opinions that she can, the painting must be done with the correct intentions.[75] The Mishnah Berurah writes that the woman "knows how to do it with the proper intentions in the same way as a man knows."[76] However, after a discussion on the subject, Rabbi Yosef David Weiss (20th century) concludes: "It is better to prevent a woman from blackening the retzuot."[77]

In almost every mitzvah that is time bound, a Canaanite slave is like a woman, and this includes tefillin, and therefore he is exempt from putting on tefillin[78] and thus there is the same halacha in the matter of blackening the retzuot.

A person who is "half slave and half a free man" must fulfill the time bound mitzvot, and therefore must put on tefillin, and it follows that he can, by almost all opinions, blacken the tefillin. Why "almost"? The reason is that there is an opinion of Rabbi Akiva Eiger (18th-19th centuries) that a person who is half a slave and half a free man is exempt from reading the shema and is therefore exempt from putting on tefillin,[79] and according to this opinion, he will have the same halachot in this matter as for a woman.

In connection with the blackening of the retzuot by a non-Jew, the "Beit Yosef" wrote: "It is good that a Jew should blacken them with the correct intentions and it should not be done by a non-Jew."[80] This shows that the "Beit Yosef" holds that even those who do not have the obligation to put on tefillin can blacken them, however, in using the word "good" he prefers that it is done by a Jew. The Mishnah Berurah explains the "Beit Yosef" who holds that even blackening the retzuot does not have to be done with the correct intention.[81] The Rema completely disagrees with the "Beit Yosef" and writes: "Even b'diavad [blackening by a non-Jew] will make the tefillin possul."[82] It is not clear from this whether the Rema believes that the blackening of the retzuot is an action that can be done by someone who is not obligated by the mitzvah of tefillin, provided that the action is done with the right intention, and that there is a fear that a non-Jew will not do it with the right intention.

A child who has not reached the age of barmitzvah does not have the requirement of observing the mitzvot, but does so only for the sake of training. According to the opinion that blackening the retzuot does not depend on the fulfillment of the tefillin mitzvah, and only that it has to be done with the correct intentions, there is a possibility that a child will do this with the correct intention. However, there are differences of opinion as to whether the child is allowed to blacken whilst a man stands behind him in order to make sure that he does the blackening with the correct intention.[83]

According to the opinion that only those who are obligated by the mitzvot of tefillin can do the blackening, it seems that already on the day the child reaches the age of barmitzvah, he can theoretically do the blackening. Why write "theoretically"? The reason is that there are two conditions for defining a child becoming an adult. It is written in the Shulchan Aruch: "He is forever a child until he grows two hairs after he has reached the age of thirteen [years] and one day."[84] The Mishnah Berurah wrote about this: "Since they reached the age of barmitzvah one has the presumption that they have grown two hairs" but he continued to limit the presumption: "And in any case we do not rely on this presumption completely, but we treated it as a doubt, and therefore we always rule in a strict manner for a mitzvah which is from the Torah, and rule leniently in a Rabbinical mitzvah.[85] Today, in the case of mitzvot from the Torah, one does not check whether he has grown two hairs, but one waits until one has a beard.[86]

On this topic, Rabbi Yosef Shalom Eliashiv (20th-21st centuries) was asked a question: "What is the ruling on a barmitzvah boy if we do not know whether or not he has grown two hairs, should he be allowed to blacken retzuot?" The Rabbi was strict about this and answered: "Because this is a doubt in a Torah mitzvah, one should be strict about it."[87] There is a comment to be made on this: "One must be careful about a very common occurrence, in barmitzvah boys whose colour on their retzuot of their tefillin has faded, and they therefore want to paint the retzuot; in such a situation they should give them to a grown up to do, and needless to add with the correct intentions, in order that the tefillin will be kosher without any doubt.[88]

In our time, when electrical devices are easily available, and a person just presses a button to do the actions, there are discussions among the Poskim if this is in the framework of correct intentions, and if so, is it possible to paint the retzuot in this way. It is true that although some allow it, it is written that one should be strict.[89] Even though some allow it while a person actually presses the buttons with their hands, today it is possible to give the instructions through the voice without the need for human physical contact.[90] If we allow the use of sound, one needs to investigate whether one can replace the electrical device with a monkey that learned to do the operation of printing the retzuot after hearing instructions from a human? [In connection with the question of using monkeys in the performing of mitzvot, Rabbi Moshe Sofer[91] (18th-19th centuries) known as the "Chatam Sofer," and also Rabbi Yitzchak Zilberstein[92] (20th-21st centuries) in his book "Chashukei Chemed" ruled that a monkey can be a courier to distribute the mishloach manot on Purim.]

Should the retzuot be blackened on both sides?

The retzuot have two sides. One of them is adjacent to the hair of the animal and is smooth; the other side, is adjacent to the flesh of the animal and is not smooth.

From Rashi's commentary on the Gemara[93] one learns that the black colour is painted on the outer side of the retzua, and the inside may be painted with any colour that one desires, for example, green, blue, or white, but with the exception of the colour red[94] the reason being that people seeing it might say that the source of the red is blood.[95]

A question to be asked is, is it possible to choose which side to paint? In the commentary on the Talmud by Rabbenu Gershom Meor Hagolah, (10th-11th centuries) it states: "According to the

Halacha l'Moshe miSinai, the black of a retzua must be visible from the outside.[96] The Mishnah Berurah rules that if the blackening is done the other way round namely, on the inside, this will not be correct and one will therefore have to then blacken them on the outside.[97]

However, in the Siddur "Beit Oved" written according to the Sefaradi custom, states that "it is proper to blacken the retzuot both on the inside and on the outside." However, it then continues with a warning: "Even though they are black on the inside, one must be careful that the outer side of the retzuot is always facing the outside."[98]

The question arises whether there is a restriction specifically to *only paint* on the inside of the retzua, or is one also allowed to make drawings or write letters on it. It could be a practicality to write (but not in red!) the name or identifying information of the owner of the tefillin, especially, but not necessarily, on the part of the retzua which is in excess of its minimum length. This is particularly recommended for high school students who all wear tefillin in the school's Synagogue, and it is there that exchanges of tefillin can occur, especially on Rosh Chodesh when everyone hastily removes their tefillin before the mussaf service and leaves them together on the tables.[99]

Although there are differences of opinion as to whether painting the batim of the tefillin black is a Halacha l'Moshe miSinai,[100] today it is customary to do so. Some interpret the Gemara[101] that the colouring of the inner side of the retzua should be the same as the colour of the batim.[102] Most of the time the inner side of the retzua is a colour close to white and according to this, the batim of the tefillin should be white in colour! On this the Tosafot writes: "Some make the batim of the tefillin from white parchment and they only require black for the retzuot."[103] (It is not known what percentage of the tefillin batim during the period of the Tosafot were white in colour!) Also, it is stated below that Rabbenu Manoah saw many batim which were white.

Incidentally, there are different opinions as to whether according to the Halacha l'Moshe miSinai, in addition to the square requirement for the titora and the shape of the gidim stitches in closing the batim, the batim should also be square. The Tosafot mentions that in the past that on the hand tefillin, the bayit was square only in the location of the titora.[104] In the Cairo Genizah they found tefillin whose batim were in the shape of a cylinder.[105] Furthermore, the Mordechi also mentions that "they made the hand tefillin in a mold which was circular at the top." However, he did not agree that this was correct.[106]

Do the sides of the retzuot need to be blackened?

In addition to the upper side of the retzuot that are visible, the two sides of the retzuot are also visible, and therefore one may well ask whether it is necessary to blacken them. The "Kesat haSofer" discussed this issue and writes: "And it seems to me that one should be particular to also blacken the sides, namely the place where the retzuot were cut."[107] In his comments in "Lishkat haSofer" he writes on this: "But the place of the cut that is visible from the outside (especially when the skin is thick and one blackens the entire skin and only afterwards cuts it for the retzuot), it seems to me that one certainly has to be particular to blacken it there, and perhaps it is included in the Halacha l'Moshe miSinai, since one can see it from the outside." (words in parentheses are in the original).[108] Likewise, in the tefillin of Rabbi Diskin, the two sides were "black as a crow."[109] Also in the tefillin of Rabbi Menachem Mendel Schneerson of Lubavitch, the retzuot were "painted also on their sides but not on the inner side."[110]

On the other hand, after a short discussion on the subject, Rabbi David Morgenstern and Rabbi Eliyahu Gitman, concluded: "In practice, it is not customary, and one does not have to be particular to paint the thickness of the sides of the retzuot."[111] A number of Rabbis of our times followed this opinion, and they include Rabbi Yitzchak Zev Soloveitchik (the GRY'Z) who "was not particular on this [to blacken the sides of the retzuot]."[112] Also the "Chazon Ish" and Rabbi Yaakov Yisrael Kanievsky (the "Steipler") "did not blacken the sides of the retzuot" of their tefillin. The proof for this was that the "Baruch She'amar" did not mention this subject.[113]

Does one blacken also the inner side of the retzuot?

The earliest source whether the inside of the retzuot must also be black is the Rambam. He wrote: "The outside surface of the retzuot of both the head and the hand tefillin must be black, and this is Halacha l'Moshe miSinai. In contrast, with regard to the inner side, since it faces the inside, it is kosher if it is green or white. One should not make this [side of the retzuot] red, since it will be embarrassing for him if the inner side of the twists that it becomes the outer side. The inner side of the retzuot should be the same colour as the batim; namely, if they are green, they should be green; if they are white, they should be white. It is attractive for the tefillin, the batim and the entire retzua to be entirely black."[114]

How should one understand this Rambam, namely, when the bayit is black, must the inner side of the retzua be black, or is it just for attractiveness?

Rabbi David Zimra (15th-16th centuries), known as the "Radvaz" discussed at length the Rambam's words and wrote: "And the Jews were found to observe the mitzvah of tefillin, even though they do not observe the attractiveness of the retzuot." However, he concludes: "But in the opinion of the Rambam, one who wants to enhance the mitzvah should also paint black the inner side of the retzua."[115]

There is a commentary on the Rambam's laws of tefillin written by Rabbenu Manoah ben Yaakov who lived in Narbona at the end of the thirteenth century. This commentary was lost for hundreds of years, but in the twentieth century it was discovered in the Cairo Genizah. His commentary of this halacha in the Rambam includes: "And I saw many tefillin whose bayit was white, that is, the skin was peeled off, but the inner side of the retzua was not [the colour of] the bayit, and even though it was therefore not possul as was written by Rashi, nevertheless it did not accord with the beautification of the tefillin, since the beautification of the tefillin is that it should be entirely black."[116]

During the time of the Rambam, lived Rabbi Shimshon ben Avraham from Sens, one of the Tosafists (Rabbis of the Tosafot era). His comments on also blackening the inner side of the retzuot were brought by Rabbi Yitzchak ben Moshe of Vienna (13th century) in his book "Or Zarua": "And now that it is customary to blacken the batim, it is a mitzvah to blacken the retzuot both inside and outside."[117] The "Darchei Moshe (Rema) quotes the words of the "Or Zarua" but adds "but this is not the custom."[118]

Rabbi Zedekiah ben Avraham Anaw Harofeh, who lived in Rome in the thirteenth century, wrote in his book "Shibbolei haLeket": "And there are those who require [the retzuot] to be black on the outside and inside," and we can see from the use of the words "and there are those," that he gives no identification as to who is saying it.[119]

On this subject, Rabbi Chaim Vital, wrote: "They should also be blackened on the inside and outside."[120] Rabbi Yosef Liberman in his book "Mishnat Yosef" interprets the words of Rabbi Vital that also the inside of the retzuot should be "black as a crow."[121] From what is written in the "Shaar haKavanot" one might think that this was universal among the Hasidim. However, this is not the case! Rabbi Yosef Chaim, (19th-20th centuries) the "Ben Ish Chai" wrote: "[The Hasidim] were not accustomed to do this. And I asked about the custom of the Hasidim in Eretz Yisrael and they answered me that there were those who were particular to blacken both the inside and the outside, and there were those who are not particular about it."[122] In similar language, Rabbi Chaim Elazer Spira (19th-20th centuries), the Gaon of Munkatch, wrote: "And we have never seen or heard of anyone who was accustomed to be strict according to the words of the "Shaar haKavanot (the Ari)."[123]

In the middle of the nineteenth century, Rabbi Mordechai Ze'ev Ettinger and Rabbi Yosef Shaul Nathansohn, in their commentary to "Elef Hamagen" on their book "Magen Gibborim" use the expression that "it is proper that they be black also on the inside," and their reason is "for a decorative reason."[124]

Although a number of Poskim bring the Rambam or the "Or Zarua," they conclude "And it is not customary" or a similar wording. Amongst them are the "Beit Yosef" on the Tur, the "Darchei Moshe" (Rema), Rabbi Chaim Benvenist

(17th century) the author of "Knesset haGedolah," and the Mishnah Berurah.[125]

The question arises, were in practice, before the period of the Rishonim (great Rabbis who lived approximately between the eleventh and fifteenth centuries) tefillin actually painted on the inside? In the Gemara, Rabbi Huna's act is given: "One day the retzua of his tefillin was (accidently) reversed whereupon he sat fasting forty days."[126] Rashi interpreted this that "the black inside,"[127] namely that after the reversal, the black side of the retzua was inside. It seems that there is not even a hint in the Gemara of blackening the inner side of the retzua. However, there is an innovation in the writings of Rabbi Yishmael Hakohen (18th-19th centuries) in his book "Zera Emet." There he writes: "Certainly the retzuot of Rabbi Huna's tefillin were black on both sides as it should be done lechatchila." If this was the case, why did Rabbi Huna fast? The "Zera Emet" continued to explain: "And if so, how can one say that the retzua was reversed, since they were black on both sides, but surely one has to say the outer side should be the side of the hair and not the flesh, so even though the retzuot of Rabbi Huna were black on both sides, it is correct to say that they had been reversed."[128]

What are the opinions of contemporary Rabbis about the blackening of the retzuot on both sides?

In our times, questions were sent to the Poskim as to whether there are advantages in painting the retzuot in addition on the inside (and this was before the era when the black colour was absorbed throughout the entire thickness of the retzua). Here are some examples of these questions and how the Poskim answered them.

Rabbi Eliyahu Katz, head of the Beit Din in Kiryat Yoel DeSatmar in Bnei Brak, who discussed the issue at length, wrote an article about it in 5755, and came to "a practical conclusion" namely: "One can learn from all this that there are several good reasons that if it is possible then one should be meticulous to use such retzuot [that have been blackened on both sides]."[129]

Rabbi Shmuel Halevi Wosner, the Head of the Bet Din of Zichron Meir in Bnei Brak wrote a responsum on the subject in the year 5755, and it appears in one of the volumes of his "Shevet Halevi" responsa. He wrote that he did not agree with the conclusion reached by Rabbi Eliyahu Katz, and also wrote: "The Great Poskim of the generations, did not want to accustom themselves in this even for the purpose of enhancement." However he concluded: "If individuals want to do so, a blessing would come upon them, but I saw no need to rule on the matter."[130]

In the book "Shelat Rav" about questions sent to Rabbi Yosef Chaim Kanievsky it states: "I heard from an ultra-Orthodox Jew who was a senior doctor in the field of the disease (cancer?) who claimed that the paint on the retzuot contains a substance that causes the disease and therefore one should not paint on the inside [of the retzuot] since then the paint will touch the skin. Is it right because of this to give up the enhancement to use retzuot blackened on both sides, or should one follow the principle a 'Shomer Mitzvah' (a Jew who observes the commandments) will not come to harm?" Rabbi Kanievsky's answer was short and sharp: "A lie and an untruth, but the enhancement of the mitzvah was not mentioned in the Gemara, and therefore it should not be done."[131]

Rabbi Meir Halevi Soloveitchik, head of the Brisk Yeshiva in Jerusalem, was not in favor of retzuot that were black on both sides. One can see this on the occasion that once the Rabbi went to buy tefillin for his grandchildren. The seller offered the Rabbi retzuot which were "black on both sides." The Rabbi "did not offer the shopkeeper halachic arguments but only said to him (in Yiddish) And what is wrong with older [Tefillin]?"[132]

Rabbi Eliashiv was asked on the subject: "Is there any enhancement in buying the retzuot that are black on both sides." The Rabbi replied: "It is against the custom, and it should not be done." He was then asked: "Is there a drawback in this, or is there just no reason," and the Rabbi answered: "To wear them because of a concern in the halacha, one should not do, but if you wear them so that the black will not peel off, it is allowed."[133]

They say in the name of Rabbi Aharon Yehuda Leib Shteinman: "There is a meticulousness about this because the colour does not come off easily."[134] This statement seems to have been be made in order to be safe.

It seems that there could even be a halachic problem in painting both sides of the retzuot. So much so that Rabbi Yitzhak Ze'ev Soloveitchik, did not use tefillin in which the retzuot were blackened on both sides, and the reason: "There is an apprehension that any spoiled ink that dries, will cause a physical separation [between the bayit and the flesh] and therefore according to the halacha and the minhag of all the Jews is to blacken just one side."[135]

What do the Poskim say about colouring that is absorbed throughout the thickness of the retzua?

For hundreds of years, the method of painting on the outside and inside of the retzua was limited to surface painting. About ten years ago, a new method was developed in which "the dyeing absorbed the colour throughout the retzua from one side to the other."[136]

Rabbi Eliashiv wrote about this: "New retzuot have recently been developed, which are painted on both sides, and many prefer to buy particularly these retzuot, because the colour stays on it much better, and the production method is that the retzuot are dyed by immersing them in the dye and the dye penetrates to inside the retzuot. By this method, even if one peels off the outer layer of the colour, there is still a black colour remaining from the colour which is under this layer,"[137] and Rabbi Eliashiv continues in his answer to the question that was asked: "If there is an enhancement in blackening the tefillin retzuot on two sides, and he replied that this is a proper and beautiful thing, but whoever has retzuot blackened on one side, even lechatchila one does not need to change his retzuot for retzuot which are blackened on both sides. Since the custom of the Jewish people is just to blacken on one side, and there is no fear that it will grind a bit, since only a physical separation will make it possul, but this is just grinding and does not make it possul, and he added that I wear tefillin which are blackened only on one side."[138]

Rabbi Ben-Zion Wozner (son of Rabbi Shmuel Wozner) added to his father's words but even in a stronger language, and wrote: "One should not blacken tefillin (on both sides) for several reasons." He gives six reasons that include "the intention of the Halacha l'Moshe miSinai was to blacken only on the outside," "one should not add to the Halacha l'Moshe miSinai," and "one should not come to transgress by reversing the retzua" and he added another reason to prohibit, (which Rabbi Kanievsky defined as "a lie and an untruth"): "We have also heard that the doctors warn to stay away from these retzuot, whose inner colour is soaked in chemicals which are absorbed into the body by the sweat, causing serious health problems."[139]

In contrast to the opinions that suggest that such retzuot should not be used, the "Kollel Halacha Rehovot" wrote: "It is worth purchasing such retzuot, so that even if the black paint on the outside peels off, the retzua will remain kosher, because underneath the shiny colour, everything is also black."[140]

An incident that occurred with two pairs of head tefillin whose retzuot had been dyed throughout their thickness. These retzuot for the two pairs of tefillin were purchased from the same seller (it is possible that more than two pairs were purchased, but it is not known!). After several months, and precisely at that time, the colour of the retzuot in the area where it encircles the head, changed its colour to brown. Because there was a black colour under the layer of the brown colour, the question is, is there a problem with the kashrut of these two pairs of tefillin, and was it necessary to blacken over the brown colour? It is possible that what is written in the "Piskei Teshuva" gives an answer, namely that "there is nothing to possul the tefillin due to the fact that the kosher bottom colour is covered."[141] However, further study of this subject needs to be done.

Appendix for vegetarians

Some vegetarians are not even prepared to benefit from animal products.[142] Regarding the colouring material for painting the retzuot and the batim there is no problem. Although it is permissible to use materials from clean animals, it is not required, and the colour can be made from completely natural materials. On the other hand, the production of tefillin can only be made from clean animals,[143] and there one has to compromise in order to observe the mitzvah of tefillin. This is done by using tefillin prepared only from animals that have died naturally! According to the language of the Gemara,[144] it is better to use animals that died naturally, but the Poskim do not quote this Gemara as the halacha. Rabbi David Rosen, the former Chief Rabbi of Ireland, wrote that it is a "hidur mitzvah" (an enhancement of the mitzvah) to use such tefillin. In fact, today one can find such tefillin in Eretz Yisrael.[145]

References

Abbreviations
SA = Shulchan Aruch
OC = Orach Chaim
YD = Yoreh De'ah
MB = Mishnah Berurah
(1) Maimonides, Rambam, Mishneh Torah, hilchot tefillin chap.3 halacha 1
(2) Talmud Bavli, Shabbat 28b
(3) Ibid.
(4) Rabbi Yechezkel Landau, *Noda biYehudah,* second edition, (Halacha Berura: New York), OC responsum 3
(5) Talmud Bavli, Shabbat 108a
(6) Rabbi Shimon Greenfeld, *Shu't Maharshag,* (Vranov, 5691), vol.1, OC chap.24
(7) Rabbi Aharon Gedalia Hoffman, *Gidulei Aharon al Kitzur Shulchan Aruch,* (Bnei Brak, 5728), hilchot tefillin, chap.15
(8) Rabbi Yonah Lansdorfer, *Bnei Yonah,* (Prague, 5563), chap.278
(9) Rabbi Shlomo Ganzfried, *Keset haSofer,* (Ungvar, 5595), second edition, part 1 chap.17 par.1
(10) Rabbi Uri Feivel Halevi Schreier, *Peirush Gedolei haKodesh* on the book Da'at Kedoshim by Rabbi Avraham David from Butschatsch, (Lvov, 5656), YD chap.278 par.2
(11) "Peninim m'Shuchan Rabbenu Hagaon Rabbi Avraham Genichovsky," parashat Bo 5776, litfor im gid hanashe, (Internet)
(12) *Gedolei haKodesh* on *Da'at Kedoshim*, op. cit., YD chap.278 par.2
(13) *Talmudic Encyclopedia*, (Talmudic Encyclopedia Publishing Company: Jerusalem), vol.3, Ben Pekuah, pp.370-71
(14) *Noda biYehudah,* op. cit., second edition, OC responsum 3
(15) *Noda biYehudah,* op. cit., first edition, OC responsum 1

(16) *Din*, Beit Hahora'a Hamercazi shel al yedai Machon Yerushalayim l'Dayanut (henceforth: din), tzeva l'retzuot tefillin. 11 Tammuz 5779, (Internet)

(17) Rabbi Shlomo Zalman Auerbach, *Halichot Shlomo*, (Jerusalem), hilchot tefillah, chap.4 par.29 and footnotes 42-43

(18) *Din*, kashrut tzeva l'retzuot, 14 Tammuz 5779, (Internet)

(19) Rabbi Shabti Ber, *Beer Esek*, (Venice, 5434), responsum 109 (end of responsum)

(20) Rabbi Shmuel Avuhav, *Dvar Shmuel*, (Venice, 5462), responsum 162

(21) Rabbi Elazar ben David Fleckeles, *Teshuva m'Ahava*, (Prague, 5575), part 3 YD responsum 391, first words: shom gilayon shom b'hagahot

(22) *Din*, tzeva l'retzuot tefillin, 26 Tammuz 5779, (Internet)

(23) *Noda biYehudah,* second edition, op. cit., responsum 3

(24) Rabbi Yitzchak Dov Halevi (Seligmann Baer) Bamberger, *Melechet Shamayim*, second edition, (Hannover, 5620), klal 20, chochma, par.2; *Keset haSofer*, op. cit. brings a similar wording, chap.23 par.2

(25) Rabbi Akiva Sofer, *Shu't Daat Sofer*, (Jerusalem, 5725), YD chap.117

(26) *Din*, tzeva kasher l'retzuot tefillin, 24 Tevet 5776, (Internet)

(27) *mi yodea*, Kosher shoe-polish to blacken tefillin; What's inside Kiwi Shoe Polish? (Internet)

(28) e.g. Madrich hakashrut of B'datz Eda Charedis, Jerusalem, 5779, vol.70, part 1, p.125, tzeva l'tefillin

(29) *Din*, tzeva l'hashcharat retzuot hatefillin ..., 24 Nissan 5772, (Internet)

(30) *Behadrei Haredim*, mahi ha'alut shel tzviat batei tefillin + retzuot + bedika? 11 July 2013, (Internet)

(31) *Din*, tzeva l'tefillin, 28 Av 5772, (Internet)

(32) Rabbi Chaim Vital, *Shaar haKavanot*, (Jerusalem, 5662), inyan tefillin, lecture 4

(33) Rabbi Avraham ben Moshe from Sinsheim, *Baruch She'amar,* (Shklow, 5564), p.4b

(34) *Piskei Teshuvot,* edited by Rabbi Simcha Rabinowitz, (Jerusalem, 5767), vol.1, chap.33 par.4

(35) MB OC chap.32 par.3 Biur Halacha, first words: yichtavem b'dyo shachor

(36) *Baruch She'amar*, op. cit., p.4b

(37) Rabbi Avraham Halevi Horowitz, *Orchot Rabbenu* (Steipler), (Bnei Brak, 5751), OC part 1, lo mashchir tzidei haretzuot, p.36

(38) SA OC chap.27 par.8

(39) SA OC chap.27 par.11

(40) SA OC chap.27 par.8, chap.33 par.5

(41) SA OC chap.33 par.5

(42) Rabbi Meir Leibush (Malbim), *Artzot haChaim* on SA OC, (Warsaw, 5620), part 1 chap.32 par.83

(43) MB OC chap.33 par.3, Biur Halacha, first words: retzuot shechorot

(44) Rabbi Yosef Chaim Sonnenfeld, *Shu't Salmat Chaim*, (Jerusalem, 5767), chap.40

(45) Rabbi Avraham Aharon ben Shalom Broda, *Lishkat haSofer*, (Vilna, 5629), principle 15, par.1

(46)   Rabbi Moshe Sternbuch, *Teshuvot V'hanhagot,* (Jerusalem, 5762), part 4, chap.9
(47)   MB OC chap.33 par.3, Biur Halacha first words: haretzuot hashechorot
(48)   Rabbi Yaakov Meir Stern, *Mishnat Hasofer,* (Bnei Brak, 5752), chap.23 par.7
(49)   *Halichot Shlomo,* op. cit., chap.4 par.28
(50)   *Shu't Salmat Chaim,* op. cit., chap.40
(51)   *Shu't Maharshag,* op. cit., chap.7
(52)   Rabbi Avraham David from Butschatsch, *Da'at Kedoshim,* (Lvov, 5656), chap.33 par.3
(53)   *Halichot Shlomo,* op. cit., chap.4 footnote 41
(54)   Rabbi Sternbuch, op. cit., (Jerusalem, 5754), part 2 par.22
(55)   "hizharu m'retzuot tefillin m'zuyafot" *Arutz 7,* 12 Av 5775, (Internet)
(56)   research by the newspaper "Yated Neeman al revevot shemanichim tefillin lo kasherot," *Arutz 7,* 27 Tammuz 5775
(57)   "hizharu m'retzuot ..." op. cit.
(58)   *Forum Otzar Hachochma,* inyan haretzuot hapesulot, August 2015, (Internet)
(59)   SA OC chap.33 par.4 and Rema; MB OC chap.33 par.22
(60)   Rabbi Avraham Gombiner, *Magen Avraham* on SA OC chap.33 par.7
(61)   MB OC chap.33 par.24
(62)   *Piskei Teshuvot,* op. cit., vol.1, chap.33 par.24
(63)   MB OC chap.33 par.4, Biur Halacha, first word: passul
(64)   Rabbi Yosef ben Meir Teomim, *Peri Megadim,* (Warsaw, 5649), Eshel Avraham, chap.33 par.7
(65)   Rabbi Sternbuch, op. cit., part 2, chap.22
(66)   Torah Devarim chap.6 verses 8-9
(67)   SA OC chap.39 par.1
(68)   SA OC chap.39 par.2
(69)   MB OC chap.39 par.2
(70)   *Pri Megadim,* Eshel Avraham, chap.39 par.6
(71)   MB OC chap.39 par.9
(72)   Rabbi Chaim Yosef David Weiss, *Vayaan David,* (Jerusalem, 5753), OC chap.10, fourth summary
(73)   MB OC chap.33 par.22
(74)   SA OC chap.38 par.3
(75)   SA OC chap.33 par.4; MB OC chap.33 par.4, Biur Halacha, first words: aval haretzuot afilu b'diavad
(76)   MB OC chap.33 par.23
(77)   *Vayaan David,* op. cit., OC chap.10, second summary
(78)   *Talmudic Encyclopedia,* op. cit., vol.9, hanachat tefillin, column 508
(79)   Rabbi Akiva Eiger, *Hagahot al Shulchan Aruch,* OC chap.70; *Talmudic Encyclopedia,* op. cit., vol.16, chetzyo eved v'chetzyo ben chorin, columns 713-14, and footnotes 614, 626
(80)   SA OC chap.33 par.4
(81)   MB OC chap.33 par.22

(82)   SA OC chap.33 par.7, Rema
(83)   MB OC chap.33 par.23
(84)   SA OC chap.55 par.9
(85)   MB OC chap.55 par.40
(86)   Words of Rabbi Avigdor Nebenzahl brought in a 25-minute shiur of Rabbi Yaakov Teller on 31 December 2017 on YU Torah Online. Title of the shiur: Shabbos Leining: Katan?
(87)   Rabbi Yosef Shalom Elyashiv, *Vayishma Moshe*, (Kiryat Belz: Bet Shemesh), part 5, hilchot tefillin, OC, l'hashchir retzuot al y'dai yeled barmitzvah, p.14
(88)   Ibid., footnote 23
(89)   *Piskei Teshuvot*, op. cit., chap.32 par.59
(90)   First Voice-Operated Assistant for Machine Tools.... 15 August 2018, (Internet)
(91)   Rabbi Moshe Sofer, *Chidushei Chatam Sofer al Masechet Gittin*, Gittin 22b
(92)   Rabbi Yitzchak Zilberstein, *Chashukei Chemed al Masechet Megillah*, (Jerusalem, 5767), Megillah 7a
(93)   Rashi on Masechet Megillah, first words: v'noiyaihen l'bar
(94)   Talmud Bavli, Menachot 35a
(95)   SA OC chap.33 par.3
(96)   Perush Rabbenu Gershom Me'or Hagolah, *Kovetz Rishonim l'Masechet Moed Katan*, (Machon HaTalmud HaYisraeli Hashalem: Jerusalem), Moed Katan 25a, first words: ithapicha lei retzua d'tefillin
(97)   MB OC chap.33 par.20
(98)   *Beit Oved, seder tefillah limei chol shel kol hashana k'minhag Sefaradim*, (Livorno), dinei makom v'ofen hanachat tefillin u'birchatan, par.16, p.46 (91)
(99)   SA OC chap.25 par.13
(100) MB OC chap.32 par.184
(101) Talmud Bavli, Menachot 35a
(102) Rambam, Mishneh Torah, hilchot tefillin, chap.3 halacha 14; Rabbi Yitzchak ben Moshe of Vienna, *Or Zarua*, (Zhitomer, 5622), part 1, hilchot tefillin, chap.564. retzuot shechorot
(103) Tosafot on Menachot 35a first words: rezuot shechorot
(104) Tosafot on Menachot 35a first words: tefillin merubaot
(105) The Jewish Encyclopedia, (Funk & Wagnalls: New York, 1901-1906), vol.10, Phylacteries, p.25
(106) Mordechi on Talmud Bavli, Masechet Menachot, Oz Vehadar edition, halachot ketanot, Menachot 35a
(107) *Keset haSofer*, op. cit., chap.23 par.2
(108) *Lishkat haSofer al Keset haSofer*, op. cit. chap.23 par.2
(109) Rabbi Sternbuch, op. cit., part 2 par.22
(110) *Ma'aseh Melech*, leket minhagim utenuot kodesh of the Lubavitch Rebbe, (Petach Tikva, 5773), p.7

(111) Rabbi David Aryeh Morgenstern and Rabbi Eliyahu Reuven Gitman, *Zichron Eliyahu*, (Jerusalem, 5749), chap.20, shachrut hatefillin, par.2

(112) Rabbi Sternbuch, op. cit., part 2, par.22

(113) Rabbi Avraham Halevi Horowitz, op. cit.

(114) Maimonides, Rambam, Mishneh Torah, hilchot tefillin, chap.3 halacha 14

(115) Rabbi David ben Shhlomo ibn Zimra (Radbaz), *Shu't Haradbaz*, (Jerusalem, 5732), part 5, responsum 1426

(116) Rabbi Elazar Horowitz, Perush al hilchot tefillin l'Rambam mitoch Sefer haMenuchah l'Rabbenu Manoach m'Narbona, *Hadarom - Kovets Torani*, (New York), no.40, Tishrei 5735, pp.98-99

(117) *Or Zarua*, op. cit., chap.564, retzuot shechorot

(118) Darchei Moshe (Rema) on Tur OC chap.33 par.2

(119) Rabbi Zedekiah ben Avraham Anaw Harofeh, *Shibbolei haLeket*, (Shlomo Buber from Lvov: Vilna 5647 and later reprinted in Jerusalem 5746), inyan tefillin, p.192 (384)

(120) *Shaar haKavanot*, op. cit., fourth lecture

(121) Rabbi Yosef Liberman, *Mishnat Yosef*, (Jerusalem, 5772), part 10, chap.10, hashcharat retzuot hatefillin mishnei hatzdadim, par.7

(122) Rabbi Yosef Chaim, *Ben Ish Chai*, (Jerusalem, 5717), first year parashat Chayei Sarah, par.4

(123) Rabbi Chaim Elazar Spira (Gaon from Munkatch), *Ot Chaim v'Shalom*, (Brooklyn, New York, 5746), chap.33 par.2

(124) Rabbi Mordechai Ze'ev Ettinger, *Magen Gibborim* on the Tur and SA OC, (Zolkieve, 5599), part 1, chap 33 par.33, Elef Hamagen, par.3

(125) Beit Yosef on Tur OC chap.33, first words: v'katav haRambam; Darchei Moshe (Rema) on Tur OC chap.33, par.2; MB OC chap.33 par.21; Rabbi Chaim Benvenist, *Knesset haGedolah*, (Livorno, 5418), hilchot tefillin Beit Yosef, chap.33; Rabbi Yitzchak Baruch, *Kol Yaakov*, (Jerusalem, 5670), chap.23 par.13

(126) Talmud Bavli, Moed Katan 25a

(127) Rashi on Moed Katan 25a, first words: ithapicha lei retzua d'tefillin

(128) Rabbi Yishmael Hakohen, *Zera Emet,* (Reggio), OC chap.32

(129) Rabbi Eliyahu Katz, *Rosh Eliyahu*, (Bnei Brak, 5768), part 1, OC chap.7 p.33

(130) Rabbi Shmuel Halevi Wosner, *Shu't Shevet Halevi*, (Bnei Brak, 5762), part 9, chap.16

(131) Rabbi Yosef Chaim Kanievsky, *Shelat Rav*, (Kiryat Sefer, 5772), part 2, chap.2, retzuot im tzeva du tzdadi, responsum 18, (p.177)

(132) Rabbi Meir Halevi Soloveitchik, *Mishnat Rabbenu Meir Halevi*, (Eretz Yisrael, 5778), halichot v'hanhagot, retzuot shechorot mishnei hatzdadim, p.583

(133) *Vayishma Moshe*, op. cit., footnote 24

(134) Rabbi Meir Arava, *Meir Oz*, (Bnei Brak, 5768), chap.33 par.3 sub-par.3 (end), p.461

(135) Rabbi Sternbuch, op. cit., par.22

(136) Rabbi Azaria Ariel, tzeva gav retzuot tefillin. *Atar Yeshiva*, she'elot uteshuvot, (Internet)
(137) *Vayishma Moshe*, op. cit., shachrut heretzuot mishnei hatzdad
(138) Ibid.
(139) Parashat Lech Lecha, retzuot hatefillin, *Hilchata*, (Kollel Halacha Rehovot, Machon l'Pirsumim Hilchatiim), halacha b'parasha, year 1 – issue no.2, 7 Marcheshvan 5778, (Internet)
(140) Ibid.
(141) *Piskei Teshuvot*, op. cit., chap.33. par.7, footnote 48
(142) Go Vegan Scotland - Non-Food Products, (Internet)
(143) SA OC chap.32 par.37
(144) Talmud Bavli, Shabbat 108a
(145) Be the Change You Want to See In the World – Non-leather tefillin, (Internet)

DOES AN ORPHAN START PUTTING ON TEFILLIN AT THE AGE OF TWELVE?

The Gemara[1] gives details from which age it is appropriate to educate a minor to observe various mitzvot. Regarding tefillin it is written: "A minor who knows how to take care of tefillin, his father should buy tefillin for him."[2] The Shulchan Aruch explains that "he should not sleep nor pass wind whilst wearing them, and also that he should not enter the toilet whilst wearing them."[3] Since today it is customary wear tefillin only during shacharit[4] (morning service) the fear of sleeping or entering the toilet with tefillin is infrequent. However, this is not so with passing wind, since the researchers have come to the conclusion that everybody passes wind between eight and twenty times a day.[5] Therefore there is a fear that a person will pass wind whilst wearing tefillin. For this reason, they ruled that one must take great care in this matter, so much so that those who cannot keep themselves from passing wind should not pray the amidah, but only recite the "shema" of shacharit (morning service) together with its berachot, "but without wearing tefillin."[6] The conclusion from this is that since we are afraid that a child will pass wind whilst he is wearing tefillin, it has been ruled that he should only start putting on tefillin close to his age of barmitzvah, or even on the actual day he reaches his barmitzvah age. This is unlike other mitzvot such as tzitzit, lulav and sukkah where the child begins to observe them at a much younger age before his barmitzvah.

In fact, there are different opinions regarding the actual age that the child should start putting on tefillin prior to reaching the age of barmitzvah. Rabbi Ovadia Yosef stated that the Sefaradim would do so a year or two before the age of barmitzvah,[7] but in practice the Sefaradim do not do so that far ahead.[8] Rabbi Yoel Sirkis (16th-17th centuries) known as the "Bach" writes that such a period is about one year.[9] The Ashkenazi Poskim (Rabbinical arbiters) write just two or three months[10] or even just one month before his barmitzvah.[11] The opinion of Rabbi Yitzchak Meri,[12] (12th century) known as the "Itur," (and the Rema[13] also holds this opinion), is that the child should start putting on tefillin only on the actual day he reaches the age of barmitzvah, and this is the opinion followed by the Hasidim.[14]

However, the Gemara as well as the Poskim from the time of the Rishonim (great Rabbis who lived approximately between the eleventh and fifteenth centuries), do not make a distinction between the age at which children begin to put on tefillin when their parents are alive, and those who are orphaned. Only in the period of the Acharonim (great Rabbis who lived from about the sixteenth century) are there opinions that a child who is an orphan should start to put on tefillin at a younger age, for example, at the age of twelve.

The period when this custom began is unknown. The earliest source which we have is from about one hundred and sixty years ago by Rabbi Shlomo Kluger. He wrote about this in his "Sefer Sta'm" published in Lemberg in 5616 (1856). One of the responsa in his book is about an orphan child, and it was not known whether his age was twelve or thirteen. There he wrote: "The world says that an orphan should put on [tefillin] before this [his barmitzvah]"[15] Also in a book written by Michael Hilton entitled "Bar Mitzvah – A History" it states: "In the middle of the eighteenth century, it is recorded that it was the custom in Lvov [Lemberg] that an orphan starts putting on tefillin from the age of twelve."[16]

This was not limited to Lemberg, but also in other parts of the world this was the custom. In the book "Mataamim Hachadash" written by Rabbi Yitzchak Lipiatz miShedlitz and published in 5664 (1904), it is written: "The orphan begins to put on tefillin one year before he reaches thirteen years. This is the custom among the Sefaradim in Jerusalem."[17] There are several sources that refer to the putting on of tefillin by an orphan at the age of twelve, and they contain words such as "a common prevalent custom in many places."[18]

Several questions can be asked on this topic:

- Where is the source that states specifically that an orphan should put on tefillin at an age younger than barmitzvah?
- On this subject, is there a difference between a child orphaned from both parents, or only from the father or only from the mother?
- What are the reasons to advance the age of putting on tefillin for an orphan?
- At what age does the orphan start putting on tefillin?

In the Talmud Bavli and the Talmud Yerushalmi there is no direct source indicating a different age between a child whose parents are alive and that of an orphaned child regarding the age of starting to put on tefillin. However, there are Poskim who write that there is a support from Masechet (tractate) Gittin on the subject of a guardian for an orphan.[19] There it is written that a guardian buys all kinds of holy and mitzvah artifacts, including tefillin, for an orphan. According to the halacha, as soon as the orphan reaches the age of barmitzvah, the guardian stops acting for him and transfers all these objects to him.[20] Namely, the guardian buys tefillin whilst the orphan is still a minor, and it follows that the orphan starts putting on tefillin before the age of barmitzvah. (The commentary of the "Penai Moshe" on the Talmud Yerushalmi in tractate Terumot confirms that the Mishnah in Gittin is referring to a minor orphan.[21]) However, there are those who reject this conclusion and write that it is impossible to wait for the actual day when the child reaches the age of barmitzvah, and only then does he buy his tefillin by himself. From the practical point of view, the tefillin should be bought before this day, and only the guardian is able to do this because the orphan is still an infant.[22]

In addition to learning from the above Gemara in Gittin, there are a number of Poskim who offer different reasons for this custom, and many point out that the early putting on of tefillin brings merit to his departed parent(s).[23]

However, there are those who reject this reason, among them being the seventh Rebbe of Lubavitch, Rabbi Menachem

Schneerson. He writes in one of his letters: "Some clarification is required [for one might ask]: What merit is there for the departed [if his son] performs an act that is not appropriate for him, not even as part of his training [in Jewish observance]? The recitation of [the Mourners'] Kaddish [for a departed parent] is, by contrast, appropriate for a minor. Moreover, [when the minor] recites Kaddish, men who are obligated [in the observance of mitzvot] respond and hear how [the child] justifies [G-d's] judgment. The question can be resolved, albeit with difficulty."[24]

Another reason for this custom is because the orphan does not have a father to instruct him regarding tefillin, and he therefore needs guidance from other people, which means that it depends on strangers, and for that reason the time to start putting on tefillin is advanced.[25]

However, Rabbi Hillel Posek (19th-20th centuries) rejects this reason and writes: "In my opinion, this point seems to be very weak. Moreover, since there is no one to watch over the boy, the tefillin should not be handed over to him before his barmitzvah ... an orphan before his barmitzvah, has no obligation to put on tefillin."[26] However, the Lubavitcher Rebbe writes: "On the surface [this explanation is also somewhat problematic], because training [in the observance of tefillin] is not so difficult that it requires extra effort." [27]

A question which arises is whether there is a difference if a boy is an orphan due to the death of his father or due to the death of his mother? The answer is that there is an obligation on the father (and not the mother) to educate his son to observe the mitzvot.[28] Therefore, before the son reaches the age of barmitzvah, the father teaches him "to concentrate properly when wearing tefillin, and he stands over him and warns him against passing wind, and should he pass wind, he will take the tefillin from him."[29] But what happens if the father has already died? In the words of Rabbi Yisrael Chaim Friedman known as the "Maharich," (19th-20th centuries): "Since he has only strangers to teach him, and this could be today one person and the next day someone else, the custom is to start a long time before his barmitzvah, and during this time he will also learn the laws concerning tefillin from other people."[30] However, one can also make an opposite claim, namely the father would be more careful than anyone else that the child does not pass wind when wearing the tefillin.[31] It thus follows that it is better for the orphan to put on the tefillin closer to his barmitzvah than for a child whose father is still alive.

We can learn from this that if only the mother has passed away, the child will start putting on tefillin like a child with two parents, because the mother does not have to educate her son in the mitzvah of tefillin.[32] On the other hand, according to Rabbi Shmuel Pinchas Galbard who writes that the advancing to the age of twelve is "in order to give merit to his father *or to his mother* in the next world"[33] (emphasis added), namely, there is no difference in which of the parents had passed away.

There are different opinions among the Poskim of how long (if at all) before an orphan reaches the age of barmitzvah should he start putting on tefillin? Above was stated the custom of an entire year. Although this custom is prevalent in several places, many Poskim reject it for various reasons.

One of the reasons is that the child will pass wind whilst wearing tefillin, namely, he will not take care over the holiness of the tefillin. Rabbi Natan Gestetner (20th-21st centuries) in his book "Lehorot Natan" writes regarding this period of a year: "One should be very concerned for the respect to be given to tefillin, and thus advance starting to put them on only by a month." [34]

There are some that hold that the day to start putting on tefillin is precisely on the day the child reaches barmitzvah age. Rabbi Shalom Rokeach (the Sar Shalom, the first Belzer Rebbe) writes: "It is customary for my children, as well to those who ask, to begin precisely on his birthday when he enters his fourteenth year, without any distinction between those who have a father and a mother and those who are orphans, and he did not know of any source for those who say that there should be a distinction between an orphan and someone else."[35] He explains his opinion that the first time one puts on tefillin, one will do so with great enthusiasm, and it is thus better to do this on the day when one has the obligation from the Torah to perform this mitzvah.[36]

There are a number of Poskim who reject outright the difference of the starting date of putting on tefillin between a child with parents and on the other hand an orphan, arguing that there is no hint or basis for this custom.[37] The "Aruch Hashalchan" writes strongly against this custom. "And one should know it is customary among the multitudes that an orphan would start putting on tefillin a year earlier, and I do not know any reason for this, and it is not right to do so."[38]

Conclusion: In view of the different opinions regarding the age of an orphan should start putting on tefillin, one should ask a Posek for a ruling on the subject.

Appendix

Does an orphan reach age of barmitzvah at the age of twelve?

It is written in the Masechet Avot, "Thirteen years old for fulfilling the mitzvot,"[39] and the age of thirteen for a barmitzvah is mentioned many times by the Poskim, and they do not differentiate between a child with two parents and a child who is an orphan. On the other hand, it is mentioned in several places that orphaned children celebrated their barmitzvah a year earlier, namely when they reached the age of twelve.

Here are some items of testimony:

a)   In the year 5771 (2011), in the forum of the "Aishdat Society," Yoel C. Salomon related about an "odd bar-mitzvah custom in [the city of] Selish" (today located in Ukraine) that took place in about 5683 (1923). Yoel said that he spoke with his grandfather who grew up in the city of Selish. The grandfather was an orphan because his father died during the years of the First World War, and "therefore had his bar-mitzvah at the age of 12. He apparently was given to believe this was the rule for yesomim (orphans)."[40] Selish was a community with a history and had notable Rabbis, and at the time of this event the Rabbis of the city were Rabbi Pinchas Chaim Klein and Rabbi Yosef Nehemiah Kornitzer,[41] and it is thus difficult to accept that the Rabbis of this city agreed to a twelve-year-old barmitzvah.

b) At about the same time, a man named "Chesky" wrote to the forum of the "Aishdat Society": "My grandfather's memory was that he was told a yossom (orphan) has his bar mitzvah at age 12."[42] It is not written in which city the grandfather grew up.

c) The Yiddish Book center in the USA interviewed many people. One of them was Paul Kaye, who was born in 1967 in New York. "He lost his father at a young age, he was considered an orphan and became a bar mitzvah at the age of twelve," that is, in 1979.[43] He did not say where the barmitzvah was celebrated.

d) In a letter to the editors of the "Aishdat Society" in the year 5777 (2017), David Havin wrote: "My maternal grandfather, who lived in Lithuania until his late teens, told me on a number of occasions that he celebrated his bar mitzvah when he was 12 years old, as his father had died and he was considered to be an orphan." David Havin went on to say that he asked several Rabbis, but they all said they "had never heard of it." At a later date, he found this topic in a book about "The Life of Reb Aryeh Levin."[44] (see below).

e) In the book "A Tzadik in our Time – the Life of Rabbi Aryeh Levin" it is written under the heading "In honor of the mother": "When a certain orphan boy became bar-mitzvah (on his twelfth birthday rather than his thirteenth, because he was an orphan), Reb Aryeh went to the family's celebration of the happy event."[45]

f) This topic also appeared in "The story of the son of the barmitzvah. Chapter 6." It says there: "In 1918, a boy was already twelve years old. At this age, the barmitzvah is celebrated for an orphan child..."[46]

The question arises, how is it possible that in several places in the world an orphan celebrates his barmitzvah at the age of twelve? Obviously, there is no halachic source for this. It is possible that the celebration of the barmitzvah at the age of twelve is rooted in the custom in which a child who is an orphan begins to put on tefillin at the age of twelve and then, at the same time, his barmitzvah is celebrated.[47]

Perhaps one can make another suggestion. A boy before the age of barmitzvah can be called up to the Torah for Maftir.[48] Likewise, a minor can be included amongst the seven people called up to the Torah on the morning of Shabbat,[49] although in practice this is not customary.[50] It is possible that a twelve-year-old orphan boy was called up to the Torah for maftir, or as one of the seven people called up on Shabbat morning, and because of this, the child and the public thought it was his barmitzvah.

References

Abbreviations
SA = Shulchan Aruch
OC = Orach Chaim
CM = Choshen Mishpat
MB = Mishnah Berurah
(1)   Talmud Bavli Sukkah 42a; Arachin 2b
(2)   Ibid., Ibid.
(3)   SA OC chap.37 par.3 and Rema
(4)   SA OC chap.37 par.2
(5)   e.g. Michael D. Levitt et al, "Evaluation of an extremely flatulent patient: Case report and proposed diagnostic and therapeutic approach", *American Journal of Gastroenterology* (1998), vol.93, pp.2276-2281
(6)   SA OC chap.38 par.2; MB chap.38 par.7
(7)   Rabbi Yitzchak Yosef, *Kitzur Shulchan Aruch Yalkut Yosef*, (Chazon Ovadia Institute: Jerusalem 5757), chap.37 Chinuch Katan b'Mitzvot Tefillin, par.1
(8)   Information from Rabbi Yitzchak Roderig, Kiryat Arba-Hevron
(9)   Rabbi Yoel Sirkis (Bach) on Tur OC chap.37, first word: katan
(10)   MB SA OC chap.37 par.12
(11)   Aruch Hashulchan OC chap.37 par.4
(12)   Rabbi Yitzchak ben Abba Meri, *Sefer Haitur*, (Vilna, 5634), sha'ar rishon, hilchot tefillin, p.61b
(13)   SA OC chap.37 par.3, Rema
(14)   Rabbi Yekutiel Yehuda Rosenberger, *She'elot uTeshuvot Torat Yekutiel* on SA OC, (Jerusalem, 5740), chap.66, p.109
(15)   Rabbi Shlomo Kluger, *Sefer Sta'm*, vol.2 (Lemberg, 5616), hilchot tefillin, chap.36 responsum 30, p.55b
(16)   Michael Hilton, *Bar Mitzvah a History*, (University of Nebraska, 2014), p.138
(17)   Rabbi Yitzchak Lifiatz miShedlitz, *Sefer Mataamim Hachadash*, (Warsaw, 5664), tefillin, p.46

(18) e.g. Rabbi Tzvi Yechezkel Michelson, *Pinot Habayit*, (Pieterkov, 5685), chap.85, p.95; Rabbi Simcha Rabinowitz, *Piskei Teshuvot*, (Jerusalem, 5767), hilchot tefillin, p.349; Aishdas Society forum, volume 35, number 65, 23 May 2017, message 8 from Zev Sero, (Internet); Mark Oppenheimer, Thirteen and a Day, (Farrar, Straus and Giroux: New York: 2005)

(19) Talmud Bavli Gittin 52a; SA CM chap.290 par.15

(20) Rambam (Maimonides) Mishneh Torah, hilchot nachalot chap.11 halacha 5; SA CM chap.290 par.16

(21) Talmud Yerushalmi, Terumot chap.1 halacha 1

(22) Rabbi Gavriel Zinner, *Nitei Gavriel*, (Jerusalem, Iyar 5762), Hilchot v'Halichot Barmitzvah, chap.33, p.226

(23) e.g. *Pinot Habayit*, op. cit.; *Torat Yekutiel*, op. cit., p.108; Rabbi Yaakov Yosef (Zeida) Shapira, *Atzmot Yosef al Hilchot Treefot ha'Atzamot*, (Husiatyn, 5665), Chidushei halachot, halacha 13, p.134; Rabbi Gershon Stern, *Yalkut haGershuni*, (Munkatch, 5661-5664), vol.1 on SA OC, chap.37, p.10; Rabbi Avraham Yitzchak Sperling, *Ta'ame haMinhagim u'Mekorei haDinim*, (Eshkol: Jerusalem, date not known); *Kitzur Shulchan Aruch Yalkut Yosef*, op. cit.

(24) Rabbi Menachem Mendel Schneerson, Rebbe of Lubavitch, *Igrot Kodesh*, vol.3, letter 503 dated 15 Tammuz 5709

(25) Rabbi Hillel Posek, *Hallel Omer*, (Tel-Aviv, 5716), responsum 18, p.10; Rabbi Yisrael Chaim Friedman, *Likutei Maharich*, vol.1, (Meir Leib Hirsch: Satmar (Sato-Mare) Romania, 5692), seder mitzvot tefillin, p.26b

(26) *Hallel Omer*, op. cit.

(27) *Igrot Kodesh*, op. cit.

(28) *Atzmot Yosef*, op. cit.

(29) *Torat Yekutiel*, op. cit. p.109

(30) *Likutei Maharich*, op. cit.

(31) *Hallel Omer*, op. cit.; *Torat Yekutiel*, op. cit.

(32) *Atzmot Yosef*, op. cit.

(33) Rabbi Shmuel Pinchas Gelbard, *Otzar Ta'amei Haminhagim*, (Petach Tikva, 5756), sha'ar 3, tefillin, p.47

(34) Rabbi Natan Gestetner, *She'elot uTeshuvot Lihorot Natan*, vol.2 (Bnei Brak, 5737), chalek Orach Chaim, chap.9, p.15; *Sefer Sta'm*, op. cit.

(35) Rabbi Dov Berish Hakohen Rappaport, *Derech haMelech*, ("Shem Olam" Institute: Bnei Brak, 5763), p.29

(36) Rabbi Yechiel Michel Hibner, *Hadrat Kodesh*, (Druck von Kohn & Klein: Munkatch, 5660), par.15, p.13

(37) *Hallel Omer*, op. cit., p.11; *Sefer Sta'm*, op. cit., *Atzmot Yosef*, op. cit.

(38) Aruch Hashulchan, OC chap.37 par.4

(39) Mishnah, Avot chap.5 mishnah 21

(40) Aishdas Society forum, letter from Joel C. Salomon, about 1971, "Odd bar-mitzvah custom in Selish ca.1923," (Internet)

(41) Kehilat Yehudai Selish , data base – Museum of the Diaspora Tel-Aviv, (Internet)

(42) Aishdas Society forum, letter from Chesky, about 1971, (Internet)
(43) Yiddish Book Center, Wexler Oral History Project, A Bar Mitzvah at Age Twelve, (Internet)
(44) Aishdas Society forum, volume 35, number 65, 20 May 2017, Message 4 from David Havin, (Internet)
(45) Simcha Raz, A tzaddik in our time: The life of Rabbi Aryeh Levin, trans. Charles Wengrov, (Feldheim Publishing: Jerusalem, 1976, second edition), In honor of the Mother, p.95. One should not think that this is a writing error, since this also appears on page 92 in an edition of this book that was published in the year 2008.
(46) The story of a Barmitzvah boy, chap.6, (Internet)
(47) Michael Hilton, op. cit.
(48) MB OC chap.282 par.12
(49) SA OC chap.282 par.3
(50) MB OC chap.282 par.12

TWO-HEADED SIAMESE TWINS
ASPECTS IN JEWISH LAW

------ ‖⟨◉⟩‖ ------

Introduction

It is written in the Gemara: "Pelemo enquired of Rebbe (Rabbi Yehudah Hanasi), if a man has two heads on which of them does he put on tefillin?"[1]

A person who has two heads is one example of what is known today as "Siamese twins," namely, that is two people who are born joined together in a certain part of their body. There are endless ways that they can be connected. For example: in the chest, the navel, the buttocks, the head, and indeed many more places. The example given in Pelemo's question is, twins born with one body but with two heads. In the medical literature they are known as dicephalus (from the Latin word di - two, cephalus - heads.)

Even in the case of dicephalus twins, there are different types. Sometimes they are attached from the shoulder down, from the chest down, from the navel down, and it is even possible that their heads are fused together. The question which arises, is what type is our Gemara referring to? The Gemara states: "He who has two heads, on which of them does he put on tefillin." From the language of the Gemara that asks *on which head* does he put on tefillin, it is possible to exclude the type in which the two heads are fused together.

According to the wording of the question, it seems that there are duplications which are limited to *just* the heads of the twins, namely, the twins are joined from the shoulder down. Another proof can be brought from the fact that Pelemo does not ask on which hand he will put on tefillin. Twins born conjoined from the chest down have four hands. However, those connected from the shoulder down have in most cases only two hands, and are called in Latin dibracius (two hands). Cases are known where twins have three or four hands, but the middle hands are attached or stuck behind the back and are therefore useless.

From all of the above it is possible to come to the conclusion that Pelemo's question is for twins connected from the shoulder down.

One could note that the phenomenon of twins with two heads is not limited to humans. The same phenomenon has also been found in animals, especially snakes.[2] Such snakes are mentioned in the Zohar, which makes a comparison with the descendants of Cain, as it is written: "And they all realised that they were the descendants of Cain, and they therefore had two heads like two snakes..."[3]

Responses to the question of Pelemo

Rebbe (who obviously thought that Pelemo was asking a silly question) answered him: "Either get out or be excommunicated."[4] Some Rishonim (great Rabbis who lived approximately between the eleventh and fifteenth centuries) learn from the question of Pelemo that one of the reasons for a person being excommunicated is that he asks an impossible question. One can learn this from the words of the Shulchan Aruch who gives this as a reason for excommunication, namely: "Asking a thing that is impossible."[5] In his commentary to the Tur, the Beit Yosef gives as a source for this halacha the answer of Rebbe.[6]

We should note that Pelemo was indeed a scholar and was a Tanna of the fifth generation,[7] and he is mentioned in several places in the Talmud, and sometimes the halacha is according to his opinion. For example:

- In Masechet Berachot, Pelemo states that in Bircat Hamazon (Grace after Meals), one must mention "brit" (covenant) before mentioning Torah.[8]
- In Masechet Pesachim, in connection with a certain aspect in the search for leaven, Pelemo says that "he does not search it at all on account of the danger."[9] On this, Rashi comments

that he does not check a hole, lest the non-Jew suspect him of witchcraft when he sees him rummaging in the hole.
• In Masechet Kiddushin it is written that Pelemo would say "An arrow in Satan's eyes."[10]

It is almost certain that Pelemo had heard of one or several cases of such twins, and we will see this in several examples mentioned in the Roman literature of that period.

One can therefore ask, why Rebbe reacted so harshly to Pelemo's question. It is possible that Rebbe did not hear about a case of such twins, or if he did hear, it is possible that he believed that such a child would die at a very young age, and would not reach the age when he was obligated to put on tefillin.[11]

The above-mentioned Gemara immediately continues with the case of a child born with two heads: "In the meantime a man came saying '[My wife has] given birth to a first-born child with two heads, how much must I give the kohen to redeem him." The Gemara replied that the sum was ten selaim, namely five selaim for each head.[12]

On the one hand, we saw that the Beit Yosef cites Rebbe's answer in connection with the Pelemo's question as a reason for excommunication, and on the other hand, the Beit Yosef[13] brings as the source of the halacha as cited in the Tur,[14] that according to the Gemara in Menachot, one must pay ten selaim to redeem a child with two heads! However, this contradiction can be resolved, by saying that the Beit Yosef believed that a child born with two heads could live up to thirty days, but not until the age of putting on tefillin. In the medical literature (as we will see below), it is stated that such children were born dead or died minutes or hours after birth. Until before about ten years ago (2005), there were only a few cases where these children lived for more than a few hours (see below).

A few hundred years after the Gemara, the Rambam (Maimonides) mentions in his medical book "Pirkei Moshe" cases that existed in the past, of a number of children born with two heads[15] (see details below). However, he does not specify whether or not they were joined from the shoulder down.

This book is the largest book in the Rambam's writings on the subject of medicine. He wrote this book in Arabic between the years 4947-4950 (1187-1190). Over the course of the years, this book has been translated into several languages, including Hebrew, and it is to be found in a number of manuscripts and published books.[16]

One can note that the Rambam does not include in his composition "Mishneh Torah" the halacha regarding the redemption of a son who has two heads, even though he mentions these cases in his book "Pirkei Moshe." It is possible, that he believed based on his medical knowledge, that such a child cannot live up to thirty days, the age of the redemption of the firstborn son.[17]

Two-headed twins throughout history in the general literature

In the history of the Romans several cases of children born with two heads are mentioned. Ammianus Marcellinus who lived in the fourth century wrote about a child in the suburb of Antioch who was born with two heads.[18] Antoninus Pius also wrote that after several natural and tragic events "a two-headed child was born."[19] Also Livy wrote: "In the territory of Veii a boy with two heads was born."[20] However, all these three writers do not specify whether or not these twins were joined from the shoulder down or from below the shoulder.

In his writings, Saint Augustine of Hippo, who lived in the fourth to fifth centuries, writes that "not many years ago, within living memory, a person was born in the East who had two heads, two chests, four hands, as though he were two persons, but one stomach and two feet as though he were one. And he lived long enough and the case was so well known that many people went to see the wonder."[21] However, he does not write if he reached the age of adulthood.

Two authors from the Middle Ages, namely Pierre Boaistuau[22] and Fortunio Liceti[23] tried to draw this man, and both drew him as an *elderly man joined from the shoulder down.*

44

According to Augustine's book, however, the connection of bodies seems to have been below the shoulder. Boaistuau also brings[24] from the letters of Caelius Rhodiginus (chapter 27) a picture of two people, each with two heads, and here they are both adults and joined from the shoulder down. There is a similar picture of an elderly woman with two heads, born in the early sixteenth century. However, "according to a German historical chronicle 'this horrible monster' died just an hour after birth."[25] Also, there is an old German print from the very late seventeenth century of a Turkish archer with two heads. However, some scholars believe that he did not exist at all, and "he was a product of wartime propaganda."[26]

Ambrose Parey was a famous doctor in the sixteenth century and in his book is to be found a drawing of *adult* twins with two heads on one body, which were joined from the shoulder down,[27] and this corresponds to the above stated Gemara. They lived before the middle of the sixteenth century. Parey writes of "their desire to eat, drink, sleep, speak and do anything."[28] However, he does not mention whether one head could feel pain as result of a blow received on the other head.

However, there is criticism[29] of Parey's degree of historical accuracy. Although he collected a number of cases from recognised sources, he also brings, for example, a picture of a child with wings instead of arms and also a bird with "an extra eye in the knee"[30] and he believed that they were facts! Therefore, apparently, his pictures are not realistic but were brought to give a religious (Christian) message!

In addition to this, there is great doubt whether at that time a man with two heads could live to grown-up age. (This criticism can also be extended to the pictures of Boaistuau and Liceti.) Even about a hundred years ago, a qualified doctor wrote about them: "Had they been scientific teratologists they would have known that such strong matured beings were a physical impossibility in the human race."[31]

About two hundred years after the publication of Parey's book, a book was published by Barton Cooke Hirst and George A. Piersol. In that book there is a chapter that brings a number of two-headed twins,[32] but it seems that they are joined below the shoulder. This chapter also includes a *photograph* of the twins from the Tocci family[33] who were born in 1877. However, we can see from the picture that their joined body starts from below the chest and therefore they are not relevant to our subject. Also in an article in a medical journal it is mentioned in the discussion of this Tocci case that their joined body is "beginning on a level of their sixth ribs."[34]

In a book written by George M. Gould and Walter L. Pyle and published in 1896, there are several cases of twins with two heads, born between the period of the Romans and the nineteenth century.[35] However, according to the explanations and pictures, they were all joined from the chest or even from below the navel. Therefore, they are not the same as the case mentioned in the Gemara. It would seem that all the twins mentioned in this book lived for several years. It is very possible that those born at that period joined from the shoulder down were stillborn or died immediately after birth, and therefore this is why they are not mentioned in this book.

However, most of the literature on this topic is from medical journals from the twentieth century, and especially in the last fifty years (namely prior to 2005). In these writings, a large number of cases were cited along with pictures of conjoined twins *from the shoulder down*. Most of them have two hands. Few of them have three or four hands, the middle one(s) of which are wedged behind the back and therefore probably unusable. They were born in various countries in the world, amongst others in Japan,[36] the United States,[37] Britain,[38] Australia,[39] Korea,[40] the Persian Gulf,[41] Singapore,[42] and even Israel.[43]

Most of them were stillborn or died immediately after birth. However, there were twins who were born in 1937 in Moscow and lived for over a year,[44] and another pair who lived for eleven days.[45] Due to the development of knowledge and more advanced medical equipment, a number of such twins born in the 1990s continued to live even to this day (2005).[46] An example are the twins from the Hensel family, who (in 2005) were still joined together, and had then reached the age of fifteen. The twins from the Holton family were separated at the age of three. One of them died after a few days from the surgery, but the second was still alive in 2005 and had then reached the age of seventeen.[47]

What causes the formation of Siamese twins, including the dicephalus type?

An internal reason is that it "probably arises from the incomplete separation of a single fertilised egg into two parts."[48] However, there are also external reasons. In the medical books it states that external causes can cause the birth of such twins. In the medical book "Pirkei Moshe" written by the Rambam, it is written: "The translator [Hunain ibn Yitzchak] explained that in the book 'Ailments of the Ischial Region' which is really the [book entitled] 'Ailments of Women' [De doloribus mulierum] of Hippocrates which he [Hunain] translated from the Greek into Arabic and which he commented upon exclusive of Galen, there are many strange cases described in addition to his own commentary. Among those that he quotes is the story of the great eclipse of the sun that occurred over Sicily and in that year, women gave birth to abnormally-shaped babies that had two heads. Other women had their regular menstrual flow but through the mouth through vomiting."[49] This corresponds to what is written in the literature of the Romans during the time of Hippocrates, namely: "a comet was seen"[50] and there were strange phenomena in the sky.[51]

Although according to scientists there is no physical danger to humans while a comet and the like are visible in the sky, it is possible that there was a spiritual danger at that time, for example, from the fear that the comet would reach the earth.

In our time, we find in the medical journals several reasons why the phenomenon that the number of children born with two heads is increasing. Among the reasons are the atomic reactor explosion in Chernobyl in 1986,[52] and in the war between Iraq and Iran during the years 1980-1988 "where many chemical weapons were known to have been used."[53] An answer with statistics is to be found in a medical article published in the city of Hiroshima in Japan. There one can see that the number of children

47

born with two heads in Japan has increased significantly since the time of the atomic bombing of Japan in 1945. From 1896-1937, that is, a period of forty-two years, there were only seven cases of two-headed twins in Japan; However, from 1951–1973, namely a period of twenty-two years, *(after the atomic bombing of Japan)* there were fourteen cases in that country.[54]

In addition to all this, teratology substances[55] can cause the birth of conjoined twins.[56]

Dicephalus twins - one person or two persons?

The Tosafot writes that in this world there is no such thing. However, in contrast, it does appear in the Midrash. There it states that Ashmedai brought out from under the ground before King Shlomo a man who had two heads, and had married a wife and begat sons like him with two heads, and also had sons like his wife who had just one head, and when they came to divide their father's property the one with two heads asked a double portion and they came to discuss before Shlomo.[57] However, the Tosafot does not come to a conclusion.

In contrast, Rabbi Bezalel Ashkenazi (16th century), the author of the "Shita Mekubetzet" continues with this Aggadah: Shlomo in his wisdom boiled water and covered one of the heads and poured the boiling water on the other head, and because of the pain from the boiling water both heads cried out. Shlomo said that one can learn from this that it is just one person with two heads.[58]

This Midrash is also mentioned in several other places but with a number of differences.[59] One of the differences is that the passage dealing with one of the heads being covered[60] does not appear in it, namely that one head actually saw the pouring of water on the *other head*.[61]

Rabbi David Bleich searched the medical literature for the reality that one head can feel pain when the other head is hit. For that there must be one nervous system, but he found no trace for such a fact.[62] It should be noted that in almost all the examples that Rabbi Bleich brought in his article, the twins were only connected from the middle of the body down.

48

To this day, the most relevant example on this matter is the Hensel family twins who are alive today (2005). They are certainly connected like the boy mentioned in the Gemara. An article while they were six years old appeared in "Time" newspaper. There it states that if one of the twins is tickled on the side of the body from head to toe, the other will not feel it "except along a narrow region on their back where they seem to share sensation." However, "they coordinate upper-body motion like clapping hands."[63] They have three lungs and were able to walk at the age of fifteen months. They share a bloodstream. If one takes medicine, it can heal the other's ear infection. Sometimes both twins think the same thing. Doctor Benjamin Carson from Johns Hopkins Hospital suggests "given the fact that they have shared organs, it is almost impossible for there not to be some overlapping in their autonomic nervous systems." [64]

Signs in which a person with two heads is considered as one person

One should note that it is not an academic discussion whether such twins are considered in *halacha* as one person or as two people. As we will see below, this question is important both in the matter of observing the mitzvot that are performed using the head organs, and also in the question that arises today when there is a possibility of separating the twins, and therefore if there is a certainty or even a chance that one of the heads will die, is it then permissible to operate?

However, the question which arises today, is whether dicephalus twins who have a separate nervous system (which is actually the situation today), are considered as one person or as two people. For the answer, one needs to study the following points:

1. The limitation stated in the "Shita Mekubetzet," that such a person will be considered as one person only if one head feels the pain as a result of a blow to the other head. However, this is not even hinted at in the Gemara.

2. The example that Tosafot brings is a person from another world. However, who says that a person from *another world* is identical in his nervous system to a person *in this world*?

3. The case cited in the Tosafot is an Aggadah, and one does not learn halacha from an Aggadah.[65]

4. In the ruling of Rabbi Chaim Elazar Spira who is known as the Gaon of Munkatch, on our topic regarding tefillin (see below), the limitation stated by the "Shita Mekubetzet" is not mentioned at all. Apparently, he thinks that you don't learn halacha from Aggadah. He also writes that only twins joined from the naval down count as two people.[66]

5. It is possible that in the case that appears in the Midrash, the pain that the second head felt was not a physical pain, but shock from the screams of the first head, or the fear that boiling water would also be poured on his head. According to some versions the second head actually *saw* the pouring of the boiling water on the first head.

6. An answer to a question asked by doctors to the Rafeh organisation, states: "... It is possible that even if there was no pain in both heads and their nervous system is separate, it does not necessarily mean that they are two people who can be said to be completely separate, since at least regarding their organs for breathing they are one body."[67] In fact, both twins have the same blood system, and together they have fewer than the usual four lungs.

7. Does the number of internal organs determine the number of people that these "twins" are? In the articles in the medical journal which focused on the heart organ in these twins,[68] it was found that in some cases two hearts were connected together. There were also cases of two separate hearts and cases of only one heart and sometimes it was large or abnormal. The case of a child who was born with two heads at the Hadassah Hospital in Jerusalem in January 1999, possessed only one heart. However, he died immediately after birth.[69] There are also articles about the number of lungs, livers, etc. in these twins. In many cases they are less than what is found in two humans. However, the Gemara does not

even hint at the possibility that the number of internal organs for a person with two heads will affect whether he is to be considered as two people or as one person.

8.   In the discussion held by Rabbi Moshe Feinstein, his son-in-law Rabbi Moshe Tendler, and other Rabbis in his family regarding the conjoined twins who were joined at the chest, they put the same question to Dr. C. Everett Koop "time and time again… in different ways because the answer would be so important to the rabbinical discussion that would ensue. Are the twins one baby or two babies? If the twins were only one baby with two heads, then it would be ethical to remove Baby A as an unnecessary appendage. However, if there were two babies with distinct nervous system, then it would require a more scholarly discussion."[70] Apparently, they did not rule out the possibility of seeing the twins as one person even if they had two separate nervous systems.[71]

Separating the Twins

With the development of medicine, today it is possible to do more complicated surgeries than in the past and thereby separate Siamese twins, including those twins who have two heads. There are two ways to do such a separation.

The first method is to cut along the length of the body and this will result in having two humans, however they will be missing several organs. Each will only have one leg and at least one of them will be missing several limbs, especially in the lower part of the body. Such operations were performed at a children's hospital in Great Ormond Street in London by Professor Lewis Spitz. In 1987 he separated two boys born eight months earlier in Sudan. The operation was successful.[72] However, they were only connected from the chest down, and this did not accord to the case mentioned above in the Gemara.

A few years later, in 1992, he separated the twins Katie and Eilish Holton when they were three and a half years old. However, a few days after the operation, Katie died. Katie and Eilish were joined from the shoulder down. With this type of surgery there is a chance that the separation will be successful for both twins and Katie's death was not expected.[73] According to the opinion that says that these "twins" are considered as one person, it is necessary to investigate and find out what will be the status of each one after the operation?!

The second method is to cut off one of the heads and arrange the body in such a way that the other head will be in the middle of the body. By such an operation, one of the heads is "killed" and then halachic questions arise. An example is that since only one head is "real," how can one know which of them is the "real" one?

[We should note that this is not similar to the case that came before Rabbi Moshe Feinstein with the Siamese twins who were joined at the chest. They were undoubtedly two people. One girl had a normal heart with four chambers and the other a heart with only two chambers and the fact of an incomplete heart could cause the death of the twin with the normal heart. However, in our case there was *one* person but one did not know which the real head was.]

There are not only ethical,[74] but there are also legal questions such as "killing" one of the heads, which can lead to a criminal case. When we know in advance that one of them will die during the operation, it is highly desirable that the doctor get permission from the court so that he does not face criminal charges.[75] We note that in the case in Germany, the court decided that "the twins were not two human beings but one monster from which a normal human being might be formed." [76] Such an operation was performed in Italy in 1998 where the doctor successfully cut off the head, which he said was the "secondary," and transplanted some of the organs that were in it into the "primary" head.[77]

The answer to the question of Pelemo

According to the language of the Gemara, one who has two heads is regarded as *one person*. The fact that the Gemara asks the question on which head he will place a tefillin proves that he is only one person. From the wording of Pelemo's question, namely "on which head to place the tefillin" one can see that there is no obligation to place tefillin on both heads. The question is *on which head*?

Rabbi Dov Berish Weidenfeld, (19th-20th centuries) the Chief Rabbi of Tshebin, and author of the book "Dovev Meisharim" asks the meaning of Pelemo's question: "What is the reason to give priority to one head over the other and what can one discuss about it? Maybe one could say, it could be that in the doubling of organs, one of them will be smaller than the other. [Pelemo] thus asked if being bigger is indeed the criterion, then one should place the tefillin on the larger head since this proves that this is the primary one."[78] One can note that one finds in the medical literature that there is indeed a small difference in size between the two heads.[79] In the case in Italy (quoted above) the doctor decided that one head was the main one and the other was secondary. However, the "Dovev Meisharim " does not reach a conclusion in his answer to Pelemo's question.

In contrast however, Rabbi Spira, the Gaon of Munkatch does come to a conclusion. He states that since the Gemara does not give an answer to the question of Pelemo, and because the mitzvot of tefillin is from the Torah, one should be strict and place tefillin on both heads.[80]

Other Mitzvot

Tefillin is not the only mitzvah performed by utilising parts of the head. There are many other mitzvot that are fulfilled by speaking, eating, hearing, etc. Some of these mitzvot are from the Torah and some are Rabbinical. The question is how can a person with two heads fulfill these mitzvot?

We note that today with the development of medicine, this is not an academic question! There are "twins" in the world with two heads who have been living for many years, an example being the twins of the Hensel family mentioned above. If, Heaven forbid, such twins are born in a Jewish family and reach the age of mitzvot (or even the age of training for mitzvot), one would need to find answers to these questions.

Let us look at a number of examples according to *the principles established by the Gaon of Munkatch* in connection with the head tefillin.[81]

Mitzvot that are fulfilled by speaking

Kiddush: Kiddush on the night of Shabbat is a Torah mitzvah.[82] For those who have already prayed ma'ariv (evening service) on Shabbat night, the mitzvah will be Rabbinical.[83] Rabbi Moshe Sofer known as the "Chatam Sofer" wrote that he had the intention of not fulfilling the mitzvah of kiddush whilst saying ma'ariv, in order that the kiddush on the wine would be for him a mitzvah from the Torah.[84] If a person with two heads wants to fulfill the mitzvot of kiddush as a Torah mitzvah, he has two options. One is that each of his two heads will recite the kiddush. The second option is that someone else will recite the kiddush and the two-headed person will have the intention to fulfil the mitzvah by listening using the principle of "listening is as saying it oneself." One should note that one mouth cannot recite kiddush and the second one fulfils the mitzvah by listening. The reason for this is that the mouth who recited the kiddush might not have been the "true" mouth, and thus could not make the "true" mouth fulfil the mitzvah! In contrast the morning Shabbat kiddush is Rabbinical,[85] and so one of the mouths could recite kiddush for the second mouth under the principle that "one is lenient in a doubt for a Rabbinical mitzvah."

Reciting the Shema: This is a mitzvah from the Torah,[86] so each of the mouths must read the Shema. However, regarding the question of whether to hear it from someone else, in accordance to the rule of "listening is as saying it oneself," there is a problem according to some opinions. Although the majority of the Acharonim (great Rabbis who lived from about the sixteenth century) hold that "listening is as saying it oneself," in reading the shema[87] there are other opinions that do not allow it.[88]

Berachot: Almost every berachah is Rabbinical.[89] As we wrote for the Shabbat morning kiddush, it is possible that only one of the heads need recite the berachot.

Birchot hanehenin: There are discussions among the Poskim (Rabbinical arbiters) whether one requires pleasure in the throat, or alternatively in the intestines when eating something.[90] Therefore it is best that the mouth that eats the food recites the berachot.

Mitzvot that are fulfilled by eating

Eating matzah on Seder night: On the first night of Pesach, it is a Torah mitzvah to eat matzah, as it states "In the evening you shall eat matzah."[91] In order to fulfill this mitzvah, one must eat it through the normal method of eating. It is written in the Gemara that if you wrap the matzah in something and then swallow it, one does not fulfill the mitzvah.[92] One learns from this that if a person is fed through a tube into his stomach (zonda) he does not fulfil the mitzvah.[93] In the "two-headed" situation, the question arises, in which mouth should he eat the matzah? One of the throats is not considered the real throat and eating it will be like eating it via a zonda. Therefore, because eating matzah on the first night of Pesach is from the Torah, one should eat a quantity of matzah (at least) the size of an olive via each of the two mouths. However, since the berachot before eating the matzah are Rabbinical, only one of the mouths needs to recite the berachot.

Eating maror and drinking the four cups of wine at the Seder: Because nowadays eating maror is a Rabbinical mitzvah,[94] it is possible to be lenient and just one of the mouths should eat it. The question arises, if he ate half the size of an olive via one mouth and half the size of an olive via the other mouth, would he fulfill the mitzvah? Probably not, because only one of the mouths is regarded as his mouth, and he did not eat the size of a whole olive in either of his mouths. Regarding the drinking of the four cups of wine on Seder night, which is a Rabbinical mitzvah,[95] one needs to drink all four glasses via the same mouth.

Mitzvot that are observed by hearing

Blowing the Shofar: On the first day of Rosh Hashanah, it is a Torah mitzvah.[96] That is why one must hear sounds from the shofar via the ears of both heads. If, for example, the ears are covered on one of the heads, there will be doubt as to whether or not the mitzvah has been fulfilled.

Mitzvahs that are observed by the sense of smell

The berachot on smelling: It is forbidden to enjoy a good smell until one recites the berachah beforehand.[97] That is why the mouth of the head containing the nose which smells is the one that must recite the berachah.

Epilogue

Let's pray that there will be no cases of a Jewish child being born with two heads, and all the discussions concerning the observance of the mitzvot by such people will be only theoretical!

References

Abbreviations
SA = Shulchan Aruch
OC = Orach Chaim
YD = Yoreh De'ah
(1) Talmud Bavli, Menachot 37a
(2) Rabbi Yosef Ba-Gad, *Nahalei Haeshkolot*, (1989?). p.88
(3) Prologue to the Zohar, par.154
(4) Talmud Bavli, Menachot 37a
(5) SA YD chap.334 par.47
(6) Beit Yosef on the Tur YD chap.334 first words: katav haRa'avad
(7) Rabbi Rafael Halperin, *Atlas Eytz Chaim*, vol.4, Tannaim and Amoraim (2), p.278
(8) Talmud Bavli, Berachot 48b; Tur OC chap.187
(9) Talmud Bavli, Pesachim 8b; SA OC chap.433 par.7
(10) Talmud Bavli, Kiddushin 81a; *Atlas Eytz Chaim*, op. cit.
(11) The question of whether such a child is regarded as a "treifah" is referred to in the Gemara (Menachot 37a), and at a later date by various Poskim in the course of the generations until this very today.

(12)   Talmud Bavli, Menachot 37a-37b

(13)   Beit Yosef on the Tur YD chap.305, first words: v'im yesh lo b' roshim

(14)   Tur YD chap.305

(15)   Rambam, Maimonides, *Pirkei Moshe,* (published by Rabbi Tzemach Magid: Vilna, 5648), paper 24, piska rishona, p.50; ibid., (published by Dr. Zissman Montner, Mossad Harav Kook: Jerusalem, 5742), p.302

(16)   Maimonides' Medical Writings, translated and annotated by Fred Rosner, Haifa, 1989, pp. xii-xvii

(17)   There are other cases where the Rambam used his scientific knowledge when writing his Mishneh Torah. For example, he omitted the condition that the chilazon is to be found only once in seventy (or seven) years (Mishneh Torah hilchot tzitzit, chap.2 halacha 2). Also zoologists today say that they do not know about fish that are only to be found after an interval of a number of years! Also, the Rambam omitted to mention the hare and the rabbit as chewers of the cud (ibid., hilchot ma'achlot assurot, chap.1 halacha 2). It is possible to say that the reason that the Torah writes that they are chewers of the cud is that it appears to the viewers that they are chewing the cud because of the movement of their mouths, and the Torah speaks in the language of man!

(18)   *Ammianus Marcellinus* vol.1, (William Heinemann: London, 1971), pp.542-44

(19)   *The Scriptores Historiae Augustae,* vol. 1, (William Heinemann: London, 1922), p.122

(20)   *Livy,* vol. xii, book xli, (William Heinemann: London, 1964), p.254

(21)   Saint Augustine. *The City of God,* ed. Vernon J. Bourke, (Image Books: New York, 1958), book xvi, chap.8, p.366

(22)   Pierre Boaistuau, *Histoires Prodigieuses,* (Paris, 1560, reprinted 1961), p.69

(23)   Fortunio Liceti. Picture from Internet: www.bium.univ-paris5.fr/monstres/biblio/pres-boaistuau04.htm

(24)   Boaistuau, op. cit. p.219

(25)   Jan Bondeson, "Dicephalus Conjoined Twins: A Historical Review With Emphasis on Viability," *Journal of Pediatric Surgery,* vol.36, no.9, (September), 2001, pp.1436-37.

(26)   Ibid., pp.1437-8.

(27)   *The Workes of that famous Chirurigion Ambrose Parey,* (London, 1665), p.644

(28)   Ibid.

(29)   George M. Gould and Walter L. Pyle, *Anomalies and Curiosities of Medicine,* (Bell Publishing Co: New York, 1896, republished by Julian Press, 1956), pp.164-65

(30)   Parey, op. cit., p.643

(31)   Robert P. Harris, "The Blended Tocci Brothers and their Historical Analogues," *American Journal of Obstetrics,* vol.25, 4, April 1892, p.467

(32)   Barton Cooke Hirst and George A. Piersol, *Human Monstrosities,* part iv, (Lea Brothers: Philadelphia, 1893), pp.151-57, plates xxx, xxxi.

(33)   Ibid., p.153

(34)   Harris, op. cit., p.461

(35)   Gould, op. cit., pp.183-87

(36)   Shinichi Miyabara, Naomasa Okamoto, et al., "A Study on Dicephalic Monsters, especially with regard to Embryopathology," *Hiroshima Journal of Medical Sciences*, vol.22, no.4, December 1973, pp.377-395

(37)   Joseph R. Siebert, Geoffrey A. Machin, et al., "Anatomic Findings in Dicephalic Conjoined Twins: Implications for Morphogenesis," *Teratology* vol.40, 1989. pp.305-310; Jonathan I. Groner, Douglas W. Teske, et al., "Dicephalus Dipus Dibrachius: An Unusual Case of Conjoined Twins," *Journal of Pediatric Surgery*, vol.31, no.12, (December), 1996, pp.1698-1700; B. James Reaves, "Dicephalous Monster," *American Journal of Obstetrics and Gynecology*, vol.37, 1939, pp.166-67; E. S. Golladay, G. Doyne Williams, et al., "Dicephalus Dipus Conjoined Twins: A Surgical Separation and Review of Previously Reported Cases," *Journal of Pediatric Surgery*, vol.17. no.3, (June), 1982, pp.259-64.

(38)   Calum N. McFarlane, "A Case of Double Headed Monster," *Journal of Obstetrics and Gynecology*, vol.67, 1960, pp.615-17; Stanley Nowell and R. Owen-Jones, "Dicephalic Monster," *Journal of Obstetrics and Gynecology of the British Empire*, vol.54, 1947, pp.507-09; Lewis. Spitz, E. M. Stringe, et al., "Separation of Brachio-Thoraco-Omphalo-Ischiopagus Bipus Conjoined Twins," *Journal of Pediatric Surgery*, vol.29, no.4, April 1994, pp.477-81.

(39)   Lawrence W. Alderman, "A Case of Dicephalic Monstrosity Delivered without Medical Aid," *The Medical Journal of Australia*, vol.35, 1948, pp.531-32; N. A. Beischer. and D. W. Fortune, "Double Monsters," *Obstetrics and Gynecology*, (New York), vol.32, 1968, pp.158-70.

(40)   Jeong Wook Seo, Sung Sik Shin, et al, "Cardiovascular System in Conjoined Twins: An Analysis of 14 Korean Cases," *Teratology*, vol.32, 1985, pp.151-61.

(41)   M. A. El-Gohary, "Siamese twins in the United Arab Emirates," *Pediatric Surgery Int*, vol.13, 1998, pp.154-57.

(42)   T. Wilson Roddie, "Case of Uniumbilical-Dibrachi-Dicephalic Monster," *British Medical Journal*, 10 March 1956, pp.552-54

(43)   "tinoket ba'alat shnei roshim nolda b'Yerushalayim -v'niftara" *Ma'ariv*, 4 January 1999, p.20. The medical doctor informed me (in 2005) that this case does not appear in the medical literature

(44)   Golladay, op. cit., p.259

(45)   Groner, op. cit., p.1698; *Encyclopaedia Hebraica*, (Jerusalem, 5732), vol.24 column 86 (picture)

(46)   "Abigail and Brittany Hensel," Internet: www.refernce.com/browse/wiki/Abigail_and_Brittany_Hensel; "Conjoined Twins in the World A-Z," updated November 15 2005, Internet: mypage.direct.ca/c/csamson/multiples/worldconjoined.html

(47)   "Conjoined Twins..." Internet, op. cit.

(48) "Siamese Twins," *The New Encyclopaedia Britannica*, 15th edition, vol.20, p.414.

(49) *Pirkei Moshe*, op. cit.

(50) *Scriptores*, op. cit.

(51) *Livy*, op. cit.

(52) Groner, op. cit., p.1700

(53) El-Gohary, op. cit., p.156.

(54) Miyabara, op. cit., p.385.

(55) Teratogen is something that can cause birth defects or abnormalities in a developing embryo or fetus upon exposure.

(56) Teratogen, Internet: de.wikipedia.org/wiki/Teratogen

(57) Tosafot Menachot 37a; first words: oh kum geli

(58) Rabbi Bezalel Ashkenazi, *Shita Mekubetzet*, Menachot 37a par.18

(59) Aharon Yellinek, *Bet Hamidrash* , (Jerusalem, 5698), section 4, pp.151-52; Rabbi Moshe Gaster *Sefer haMaasiyot*, (Leipzig, 5684), pp.75, 150-51

(60) *Sefer haMaasiyot*, op. cit.

(61) in Rabbi Yehudah L. Zlotnik's, "Anashim ba'alei shnai roshim" there is a comparison between the different variants, *Sinai*, (Mossad Harav Kook: Jerusalem), 5706, vol.19, pp.17-25

(62) Rabbi J. D. Bleich, Survey of Recent Halakhic Periodical Literature, "Conjoined Twins," *Tradition*, vol.31, no.1, Fall 1996, (Rabbinical Council of America: New York), p.97

(63) Claudia Wallis, "The Most Intimate Bond," *Time*, vol.147, no.13, 25 March 1996, pp.38-41.

(64) *Life*, op. cit., pp.44-56

(65) Talmud Yerushalmi, Peah chap.2 halacha 4

(66) Rabbi Chaim Elazar Spira, *Ot Chaim veShalom*, (Beregaz, 5681), chap.27 par.9

(67) Rafeh organization, Jerusalem, Question which was asked by Medical Doctors, 6 Adar 5756, pp.2-3

(68) e.g. Leon Gerlis, J. Seo, et al, "Morphology of the Cardiovascular System in Conjoined Twins: Spatial and Sequential Segmental Arrangements in 36 Cases," *Teratology*, vol.47, 1993. p.103

(69) *Ma'ariv*, op. cit.

(70) Donald C. Drake, "One Must Die So the Other Might Live," *Nursing Forum*, (New Jersey), vol.xvi, no.3, 4, 1977, pp.233-34

(71) The ruling in this case does not appear in "Igrot Moshe" of Rabbi Moshe Feinstein

(72) L. Spitz, S. Capps, et al., "Xiphoomphaloischiopagus Tripus Conjoined Twins: Successful Separation Following Abdominal Wall Expansion," *Journal of Pediatric Surgery*, vol.26, no.1, (January), 1991, pp.26-29

(73) "Siamese twins" (of the Holton family), cable television, family channel, 13 December 1992

(74) C. K. Pepper, "Ethical and Moral Considerations in the Separation of Conjoined Twins," *Birth Defects Original Article Series*, vol.iii, no.1, April 1967, Conjoined Twins, pp.128-134; J. Raffensperger,

"A Philosophical approach to conjoined twins," *Pediatric Surgery Int*, vol.12, 1997, pp.249-255.

(75)   e.g. Drake, op. cit., pp.245-46.
(76)   Rabbi Bleich, op. cit., p.117.
(77)   "tinoket shenolda im shenai rishim nutcho – v'nitzlah" *Yediot Achronot*, 2 December 1998, p.22. The medical doctor informed me that this case does not appear in the medical literature.
(78)   Rabbi Berish Weidenfeld, *Shu't Dovev Meisharim*, (Jerusalem, 5743), part 3, chap.31, (p.27)
(79)   e.g. Miyabara, op cit., p.378; McFarlane, op. cit., p.615.
(80)   *Ot Chaim veShalom*, op. cit., (p.61)
(81)   it should be stressed that things brought in this paper are only for the purpose of study and not as a halachic ruling.
(82)   *Sefer haChinuch*, mitzvah 31
(83)   Rabbi Avraham Gombiner, *Magen Avraham*, SA OC chap.271 par.1
(84)   Rabbi Moshe Sofer, *Minhagei Chatam Sofer*, (Jerusalem, 5731), chap.5, b'yom haShabbat, par.4, (p.28)
(85)   *Mishnah Berurah* OC chap.289 par.3
(86)   *Sefer haChinuch*, mitzvah 420
(87)   *Mishnah Berurah* OC chap.61 par.40
(88)   see: *Shu't Eliyahu Mizrachi*, (Jerusalem, 5698), chap.42, (p.111)
(89)   Rambam, Maimonides, *Mishneh Torah*, hilchot Berachot chap.8 halacha 12
(90)   Rabbi Avraham Sofer Avraham, *Nishmat Avraham*, (Jerusalem, 5743), Chalek Orach Chaim, p.93
(91)   *Sefer haChinuch*, mitzvah 10
(92)   Talmud Bavli, Pesachim 115b
(93)   *Nishmat Avraham*, op. cit., p.270
(94)   Rambam, Maimonides, *Mishneh Torah*, hilchot Chametz uMatzah chap.7 halacha 12
(95)   Ibid., halacha 7
(96)   *Sefer haChinuch*, mitzvah 405
(97)   SA OC chap.216 par.1

THE DATE FOR STARTING TO ASK
FOR RAIN IN THE DIASPORA

[The term "Tekufah" is continually used throughout this paper.

The solar year is divided up into four tekufot which are not of equal length, and they mark the beginning of the four seasons. They are:

Tekufah of Nissan – the Vernal Equinox which is the beginning of Spring - March 21

Tekufah of Tammuz – the Summer Solstice which is the beginning of Summer – June 21

Tekufah of Tishrei – the Autumn Equinox which is the beginning of Autumn – September 23

Tekufah of Tevet – the Winter Solstice which is the beginning of Winter – December 22

The one most relevant to this paper is the Tekufah of Tishrei

❀ ❀ ❀ ❀ ❀ ❀ ❀ ❀ ❀ ❀ ❀

Rain is mentioned or asked for in three of the berachot of the amidah, and they will be referred to in this paper as follows:

"mechaye hameitim" – here rain is mentioned (but not asked for) "mashiv haruach umorid hageshem"

"barech aleinu" – here one asks for rain "v'tein tal umatar livrachah"

"shema koleinu" – in certain circumstances rain is asked for in this berachah

❀ ❀ ❀ ❀ ❀ ❀ ❀ ❀ ❀ ❀ ❀

Introduction

It is stated in Masechet (tractate) Berachot:[1] "If one made a mistake and did not mention the 'power of rains' in the berachah

61

'mechaye hameitim' or did not ask for rain in the berachah 'barech aleinu, one has to repeat the amidah." The date on which one begins to ask for rain is given in Masechet Ta'anit[2] namely, in Eretz-Yisrael one begins to ask on the seventh day of Marcheshvan,[3] and in the Diaspora "sixty days after the tekufah," namely the tekufah of Tishrei. However, there are several ways to interpret these words:

1) Does one begin to ask on the sixtieth day after the tekufah, or on the following day? There is a discussion in the Gemara[4] on this, and the conclusion is that one begins at the ma'ariv service of the sixtieth day.

2) Does one begin to count sixty days from the day that the tekufah occurred, or from the following day? There is also a discussion on this in the Gemara. An answer can be found in Masechet Sanhedrin,[5] when the Gemara speaks of the decision to declare a year to be a leap year. Rav Yehuda believed that the day the tekufah occurred is considered as part of the previous tekufah, whereas Rav Yossi believed that the day the tekufah occurred is considered as the first day of the next tekufah. There is a rule in the Gemara[6] that when there is a dispute between "Rav Yehuda and Rav Yossi, the law goes according to Rav Yossi," thus the day that the tekufah occurs is considered as the new tekufah.

Rabbi Yitzchak ben Moshe (13th century), the author of the book "Or Zarua" uses this conclusion regarding the question of rain and he writes: [7] "Therefore on the question of asking for rain, namely 'tal umatar' (dew and rain) one starts on day sixty, in accordance with Rav Yossi". Likewise, in the books of a number of other Rishonim,[8] (great Rabbis who lived approximately between the 11th and 15th centuries) the same conclusion is reached.

3) There are those who want to count the sixty days "me'et le'et" (60 periods of 24 hours each). The source of this is Rabbi Eliezer ben Yoel Halevi ("Raviah,"12th-13th centuries) who quotes from the Talmud Yerushalmi":[9] "It is stated in the Talmud Yerushalmi that the sixty days from the tekufah of

Tishrei are calculated "ma'et la'et" (periods of 24 hours) and after one completes fifty-nine days, one begins on day sixty to pray for rain, namely, one begins to say 'v'tein tal umatar' sometimes one begins during the day and sometimes during the night." However, the "Korban Netanel"[10] who searched for this in the Talmud Yerushalmi wrote "I searched but did not find it." However a number of other Rishonim including "Hagahot Maimuniot,"[11] Rabbi Alexander Hakohen Zuslin known as the "Agudah"[12] (14th century), and Rabbi Yaakov Moelin who was known as the "Maharil"[13] (14th-15th centuries) bring this Yerushalmi. It is obvious that they had in front of them a different version, since in the course of generations a number of distortions entered the text. The Raviah even ruled according to this Yerushalmi. In addition, according to the Hanwa edition[14] (from the year 5571) the Maharil writes on this subject that the halacha did not rule according to any of the opinions, and one can therefore act according to any of them, namely from the outset one can act according to the Yerushalmi. In addition, in one of his responsa, he writes:[15] "There are places where it is customary to calculate "ma'et la'et" from the exact hour of the tekufah and therefore one should go and see what the custom in other places is."

Although there are other Rishonim who do not follow the Talmud Yerushalmi in this matter, there are Acharonim (great Rabbis who lived from about the 16th century) including Rabbi Hezekiah da Silva[16] who is known as the "Pri Chadash," (17th century) and the "Chayei Adam,"[17] who b'diavad (in retrospect) rely on the opinion of the Raviah. In addition, Rabbi Yosef Ostreich[18] the "Leket Yosher" writes: "I copied from Rabbi Yuda Ovranik who copied Rabbi Pinchas who heard from the Gaon (the Vilna Gaon?) that once he forgot to say 'v'tein tal umatar' before 60 days 'ma'et la'et' from the tekufah and he said that one can rely on the Talmud Yerushalmi on an occasion of forgetfulness."

There are also a number of discussions in the Rabbinic literature about which places in the world the law of "sixty days after the tekufah" refers to. As we will see below, the main source for these discussions is the Rambam's (Maimonides) commentary on the Mishnah, and there even exists a copy of it in Arabic in the handwriting of the Rambam.[19] Because this manuscript is an original,[20] we know that it does not contain any distortions that go into manuscripts in the course of their being copied.

Asking for Rain in Locations near to Eretz Yisrael

The first person who made a study in depth of the question regarding the border between Eretz Yisrael and the Diaspora for the asking for rain was Rabbi Yehosef Schwartz and this was about one hundred and fifty years ago. He, in his generation was one of the great researchers on the questions concerning Eretz Yisrael, and he prepared a number of maps of Eretz Yisrael. In his book entitled "Divrei Yosef"[21] he argued that the question before us involved places near to Eretz Yisrael like the city of Aleppo (Halab), Damascus, Antioch, Tarshish, Ein Tov, Mount Sinai, the land of Ammon and Moav. He held that the date for the start of asking for rain does not depend on the sanctity of the Land but on the climate, and places near to Eretz Yisrael which have a climate like that of Eretz Yisrael should start to ask for rain on the seventh day of the month of Marcheshvan. He cites evidence from the Rambam's commentary on the Mishnah:[22] "And all this refers to Eretz Yisrael and all that is similar to it." In addition, there is indirect evidence from the Hilchot Nedarim in his Mishneh Torah:[23] "... he is forbidden until it rains, provided it rains from the second phase of the preliminary rainy season. In Eretz Yisrael and in the places close to it, this is from the twenty-third day of Marcheshvan onward..."

However, there seems to be a contradiction between the above and the Rambam's Mishneh Torah in Hilchot Tefillah:[24] "[Beginning] from the seventh of Marcheshvan, one asks for rain in the blessing of prosperity (barech aleinu) [and continues to do so] as long as one mentions the rain. Where does the above apply? To Eretz Yisrael. However, in Shin'ar, Syria, Egypt and areas adjacent to or similar to these, one asks for rain sixty days after

64

the autumnal equinox [tekufah of Tishrei]." The "Divrei Yosef" explains the contradiction and claims that according to the Rambam only the places in Syria which are far from Eretz Yisrael begin as in Bavel (Babylon) "but for the parts of Syria which are close to the border of Eretz Yisrael and have a similar climate to that of Eretz Yisrael, it is obvious that the Rambam's intention is that they begin to ask on the seventh of Marcheshvan, even though they are in Syria."

The "Divrei Yosef"[25] also gives a list of places near to Eretz Yisrael (really within the future borders of Eretz Yisrael as promised to Avraham Avinu, the Patriarch Abraham), which begin the asking for rain as in Eretz Yisrael. These are: all the places that are on the shores of the Great Sea (Mediterranean) and adjacent to it, from the River of Egypt (Nile?) and Wadi El-Rish until the south to Iskendrun (Alexandretta), which is located in the region of Antioch (but does not include the city of Antioch), whose location is today in southern Turkey; on the Eastern side until the border of Tadmor in the wilderness of Palmira which is situated today in Syria. On the southern side until Yam Suf Ezion Gever (Aqaba), and Suez (but excluding the city of Suez), and Mount Sinai; also included is the southern part of Syria including the city of Damascus. However, the cities of Aleppo, Antioch and Suez ask for rain as in the Diaspora.

In addition, Rabbi Yitzchak Nissim,[26] the author of "Yayin haTov" (a former Chief Rabbi of Eretz Yisrael), wrote in the year 5675, that the day when one starts to ask for rain does not depend on the sanctity of the land but on the climate, and in "the border of the Damascus wilderness" they ask as in the Diaspora.

From the book "Milei d'Ezra" written by Rabbi Ezra Hakohen Trab[27] we see that in practice in Damascus they began to ask for rain in Marcheshvan.

Asking for Rain in the Northern Hemisphere

In the entire northern part of the world, the seasons occur at the same time as in Eretz Yisrael. However, there are large differences in the climate in the different places and their requirements for rain are not uniform. There are different interpretations on the word "golah" that appears in the Gemara in Masechet Ta'anit, regarding the date when one begins to ask for rain. Rabbi Avraham ben Yitzchak of Narbonne[28] known as the "Eshkol" (12th century) writes that there are for example - the Geonim[29] (great Rabbis who lived between about the 6th to the 11th centuries) who interpret this word specifically as "Bavel" and others - for example "Sefer haOra"[30] attributed to Rashi, who interpret it as "kol hagalut" (the entire Diaspora).

A question which arises is what is the halacha in places where the climate is not like the climate of Bavel? This problem was realised since the time of the Amoraim (great Rabbis of the period of the Gemara about 200-500 C.E.), and the Gemara[31] discusses the city of Nineveh (its location is near Mosul in present-day Iraq) that needs rain in the summer, indeed "even at the period of the tekufah of Tammuz (the summer)." The conclusion of the Gemara is that in Nineveh one asks for rain in the summer but as "individuals" in the berachah "shema koleinu."

But what about places that need rain at the same period as in Eretz Yisrael? We see in the books of the Geonim and the Rishonim about places that begin to ask for rain at the same date as in Eretz Yisrael, namely the seventh of Marcheshvan. Rabbi Amram Gaon[32] writes: "In all places who use rain water for drinking and watering as in Eretz Yisrael and Francia (the predecessor of France and Germany), they pray for dew and rain on the seventh of Marcheshvan." In an addition to the composition "Hilchot Gedolot"[33] written at the period of the Geonim it states: "The custom in all of Africa" was to begin on the seventh of Marcheshvan. Rabbi Avraham haYarchi[34] (12th-13th centuries), author of the book "Hamanhig" also saw this custom in "Narbonne and its environs (southern France)," and Rabbenu Asher,[35] the "Rosh" (13th-14th centuries) in Montpellier (a city in southern France).

In addition, the Rosh[36] held that even after Pesach, people ask for rain in the berachah "barech aleinu" and he based his reasoning on the Rambam's commentary on the Mishnah. On this, the Rambam[37] writes: "But in other lands the asking for rain will be at the appropriate time for rains in that place." However, this contradicts what the Rambam wrote in his Mishneh Torah, and in order to resolve this contradiction, the Rosh holds that the intention of the Rambam in the Mishneh Torah is only for distant islands but not for entire countries such as France or Ashkenaz (Germany). (However, this conclusion was not accepted by a number of great Rabbis. Rabbi Naftali Zvi Yehuda Berlin[38] known as the "Netziv" (19th century), explicitly rejects it and gives reason for his conclusion from the laws regarding the fasts for rain.)

The Rosh writes: "in Ashkenaz, where it is impossible for the grain to grow without rain between Pesach and Atzeret (Shavuot), rain at that period would have been a sign of blessing for them. Therefore why not both ask and make mention of it until Atzeret, and then the laws regarding not saying "morid hageshem" from Atzeret onwards will be like in Eretz Yisrael not saying it from Pesach onwards." He put forward these suggestions to Ashkenazi Rabbis "but there was not a man who challenged my words, but they said that we do not need to change the Ashkenazi custom since in any case the cessation of rain does not occur, and also on many occasions the grain is ruined because of an excess of rain." However, because the people did not accept his opinion and he did not want to make splits between the various groups of people, he retracted what he had said.

In addition, the Rabbi Yom Tov Ashevilli[39] known as the "Ritva" (13th-14th centuries) wrote at length on the subject, but came to a different conclusion: "My Rabbi would say that the opinion of the Geonim that according to the Talmud the Rabbis only fixed two dates for beginning asking for rain, one for Bavel and one for Eretz Yisrael, and the rest of the world follow either Bavel or Eretz Yisrael."

In an article on the subject, Rabbi Avraham Rapoport[40] explains the difference between the opinions of the Rosh and that of the Ritva. According to the Rosh, the principle held by the Rabbis is to ask for rain "when rain is needed in a particular place, and one asks according to what is needed in that place" and therefore one fixes the time according to the climatic conditions in that place. However according to the Ritva in the name of his teacher "times were fixed for the asking for rain in Eretz Yisrael and in Bavel, and this was the very essence of the decree of the Rabbis in this matter" and therefore one cannot fix additional times.

But the question arises. Why in places in the Diaspora that need rain before sixty days after the tekufah, does one wait until the seventh of Marcheshvan? The reason for delaying asking for rain from the end of Sukkot until the seventh of Marcheshvan was to give time for those who made the pilgrimage to Jerusalem for Sukkot to return home before the rain started.[41] Therefore what is the connection between those who made the pilgrimage and those living in the Diaspora. Why then should they not begin on the termination of Sukkot to ask for rain?

Rabbi Shlomo Zevin[42] (20th century) suggests an answer to this question): "The answer is obvious, that this date serves as a spiritual link between them and Eretz Yisrael." One can find support for the opinion of Rabbi Zevin from the customs of the Jews of Djerba[43] (an island, which is part of Tunisia and is situated close to it). On the seventh of Marcheshvan, the day that the Jews of Djerba start asking for rain they do not say vidui and tachanun. The basis of this custom is out of joy because the weather of Djerba is like that of Eretz Yisrael, and therefore they act this way out of their love for Eretz Yisrael. Rabbi Chaim Palaji[44] (19th century) ruled that this custom is permitted after he had deliberated on this matter regarding Djerba and according to him: "There is insufficient rain there, and there is need for rain from the Heavens, and there is no prohibition to advance the asking for rain by following Eretz Yisrael."

However, one can learn another reason from Rabbi Zvi Hirsch Orenstein[45] (grandson of the "Yeshuat Yaakov"). He believes that according to Rabbi Yitzchak Alfasi, known as the "Rif" (one of the early Rishonim), "the essence of the asking for rain is after the tekufah of Tishrei ... and therefore it is good to delay the asking until the seventh of Marcheshvan so that it would always be after the tekufah of Tishrei, since if the tekufah of Nissan would occur at the end of the month of Nissan, then the tekufah of Tishrei would occur on the fourth or fifth of

Marcheshvan. According to this reasoning, if the tekufah of Tishrei is at the latest on the fourth or fifth of Marcheshvan, why then does one wait until the seventh of Marcheshvan? [In fact, this question was even stronger at the time of the Gemara. The reason is that due to changes in the number of days (as will explained below) between the dates of the Festivals in relation to the sun and between the dates of the tekufah of Shmuel, the latest date for the tekufah was at the beginning of the month of Marcheshvan, namely, about four to five days earlier than today]. Albeit Rabbi Orenstein[46] answered this question himself: "Since at the time that the Temple was standing, they would they would ask for rain on the seventh day of Marcheshvan, one does not establish a new date today since it was impossible for them to ask for rain from the time they said mashiv haruach." It is possible to use the same reasoning to answer why one begins to ask precisely on the seventh of Marcheshvan in specific places in the Diaspora.

Indeed, the Shulchan Aruch and its commentaries do not accept the opinion that it is possible to start the asking for rain in certain places in the Diaspora from the seventh of Marcheshvan. The Shulchan Aruch[47] states: "And we begin to ask for rain in the Diaspora in the ma'ariv (evening) service of day sixty after the tekufah of Tishrei, and in Eretz Yisrael one starts from the evening of the seventh of Marcheshvan." On this Rabbi David Halevi Segal (17th century), known as the "Taz"[48] writes "and us in the Disapora follow Bavel." Similarly, the Mishnah Berurah[49] writes: "And we in all the Diaspora go after the exiles of Bavel."

In addition, the "Divrei Yosef"[50] writes "that there are many strips of land and territories in the world whose nature is similar to the nature and climate in Eretz Yisrael, and these include the southern part of Italy, Sicily, Malta, Karafuu. These should be considered as the "golah" and thus ask for rain there sixty days after the tekufah."

The reason why we do not start asking for rain in these places on the seventh of Marcheshvan was given by Rabbi Shimon Sofer[51] (19th century), Av Beit Din of the Hungarian city of Eger (Erlau). He writes that because the asking for rain in places in the Diaspora in the month of Marcheshvan is not for the sake of the sprouting of the seeds, but to guard them from birds and mice, and one can find other ways to guard them. Therefore it is possible to begin the asking for rain sixty days after the tekufah.

Asking for Rain in the Southern Hemisphere

The big problem regarding the timing of the asking for rain is in the southern part of the world. In this area, the summer is between Tishrei and Nissan, and winter is between Nissan and Tishrei. Thus if one asks for rain at the same time as the rest of the world, one would be asking for it only in the summer!

The first Jewish settlement in this area was in the seventeenth century in Brazil in South America,. The settlers sent a question on this subject to Rabbi Chaim Shabtai,[52] (17th-18th centuries) the "Torat Chaim." They wrote that rain from Tishrei to Nissan, namely during their summer, "is very bad for them" and that the rain "will dampen the atmosphere and cause serious illnesses." Therefore, they wanted to "change the order of prayer regarding of the asking and mentioning of rain" and instead to ask for and mention rain only from Nissan to Tishrei.

In his answer, the "Torat Chaim" quotes the Rambam's commentary on the Mishnah regarding the damage that rains can cause in certain places. According to the Rambam: [53] There are places where Marcheshvan is in the summer and the rain is not a blessing but is destructive, and therefore how can people in such a place ask for rain in Marcheshvan, and doing so is that not a lie?" From this the "Torat Chaim" learns not to mention nor ask for rains in Brazil between Tishrei and Nissan.

He also brings proofs of this from the Gemara:[54] The reason for postponing asking for rain in Bavel until sixty days after the tekufah is that "there is produce in the fields," and in Eretz Yisrael the reason for postponing the asking for rain until the seventh of Marcheshvan is not to disturb the pilgrims (due to the effects of rain) who are returning home (after coming to Jerusalem for Sukkot).

Regarding the mention of rain in the berachah of "mechaye hameitim," he believes that the reason for mentioning rain in

Bavel from the last days of Sukkot is because the rains are not a sign of a curse. However, this is not so in Brazil where the rain in their summer is bad, so therefore he ruled not to mention it. However, this raises a question that according to the Rambam, rain in lands where the summer is in Marcheshvan is destructive. It is thus clear that this is a curse. However, the Rambam rules not to ask for rain, but he does not write not to mention rain.

On this subject, in a letter sent by Rabbi Shmuel Salant[55] (19th-20th centuries) to a Rabbi in Australia he made a distinction between the asking for rain and the mentioning of rain. He wrote "Indeed in saying 'mashiv haruach umorid hageshem' my intention was to behave there as in the whole world," and only because they were accustomed not to mention it, he did not rule to mention rain in the berachah "mechaye hameitim." In addition, the "Aruch Hashulchan"[56] writes: "... when the time has arrived for rain in Eretz Yisrael then everyone should mention it." It seems that "everyone" includes settlements in the southern hemisphere that already existed at the time of the "Aruch Hashulchan" (early 20th century). In a similar opinion, Rabbi Bezalel Stern in his book "Betzel Hachochma"[57] thinks that only while the rains are harmful to the entire world, as during the days of Sukkot, one does not mention them. "But while rains are good and beneficial to most of the world, then even in places where they are harmful, they should be mentioned." He also gives his opinion according to the "Torat Chaim" on the subject: In places where the rains are not harmful, one should ask for it from sixty days after the tekufah, and according to the Gemara,[58] mentioning something at the beginning of a prayer can be classed as propitiation, but this is not so at the end. However, in the opinion of the "Torat Chaim," in Brazil where they do not ask throughout the entire year, and thus there is no question there of propitiation, and because of this one does not make mention of rain.

Regarding the question in the winter months of the southern hemisphere, the "Torat Chaim" doubts the Rosh's understanding of the Rambam's commentary on the Mishnah on the question of asking for rain in the berachah "barech aleinu" after Pesach, and claims that the Rosh's understanding is "very forced." The "Torat

Chaim" therefore rules that they are considered as "individuals" and it should be asked in the berachah "shema koleinu."

Near Brazil there is a country called Argentina. There, the Jewish settlement began only in the middle of the nineteenth century and even by the twentieth century it was small. During that century, however, this settlement grew until today it numbers close to 300,000 people. Jews who came to Buenos Aires from Damascus sent a number of questions on various matters related to prayers to Rabbi Yehuda Hakohen Trab[59] who was a Rabbi in Damascus. On the question of rain, they wrote: "The community is accustomed to ask for rain from Marcheshvan onwards, and from Nissan onwards they say "barchenu" [the text that Sefaradim say in the summer for the berachah "barech aleinu] according to the customs of Damascus. Are they acting correctly?"

In a long responsum (printed in 5684), Rabbi Trav collected and compiled all the rulings of the Poskim (Rabbinical arbiters) on this subject, and its conclusion was similar to the ruling of the "Torat Chaim." In order to obtain the agreement of Rabbi Eliyahu Yaluz[60] who was head of the Rabbinical Court in Tiberias, Rabbi Trav sent him his responsum. Rabbi Yaluz agreed, but added that in the situation that the rains were needed "for some fruit ... then it seems that they should ask for rain from the time that they need it until Pesach in the berachah 'barech aleinu'." It was four years later that Rabbi Shaul David Sitton,[61] who was the Rabbi in Buenos Aires wrote: "I have acted this way for ten years." He added that he had sent his ruling on this issue to several Rabbis and they all agreed with him.

In a letter to Rabbi Yitzchak Weiss[62] (the "Minchat Yitzchak"), written in 5735 by Rabbi A. Ben-David, he wrote about the custom of asking for rains in the "Shuva Israel" congregation in the city of Buenos Aires. The founders of this community belonged to those who had arrived there from Syria, and among their great Rabbis was Rabbi Sitton. The custom in most of the older communities regarding the question of asking for rain was according to a ruling by Rabbi Sitton. (Indeed, Rabbi Trab, who mentioned in his answer the ruling of Rabbi Sitton, wrote that the community "did not listen to his words.") But with the founding of the "Shuva Yisrael" congregation in the middle of the twentieth century, the custom there was as in the northern hemisphere. Likewise, those who had become observant, and there were a large number of them, also did so. Rabbi Ben-David was satisfied with which ever opinion they followed, and the "Minchat Yitzchak" answered that they should follow according to most of the world.

Another community, however smaller, in South America who asked the same question was that of Uruguay. In his responsum given in 5696, Rabbi Ben Zion Uziel[63] wrote that they should follow the northern hemisphere. However, about twenty-five years later, Rabbi Shmuel Halevi Wozner[64] answered with a new opinion. He wrote that the not praying for rain between Pesach and Sukkot is only relevant to the north of the world, but "in a place where the seasons have completely changed, and the days of rain are after Pesach, and the days of the harvest are after Sukkot ... the Poskim had not dealt with this and I have not found any famous Posek who says that it is then forbidden to ask for rain." However, he did not want to rule like that "until a famous Posek agrees with me." The "Minchat Yitzchak"[65] rejected this opinion.

Like the countries in South America, Australia also is in the southern hemisphere. Jews began to settle in Australia in the early nineteenth century. Mostly they came from England, and so they asked the halakhic questions from the Chief Rabbi of England, who at that time was Rabbi Shlomo Herschel,[66] and his answer was like that of the "Torat Chaim's." The same answer was given in the middle of the nineteenth century by Rabbi Yehosef Schwartz,[67] and also by the Chief Rabbi of England, Rabbi Nathan Adler[68] to the community in Hobart, Tasmania. There is a letter[69] from 1848 from the president of the Jewish community in Melbourne in which he asked the question to Rabbi Adler. We do not have his answer, but apparently, there was no change in Rabbi Adler's opinion on the subject. All the answers to the above questions were based on the fact that rain falling during the summer is harmful.

About a century ago, Rabbi Avraham Aver Hirshowitz (19th-20th centuries) arrived in Melbourne. He was a great scholar and in Melbourne he set up his own Bet Din (Rabbinical Court) (which was outside the city's official Bet Din!). However, he held[70] that in Australia there was a need for rain between Tishrei and Nissan, and therefore it was necessary both to mention and to ask for rain as was done in the whole world. He submitted questions on this subject to a number of great Rabbis throughout the world.

The first answer he received was written in Elul 5652 by the Rabbi of Kovno, Rabbi Yitzchak Elchanan Spektor.[71] In it he agreed with Rabbi Hirshowitz: "Yes, this is also my opinion."

Rabbi Shmuel Salant[72] (19th-20th centuries) in a letter to Rabbi Hirshowitz and also to Rabbi Kalonymus Zeev (known as Wolf Davis) dated Kislev 5653 wrote that in the past he had received "many questions from different people who formerly lived in Jerusalem but now lived in Melbourne and Sydney and they wrote to him stating that rain from after Sukkot until the end of Nissan will be harmful to one's health." Therefore Rabbi Salant answered them that they should not change their custom as set by the Rabbis of London. However, he then received a letter from Rabbi Hirshowitz who said that rainy days from Sukkot until Pesach are good for both one's health and the land. He answered that "obviously and without doubt it is appropriate that rains should also be asked for in the berachah 'barech aleinu'."

Apparently, Rabbi Davis was not satisfied with Rabbi Salant's answer, and on the twenty-eighth of Shevat he wrote him a letter. We do not have this letter, but from Rabbi Salant's[73] answer (from twenty-sixth of Nissan) it is clear that Rabbi Davis had emphasised that the rain between Sukkot and Pesach is harmful. For this reason, Rabbi Salant went back on his first ruling and said that the people of Australia should not change their custom. This was not the end of the affair, since Rabbi Hirshowitz appealed on this ruling, and three months later Rabbi Salant[74] wrote to him: "I do not intend to reply now, especially since the letters I received are not identical since there are some who write that rain from the month of Tishrei until Nissan is harmful to the fields and to the health, and there are those who write the opposite," and he informed him that Rabbi Adler had ruled on this question.

Therefore Rabbi Salant answered "I do not want to get involved in this matter now." In the next letter, Rabbi Salant[75] wrote to the two Rabbis that they should "prepare a letter and give just one definite opinion of the climatic conditions of this country on the matter of rain from Tishrei to Nissan." The two Rabbis did so,[76] but we have no further answers from Rabbi Salant on the subject.

We see that this debate is not primarily a halakhic debate, but a debate among naturalists on whether or not rain in Australia between Sukkot and Pesach is good or harmful? This is also the subject of an argument between Rabbi Hirshowitz and Rabbi Naftali Adler, the Chief Rabbi of London (and the son of Rabbi Nathan Adler, the previous Chief Rabbi) and details can be found in letters between them from the years 5653.[77]

Rabbi Ari Leib Rashkes[78] wrote a long responsum to Rabbi Hirshowitz on the subject. He relied on the responsum of the "Torat Chaim" and the principle of not changing customs. However, he concluded[79] with an "advice and a trick" so that the worshipper would not enter a situation that would not fulfil his obligation with his prayer. Throughout the year the worshiper would say "mashiv haruach umorid haTAL," and on the question of asking for rain he will say throughout the year "v'tein TAL livrachah," because according to some Rishonim by this wording, the worshipper does not have to not pray a second time even in the rainy seasons.

Rabbi Bezalel Stern,[80] who was an Av Beit Din in Melbourne, and served as Rabbi of the Adat Yisrael congregation in that city between the years 5716 and 5728, wrote that all the time that he was there, they would ask for rain at the same time as at the north of the world. He researched the subject and came to a conclusion that supports this, and it is on the condition that the rains between Pesach and Sukkot are not harmful. Also, Rabbi Tzvi Pesach Frank[81] (19th-20th centuries) in a responsum to a Rabbi in New Zealand ruled like this, but he did not mention in the situation where the rain is harmful.

In contrast to what was written above, it seems from a ruling by Rabbi Shmuel Moholiver[82] (19th century) that even if the rains are a curse, from the outset one should ask for dew and rain.

According to him: "If one forgot to ask for dew and rain, in a place that the rain is nothing but a curse in the winter, one does not have to pray a second time," that is, only b'diavad one does not pray a second time.

What is the law in practice today? Rabbis who lived in Brazil,[83] Argentina,[84] Australia,[85] and South Africa[86] stated that in practice these countries follow the rest of the world.

We have seen that according to many Poskim, the question of asking for rain for Jews living in the south of the world goes according to the "majority the world." Therefore, the question arises that if the majority of the Diaspora Jews emigrated to the south of the world, what would be the law?! Does the location of the Jewish Diaspora population at a particular time determine, or is the location of the population at the period of the Gemara the determinant? It is possible that the answer depends on the difference in opinions between the Rosh and the Ritva, which were discussed above?

The Tekufot of Shmuel and Rav Adda

The question to be asked is what is the meaning of the word "tekufah" in the law of the asking for rain "sixty days after the tekufah"?

The solar year is divided into four seasons. Although these four seasons are not equal in length, Shmuel and Rav Adda bar Ahavah divided this year into four equal tekufot, and each of the tekufot was called the "average tekufah."

The tekufah of Shmuel appears in the Gemara,[87] and is of length exactly 91 days and seven and a half hours. According to Rabbi Avraham ibn Ezra,[88] Shmuel knew that his figure was not exact and he simplified in a way that the people of his generation would be able to understand it.

Nevertheless, Rav Adda gave the length for a tekufah that was more accurate. According to him, its length is 91 days 7 hours 519 parts 31 moments. (There are 1080 parts per hour, and 76 moments per part). This figure is not mentioned in the Talmud, but Rabbi Avraham Bar Chyiah[89] (who lived in the 11th century), saw this figure in the "Baraita d'Rav Adda." Also, Rabbenu Ovadia[90] (who lived in the fourteenth century in Egypt), in his commentary to Hilchot Kiddush Hachodesh of the Rambam's Mishneh Torah mentions this Baraita. However, for centuries this Baraita has not been extant.

Because the Hebrew calendar is based on the molad (mean conjunction – the "birth of the New Moon"), the difference between the time of the molad and the tekufah of Rav Adda repeats itself every nineteen years. [The reason: nineteen solar years (called the "small cycle") according to Rav Adda is exactly equal to the 235 lunar months - that is, 235 times the average time between two molads. (In 19 years there are 235 months - that is, 12 years with 12 months, and 7 years with 13 months.)] It follows that in relation to the Hebrew calendar, the tekufah of Rav Adda has not moved from the time of the Creation of the World. Therefore, if the tekufah in the law of asking for rain is according to Rav Adda, the date in the Hebrew calendar that one begins the asking for rain, will be repeated until the end of the world (within a day or two) after every 19 years.

In contrast, the tekufah of Shmuel is a little longer, and after a cycle of 19 years it will occur one hour and 485 parts later. Over a period of time, these "residues" accumulate, and the date of the beginning of asking for rain is slowly approaching Pesach!! It is possible to calculate when the date will reach Pesach and there will be no opportunity to request rain in the Diaspora!!

According to the Gregorian calendar (the calendar we use today) there are 28 days in the month of February in the centuries that are not wholly divisible by four, (namely, for the years 1700, 1800, and 1900, there were 28 days in February, and for the year 2000 there was 29 days.) One can see from this, that after 400 years, the date in the Gregorian calendar for starting to ask for rain will be three days later and in the course of 1,000 years will be 7.8 days later. [The reason for 7.8 days instead of 7.5 days is because the Gregorian year is 26 seconds longer than the solar year, and in the course of 3,323 years it accumulates to one day, which is equivalent to a correction of 0.3 days per 1,000 years]. However, the astronomical solar year is about 7 minutes shorter than Rav Adda's solar year - that is, close to 5 days in every 1000 years. Because the Hebrew calendar is based on the length of the year according to Rav Adda, Pesach occurs about 4.6 days later in relation to the sun, as compared to about 1000 years ago. Therefore the beginning of the asking for rain approaches Pesach by 3.2 days every thousand years.[91] This change was calculated on a computer by Charles Elkin, who came to the conclusion that in about 42,000 years the date of the beginning of the asking for rain will be Pesach!! [92]

Regarding the accumulation of hours over a period of time due to this slight inaccuracy, Rabbenu Ovadia[93] writes that before the tekufah of Nissan will occur in Iyar, "a 'great person' will come and let us know the law regarding this late occurrence." (However, Rashi[94] accepts the possibility that sometimes the tekufah of Nissan will occur in the month of Iyar); and the grandson of the "Yeshuat Yaakov"[95] brings the Gemara[96] that the duration of the world is six thousand years and during this time "the accumulation of this inaccuracy [i.e.one hour and 485 parts in every 19 years] will not exceed nineteen days."

The question to be asked is which tekufah does one use for the asking for rain, the tekufah of Shmuel or the tekufah of Rav Adda? On this, Rabbi Avraham bar Chyiah[97] in the eleventh century wrote that one begins to ask for rain sixty days after the tekufah of Tishrei in accordance with the tekufah of Shmuel. Furthermore, in the fourteenth century the "Tashbatz"[98] wrote: "And I heard from my holy teacher the late Rabbi Vidal Efraim (a pupil of Rabbi Nissim of Geronda the "Ran") that there is a tradition from the Geonim that we calculate sixty days from the tekufah of Shmuel and not from the tekufah of Rav Adda."

The Time of the Tekufah – its Location

One can see from the yearly calendars that the tekufah occurs at a certain time. To what place in the world does this time refer?

It seems that the first to mention this subject was Rabbi Avraham ibn Ezra,[99] in the middle of the twelfth century, and he wrote that the time of the tekufah of Shmuel refers to the clock in Jerusalem.

This question also appears in other books of the Rishonim, but in connection with the molad. Therefore the question arises whether it is the same place as for the tekufah? Rabbenu Yitzchak haYisraeli,[100] a student of the Rosh, in his book "Yesod Olam" which was written in the early fourteenth century, answers this in the affirmative: "At a certain place from the equator our ancestors established both the molad and the tekufah." Likewise, today, in the yearly calendar brought out by of Rabbi Yechiel Tucazinsky[101] it is clearly stated that the molad and the tekufah were established in the same place in the world, (in his opinion Jerusalem): "The average molad (and also the average tekufah) have been since the period of the Creation of the World, namely from the molad "baharad" (an abbreviation which gives the day of the week and the time of the first molad) until today located in Jerusalem."

The subject of the place where the molad was established is not mentioned at all by the Rambam. However, in other books of the Rishonim there is a dispute on this subject.

The places which appear in the various Rabbinical books[102] are:

1. The extreme east: 120 degrees east of Baghdad
2. Edge of the east: 90 or 114 degrees east of Jerusalem
3. Centre of Settlement ("naval of the land"): 24 degrees east of Jerusalem
4. Near to Nusaybin (then) in Bavel
5. Eretz Yisrael – Jerusalem or the eastern edge of Eretz Yisrael or Yavneh (where after the destruction of the Temple the Sanhedrin was situated)

Rabbi Azariah dei Rossi min ha'Adumim[103] (16th ecentury) in his book "Mtzaref l'Kesef" mentions only two of these places, namely Jerusalem and Nusaybin (it seems that he prefers Nusaybin), and we shall take these two locations as examples in our calculations.

Furthermore, the subject of the tekufah in connection with our clocks requires clarification. When the tekufah occurs at a specific time, the time is not according to our clocks, but is in relation to the tekufah of Nissan at the period of the Creation of the World. To translate that hour into our time we need to make the following corrections:

1) For every 15 degrees longitude around the world, the clock changes by one hour. To make it more convenient, the world is divided into 24 time zones, with an hour difference between one zone and the next. The time in a particular area is called the "official clock" (or "standard time"), and the real time is called the "local time." In Jerusalem, the difference between these two "times" is about 21 minutes, and in Nusaybin 46 minutes (assuming that Nusaybin is in the same "time zone" as Jerusalem).

2) Even though the length of daylight varies throughout the year, for the calculation of the average molad and the average tekufah, the night begins six hours (60 minutes per hour[104]) prior to midnight throughout the year.[105] However, midday, namely the time when the sun is at a maximum height, is not at 12.00 noon throughout the year, but moves from about a quarter of an hour both before and after 12:00 noon during the course of the year. To know when midday is on a particular day in the Hebrew calendar, it is necessary to utilise a solar calendar. The Gregorian calendar with a small correction (see above for details of this correction) will give this.

[To be precise, the time of midday today on a particular date on the solar year is not the same time of midday that was thousands of years ago on the same date on the solar calendar. However, in order not to complicate the calculations, we will not address this change.]

The times of the tekufot shown in our tables are according to the "time of the first tekufah" and its date was on the Tuesday night (strictly one should write night of Wednesday since the Jewish day of Wednesday begins on Tuesday night), 22 Adar in the first year of Creation of the World, [106] and so it is necessary to know when it was midday on the day of the first tekufah. The date in the Gregorian calendar together with the correction was 23 February.[107] On that day, midday according to the "local time" was 12.13.[108] Therefore we need to offset 13 minutes from the 21 minutes (mentioned above), in order to translate the time period to that of our clocks (the "official clock"). [Perhaps it is surprising that the day of the first average tekufah (23 February) did not occur on the actual day of the tekufah; and also the first average molad ("baharad") was not at the time of the actual molad. [109]

From the above, at the time when the tekufah according to "the time of the first tekufah" is (for example) 9.00, the correction that needs to be performed in Jerusalem is to deduct 8 minutes - (namely 21 minus 13 minutes) - so the tekufah according to the "official clock" is at 8.52, (and at Nusaybin is at 8.27).

From the above, the following question can be asked: In the year 5752, the tekufah of Tishrei was on the night of 30 Tishrei (October 7) at 8:52 PM in Jerusalem [or at 8:27 PM in Nusaybin]. Sixty days later was the 29 Kislev. At that time in Cairo it was already night, and therefore they began to ask for rain on the evening of 29 Kislev. However, in New York it is still the daytime of 28 Kislev. Was it necessary to start in New York asking for rain on the evening of 28 Kislev instead of the evening of 29 Kislev?

Maybe one can learn an answer from the time of the molad in Jerusalem (or in Nusaybin).[110] If, for example, the molad of Tishrei occurs at 12.30 on Thursday. Rosh Hashana will occur on Shabbat (due to the postponement of "yach adu"). In New York, however, it was still morning at that time, so the postponement of "yach adu" was not relevant. Will Rosh Hashanah in New York be on Thursday, namely two days before the Rosh Hashanah in Jerusalem?!! Obviously not! The day that Rosh Hashanah occurs is determined for the entire world according to the time of the molad in Jerusalem (or Nusaybin). It would seem that the same logic is used to determine the beginning of the question of asking for rain anywhere in the Diaspora.

We have seen above that there are Acharonim including the "Pri Chadash" and the "Chayei Adam" who rule b'diavad according to the Raviah in the name of the Talmud Yerushalmi, namely counting "ma'et la'et" to determine the start of asking for rain. According to this opinion it would appear that if the tekufah is at 8.52 in the morning in Jerusalem, in Cairo one would not repeat the shacharit amidah (provided one prayed before this time). However, in New York where it is still nighttime it would be necessary to repeat the shacharit amidah if one forgot to ask for rain. Maybe one can learn this from the law regarding the latest time for "kiddush levana" (blessing the Creator of the moon).[111] The latest time is 14 days 18 hours and 22 minutes after the time of the molad "and if the latest time is in Eretz Yisrael at the end of the night, in America it will be at the beginning of the night." [112]

So far we have made our discussion on the basis that the tekufah was established according to the opinions that its location is Jerusalem or Nusaybin. Although, Rabbi Yechiel Zilber[113] holds "that the beginning of the tekufah follows the sun from place to place," he adds regarding "danger at the beginning of the tekufah, etc.," namely that it is, "a custom not to drink water at the time of the tekufah."[114] It is possible to follow the rule that when there is a danger, one should be stricter than in the case of just a prohibition,[115] and therefore in the matter of not drinking water, different rules apply to the determination of the time of the tekufah. In connection with this custom, we see from the answer of Rabbi Menachem Mendel of Lubavitch[116] known as the "Tzemach Tzedek" (19th century) that one should be cautious and not drink water also in accordance with the calculation of the tekufah, from the beginning of that night (in addition to the customary calculation). R. Avraham Moshe Lunz[117] writes in his Hebrew calendar that one also to be careful in this matter during the tekufah of Rav Adda (in addition to the tekufah of Samuel).

Date for the Start of Asking for Rain in the Rabbinic Literature

Until near the end of the sixteenth century (and in many countries until a later period), the Julian calendar was in one in use. In this calendar, the length of the solar year was 365 and a quarter days exactly. There was a cycle of four years - in three of them there were 28 days in February, and in the fourth year 29 days. Because the length of the Julian year is exactly the length of the year according to Shmuel, the date of the tekufah of Tishrei, and therefore the date for the beginning for the asking of rain did not change in relation to the Julian calendar.

The first book in which there is a calendar date for the beginning of the asking for rain was written in the fourteenth century by Rabbi David ben Yosef Avudarham[118] (14th century): "And the sixtieth day will be the 22 day of November if during that year there were 28 days in February. But if February were to have 29 days, one would start to ask on 23 November." (At that time, the Julian calendar was still in use). However, if one studies his words carefully one can see that there is no compatibility between the words of the Avudarham and the mathematical calculation for two years out of every four years. According to a mathematical calculation that when the following February will have 29 days, sixty days after the tekufah of Tishrei will be the 23 February, and if February of that year has 29 days, sixty days after the tekufah will be 22 November. This is in fact the opposite of what the Avudarham wrote. Even if we say that at the period of Avudarham, the year began with the first of March (as there was a period when this was the case), it would not solve this problem, since the Avudarham uses the expression "that February was" namely the *past* February.

In order to resolve the problem, Rabbi Nachman Kahana[119] in his book "Orchot Chaim" writes: "I saw that a current scholar checked with the agreement of the great Rabbis of that generation including the Maharam Schick, that there was a printing error and it should have been written that if the month February has 29 days, then the previous asking for rain is on fifth of December." It is obvious that the intention of the "Orchot Chaim" who wrote in the nineteenth century (the Gregorian method was then in use, and hence a change of days of about 11 days) 5 December) refers to ma'ariv (evening service) of 4 December. Furthermore, R' Charles Elkin,[120] suggested that there was a mix up and instead of "was" one needs to write "will be." However, in the manuscripts[121] in our possession, and also in the various printings of the Avudarham,[122] no proof has been found of a printing error for this section of the Avudarham. One cannot exclude the possibility that that they had additional manuscripts which are no longer extant. The "Porat Yosef"[123] also quotes the Avudarham, but he explains that "the Avudarham's intention was the reverse."

About a hundred years after the Avudarham, Rabbi Shimon ben Tzemach[124] (the "Tashbatz") wrote a responsum on this subject. In it he gives a cycle of these four years with dates of the tekufot of Tishrei and the day on which one begins to ask for rain. This is his language (with the addition of punctuation and explanations): "The order of the four years is as follows: In the first year the tekufah of Tishrei is three hours of the day [9 o'clock in the morning]. In the second year it is at nine hours of the day [3 o'clock in the afternoon]. Both of them are on the twenty-fourth of September. [The sixtieth day after the tekufah is 22 November and this is the day that one begins to ask for rain in these years.] In the third year it is at three hours of the night [9 o'clock at night] on the twenty-fifth of September. [As is known according to Jewish practice (and also according to some other nations) the day begins at the beginning of the night; likewise, the "Tashbatz" begins on 25 September at six o'clock in the evening of 24 September, in place of the usual custom today to begin the new date at midnight.] And sixty days after the tekufah will be on the twenty-third of November. [The reason that it occurs on a day later 23 November in place of 22 November is that nine o'clock at night is already after dark and therefore it is a day later in the Jewish calendar.] In the fourth year it is nine hours into the night [three o'clock in the morning] on the twenty-fourth of September because one has already added a day to the month of February. [the reason is that after adding a adding a day in February, sixty days from the tekufah will again be 22 November]. This sequence is then repeated, [namely, the cycle of four years begins again]." In addition, the "Tashbatz" concluded his responsum that in the year 5193 one began to ask for rain on twenty-third of November, namely 23 November 1431, the following February had 29 days.

Unlike the Avudarham, the calculations of the "Tashbatz" are exactly according to the mathematical calculation. He also mentions places that were already wrong regarding the date. The "Tashbatz" wrote in the same responsum: " Rabbi Yitzchak ben Sheshet (14th century) said to me that in Saragossa in Aragon they are accustomed to start on the twenty-fourth of November asking for rain and it seems to me that they made a mistake that they say that the day of the tekufah has ended and they thus start counting after this day saying that day sixty is as if it was before sixty, and all this is not correct according to Rav Yossi (of the Gemara) that the day the tekufah then begins, and he also holds that day sixty is like after sixty." Also, the question arises whether in the year after, that February was 29 days, they started asking for rain on 25 November? It is also not clear from the language of the "Tashbatz" whether he only brings the custom in Saragossa in Aragon for most years (namely three in a cycle of four years or they made another mistake and did not delay asking for rain by one day in the year following February with 29 days.

Also the book by the Maharil"[125] contains the date for the beginning of asking for rain. In this book appears the customs of the Maharil which was written by his disciple Zalman after the death of the Maharil in the year 5127. However, in this book there is only one date given for all the years: "one always asks on twenty-second of November of the secular calendar." The question arises: Is this date according to the Maharil (written by his disciple Zalman), or is it a later addition. A perusal of the editions of this book[126] and also in some of the manuscripts[127] shows that a date in November is given. However, there is also a manuscript[128] in which no date is given, and therefore it is possible that this is a later addition. In one of the manuscripts[129] of "Sefer Hamaharil" there is an addition: "And I who wrote this book found that in the year 5232 it did not state asking for rain on twenty-second of November but on the twenty-third of November, signed Asher ben Yechiel Hakohen Rapa also known as Anshkin Yitz." November of the year 5232 occurred in the year 1571 and the following February had twenty-nine days, and therefore this behavior in the year 1571 was correct.

In the middle of the sixteenth century, the "Beit Yosef"[130] in his commentary on the Tur quoted the Avudarham. We know, however, that the Beit Yosef. saw the responsa of the "Tashbatz"[131] because he brings them in his commentary to the Tur. Therefore, the question arises: Why does he quote the Avudarham and not the "Tashbatz"? (However, if the world had followed Avudarham, they would have been wrong every two years out of the four!) In the Shulchan Aruch the Beit Yosef does not mention the dates in November.

In the eighteenth century Rabbi Shmuel ben Natan Neta Halevi Kelin[132] (18th-19th centuries), who wrote the "Machazit haShekel" quoted the Beit Yosef in the name of the Avudarham without any suggestion that the words of the Avudarham were inaccurate. The "Machazit haShekel" was born and studied in Bohemia, and was a Rosh Yeshiva in Moravia. In these two places the transition from the Julian calendar to the Gregorian calendar was in the year 1584.[133] It is therefore interesting to note that in the middle of the eighteenth century, the "Machazit Haskekel" brought dates from the Julian calendar. Maybe his reasoning was that only in the Julian calendar was it possible to write a fixed date for all time.

A date also appears in the "Kitzur Shulchan Aruch" which was written in the nineteenth century. According to the author:[134] "One begins to mention, 'tal umatar' (dew and rain) in the ma'ariv service of day sixty after the tekufah of Tishrei, and it occurs on the fourth or fifth of December." (It seems that his intention that when he wrote the fourth of December, he meant that the fourth of December starts at six o'clock in the evening of the third of December; in the nineteenth century the prayer for rain started at ma'ariv of the third of December and in a year when the following February had twenty-nine days one began on the fourth of December). However, the Kitzur Shulchan Aruch does not explain on which years one begins on ma'ariv of the third of December and in which years on the fourth of December. In his notes on the Kitzur Shulchan Aruch, Rabbi David (son of Rabbi Sraya) Davlitzki[135] adds "beginning from 1900 to 2100 – fifth of December (when the following February has twenty-eight days) or the sixth of December (when the following February has twenty-nine days).

In the edition of the Kitzur Shulchan Aruch brought out by Rabbi David Feldman,[136] there is a table of days from 5711 to 5740 of dates when one "begins to ask for dew and rain in the previous ma'ariv." A perusal of the table proves that it is consistent with the text of the "Tashbatz." In his commentary on the Kitzur Shulchan Aruch[137] he quotes Rabbi Nachman Kahana:[138] "When we will arrive at the year 5660 which in the secular calendar will be 1900, then the thirteen days between the old calculation (Julian) to the calculation which we use today (Gregorian), from that time onwards the tekufah of Tishrei will occur in a simple year (solar non-leap year) on the seventh of October and the start of asking for rain will be on the fifth of December, which on the old calculation (Julian calendar) is the twenty-second of November, and every fourth year which will be a solar leap year the tekufah of Tishrei will be on the eighth of October. Thus the day when one begins to ask for rain is on the sixth of December which is the twenty-third of November on the Julian calendar. On a first reading it seems that Rabbi Kahana is following the Avudarham. However, it is difficult to accept this, since he himself says that in the same paragraph that that there is a misprint in the Avudarham. In order to resolve this problem, one must say that Rabbi Kahana does not utilise the secular year from January to December as is customary today, but he counts it from the first of March in accordance with the old system. A proof of this is that in all this section he equates the date with "the old system." In accordance with the mathematical calculation, in a year which begins on the first of March, which has twenty-eight days in February one begins asking for rain) in the twentieth and twenty-first centuries) on fifth of December (ma'ariv of fourth of December) and in a year which has twenty-nine days in February, one begins on the sixth of December (ma'ariv of the fifth of December). All this accords with the words of Rabbi Kahana. (However, as explained above, it is not possible to use this explanation to explain the words of the Avudarham.)

87

To know in which years one has to start asking for rain on the fourth of December and which years on the fifth of December has a very practical halachic importance.[139] It is written in the Shulchan Aruch:[140] "If one does not ask for rain at a time when one is required to do so, one has to repeat the amidah (silent prayer recited at each service)." This means that if in a year when there twenty-nine days in February, one begins asking for rain on the evening of the fifth of December instead of the fourth of December, as prescribed by the Rabbis, then in the amidah for ma'ariv on the fourth of December and in the amidot for shacharit and minchah on the fifth of December, one would have to repeat the amidot.

It is also written in the Shulchan Aruch:[141] "If one asks for a rain on a day in the summer, one must repeat the amidah." Are the days between the end of Sukkot until sixty days after the tekufah considered "summer days"? According to Rabbi David ibn Zimra known as the "Radbaz,"[142] (15th-16th centuries), and the "Torat Chaim,"[143] who hold that rain during these days is not a sign of a curse, one does not have to repeat the amidah. However, the "Pri Chadash"[144] disagrees with the words of the "Radbaz." Also, the Mishnah Berurah[145] holds that if that country requires rain during those days, one should say the amidah again "but as a voluntary prayer." It seems from this that if one does not need rain, one must repeat the amidah in a regular manner. However Rabbi Yaakov Chaim Sofer[146] (19th-20th centuries) in his work "Kaf Hachaim" writes that in such a case "when he returns to repeat the amidah one should make it conditional, and add something new into the amidah." Therefore, according to a number of Poskim if in the following year there will be twenty-nine days in February, one begins asking for rain as ruled by the Rabbis, on the fourth of December instead of the fifth of December, then one has to repeat the amidah (either as a regular prayer or as a conditional one for ma'ariv on the fourth of December, and shacharit and minchah on the fifth of December.

From this we can see the importance of siddurim having precise instructions of when to start asking for rain. However, in practice only a few siddurim give the correct instructions!!

One can divide the incorrect instructions or ambiguous instructions into several groups: [147]

1) "Starting on December 4th." This would mean that every year one begins on this date.
2) "Starting on December 4th or 5th." There is no detail as to which year it is the 4th and which year it is the 5th.
3) "Beginning in ma'ariv on December 3." Why December 3 in a siddur printed in the 20th century?!
4) "Starting in ma'ariv on December 4th and in a year that has 29 days in February on December 5th". As we saw above, quite the opposite.
5) "Starting on the 5th or sometimes on the 4th of December. Refer to the Hebrew calendar." Can someone in the middle of the amidah on the 4th or 5th of December stop praying to peruse the calendar?! In addition there are errors at this subject in a number of calendars that have been published today and in previous generations. For example, in 5640 Rabbi Shmuel David Halevi Jungreiz[148] received a question regarding a discrepancy in the date of commencement of the asking for rain between the calendar published in the city of Pest in Hungary, and the calendar in the city of Pressburg (now in) Slovakia; the first gave the date as ma'ariv of 3 December whilst the second gave the date as ma'ariv of 4 December.

Summary

- The custom today in almost the entire Diaspora, including the southern hemisphere of the world, is to begin asking for rain in ma'ariv on the sixtieth day after the tekufah of Tishrei.
- "Tekufah" in the laws of asking for rain is the average length of a tekufah, and its length is according to the opinion of Shmuel.
- The location of the place of the tekufah was established (among other places) in Jerusalem or in Nusaybin, and is calculated according to the "time of the first tekufah with the Creation of the World."

- The date of the beginning of asking for rain in the 20th and 21st centuries, is in ma'ariv of 4 December, and in the years that are followed by February with 29 days, at ma'ariv of 5 December.

References

Abbreviations
SA = Shulchan Aruch
OC = Orach Chaim
(1)   Talmud Bavli, Berachot 26b
(2)   Talmud Bavli, Ta'anit 10a
(3)   It is written in the Mishnah: "On the seventh [of Marcheshvan] fifteen days after the Festival of Sukkot [one starts to ask for rain]" (Ta'anit, chapter 1, Mishnah 3). This is true according to the fixed calendar of today. However, during the period of the Mishnah, the month was consecrated according to the sighting of the new moon, and therefore it was possible that the month of Tishrei would have only twenty-nine days. Therefore, fifteen days after the Festival could be either on the seventh or the eighth of Marcheshvan. The question arises, why does the Mishnah mention only mention the seventh of Marcheshvan? The Talmud Yerushalmi (Ta'anit chap.1 halacha 3 and Pnai Moshe) discusses this question that if the month of Tishrei was deficient and only had twenty-nine days, one would have to calculate accordingly. Indeed, there were different opinions on what to do. According to Rav Shimon ben Gamliel one would have to go according to the situation of that particular year and that fifteen days after Sukkot would be the eighth of Marcheshvan. In contrast, Rebbe holds that such a situation does not merit a change, and one does as in other years, since in most years Tishrei has indeed thirty days, and one therefore starts to ask for rain on the seventh of Marcheshvan. Why in most years has Tishrei thirty days? One could suggest the following. From the time of Ezra we do not find a case where Elul had thirty days." [Beitzah 6a]. However, from a number of Mishnayot [Sheviit chap.10 Mishnah 2, Eruvin chap.3 Mishnah 7, Rosh Hashanah chap.1 Mishnah 3] we can see that the possibility of thirty days in Elul does exist The average length of a month is twenty-nine and a half days. Therefore, if Elul always has twenty-nine days, in the majority of the years Tishrei will have thirty days. The "Tosafot Yom Tov" (Ta'anit chap.1 Mishnah 3) explains according to Rav Shimon ben Gamliel and writes "Everything depends on fifteen days after the Festival, and if Tishrei is lacking a day, one would start to ask on eighth of Marcheshvan."
(4)   Talmud Bavli, Ta'anit 10a
(5)   Talmud Bavli, Sanhedrin 13a
(6)   Talmud Bavli. Eruvin 46b
(7)   Rabbi Yitzchak ben Moshe, Or Zarua, vol.2., (Zhitomir, 5622), chap.400

(8)   e.g. Rabbi Shimon ben Zemach Duran, *Tashbatz*, vol.3, (Amsterdam, 5501), responsum 123; Rabbi Zedekiah ben Avraham Harofe Anaw, *Shibbolei haLeket*, (Vilna, 5647), chap.21

(9)   Rabbi Eliezer ben Yoel Halevi, *Sefer Raviah*, (Avigdor Aptowitzer: Jerusalem, 5724), part 2, chap.848, (p.594)

(10) Korban Netanel, Rosh, Masechet Ta'anit , chap.1 par.2, (p.32)

(11) Hagahot Maimuniot, Rambam, Mishne Torah, Hilchot Tefillah, chap.2 par.10

(12) Rabbi Alexander Hakohen Zuslin, *Sefer Haagudah*, (Krakow, 5331), Masechet Ta'anit, chap.1, (p.132)

(13) Rabbi Yaakov ben Moshe Levi Moelin, *She'elot uTeshuvot Maharil*, (Machon Yerushalayim, 5740), chap.21

(14) Ibid., (Hanuva, 5571)

(15) MS Moscow – Günzburg 604/2 folio 134b

(16) Rabbi Hezekiah da Silva, *Pri Chadash*, SA OC chap.117 par.1

(17) *Chayei Adam*, section 24, par.12

(18) Rabbi Yosef ben Moshe Ostreich, *Leket Yosher*, (Rabbi Dr. J. Freimann: Berlin, 5663), OC p.21

(19) MS Sassoon 72

(20) *Mishnah im Peirush HaRambam*, translated by Rabbi Yosef Kapach, (Mosad Harav Kook: Jerusalem, 5724), introduction, p.16

(21) Rabbi Yehosef Schwartz, *Divrei Yosef*, (Jerusalem, 5722), Responsa 4

(22) *Mishnah im Peirush HaRambam*, op. cit., Seder Moed, masechet Ta'anit , chap.1 par.3

(23) Rambam, Mishneh Torah, Hilchot Nedarim, chap.10 halacha 11

(24) Rambam, Mishneh Torah, Hilchot Tefillah, chap.2 halacha 16

(25) *Divrei Yosef*, op. cit., p.30a fn.1

(26) Rabbi Yitzchak Nissim, *Yayin Hatov*, Responsa, (Jerusalem, 5707), section 1, OC chap.35

(27) Rabbi Ezra Hakohen Trab, *Milei d'Ezra*, (Jerusalem, 5684), OC chap.10, (p.12a)

(28) Rabbi Avraham ben Yitzchak of Narbonne, *Sefer haEshkol*, (Vagshal Publishing House: Jerusalem, 5744), section 1, p.36

(29) Rabbi Yom Tov ben Avraham Asevilli, ("Ritva"), *Chidushei haRitva*, Ta'anit 10a

(30) *Sefer ha-Orah*, (attributed to Rashi), ("Yahadut": Bnei Brak, 5740), section 1, chap.30

(31) Talmud Bavli, Ta'anit 14b

(32) *Seder Rav Amram ha-Shalem*, (Jerusalem, 5672), chap 39 and variant readings, (p.123)

(33) *Halachot Gedolot*, (Hevrat Mekitze Nirdamim: Berlin, 5652), p.175

(34) Rabbi Avraham ben Natan haYarchei, *Sefer haManhig*, (Rabbi Yitzchak Refoel, Mosad Harav Kook: Jerusalem, 5728), part 1, p.90

(35) *She'elot uTeshuvot l'Rabbenu Asher* (haRosh), (Jerusalem, 5741), section 4, par.10

(36) Ibid.

(37) *Mishnah im Peirush HaRambam*, op. cit.
(38) Rabbi Naftali Zvi Yehuda Berlin ("Netziv"), *Meromei Sadeh*, (Jerusalem, 5715), part 2, Ta'anit 10a, (p.184)
(39) *Chidushei haRitva,* op. cit., Ta'anit 10a
(40) Rabbi Avraham Rapoport, "b'inyan zeman sheilat geshamim b'birchat hashanim b'chul", *Noam,* annual to investigate problems in halacha, vol.11, (Machon Torah Shelemah: Jerusalem, 5728), pp.90-91
(41) Talmud Bavli, Ta'anit 10a
(42) Rabbi Shlomo Yosef Zevin, *haMoadim baHalacha,* (Avraham Tzioni: Tel-Aviv, 5724), p.151
(43) Rabbi Kalfon Moshe Hakohen, *Berit Kehunah,* (Djerba, 5701), OC part 1, ma'arechet gimel, par.2
(44) Rabbi Chaim Palagi, *Artzot haChaim,* (Jerusalem, 5637), gate 10, chap.49
(45) Rabbi Tzvi Hirsch Orenstein, *Yeshuat Yaakov,* (Lorje and Sperling: Lemberg, 5623), OC end of chap.117, p.80a
(46) Ibid., p.80b
(47) SA OC chap.117 par.1
(48) Taz, SA OC chap.117 par.1
(49) Mishnah Berurah SA OC chap.117 par.5
(50) *Divrei Yosef*, op. cit., p.30
(51) Rabbi Shimon Sofer, *Hitorerut Teshuva,* (Jerusalem, 5734), chap.180
(52) Rabbi Chaim Shabtai ("Maharchash"), *Torat Chaim,* (Saloniki (Thessaloniki), 5482), part 3, chap 3
(53) *Mishnah im Peirush haRambam*, op. cit.
(54) Talmud Bavli, Ta'anit 4b
(55) Rabbi Avraham Eber Hirschowitz, *Beit Avraham,* (Jerusalem, 5668), part 1, second responsum by Rabbi Shmuel Salant, p.36
(56) Aruch Hashulchan, OC chap.114 par.3
(57) Rabbi Bezalel Stern, *Betzel Hachochmah,* (Jerusalem, 5750), part 6, chap.85
(58) Talmud Bavli, Ta'anit 4b
(59) *Milei d'Ezra,* op. cit.
(60) Approbation by Rabbi Eliyahu Yaluz to *Milei d'Ezra,* op. cit., p.19
(61) Rabbi Shaul David Sitton, *Diber Shaul,* (Jerusalem, 5688), OC chap.3
(62) Rabbi Yitzchak Yaakov Weiss, *Minchat Yitzchak,* (Jerusalem, 5749), vol.6, chap.171
(63) Rabbi Ben-Zion Meir Chai Uziel, *Mishpatei Uziel,* secomd edition, (Jerusalem, 5707), vol.1, OC chap.6
(64) Rabbi Shmuel Halevi Wosner, *Shevet Halevi,* (Bnei Brak, 5730), OC chap.21
(65) *Minchat Yitzchak,* op. cit.
(66) *Beit Avraham,* op. cit., p.39
(67) *Divrei Yosef,* op. cit.
(68) *Beit Avraham,* op. cit., pp.37, 39
(69) Lazarus Morris Goldman, *The Jews in Victoria in the Nineteenth Century,* (Melbourne, 1954), pp.58-59

(70) *Beit Avraham*, op. cit., p.34

(71) Responsum from Rabbi Yitzchak Elchanan Spektor, *Beit Avraham*, op. cit., pp.34-35

(72) First Responsum from Rabbi Shmuel Salant, *Beit Avraham*, op. cit., pp.35-36

(73) Second Responsum from Rabbi Shmuel Salant, *Beit Avraham*, op. cit., p.36

(74) Third Responsum from Rabbi Shmuel Salant, *Beit Avraham*, op. cit., p.37

(75) Fourth Responsum from Rabbi Shmuel Salant, *Beit Avraham*, op. cit., pp.37-38

(76) *Beit Avraham*, op. cit., p.38

(77) *Beit Avraham*, op. cit., pp.38-40

(78) Responsum from Rabbi Aryeh Leib Rashkas, *Beit Avraham*, op. cit., pp.41-47

(79) Ibid., pp.46-47

(80) *Betzel Hachochmah*, op. cit.

(81) Rabbi Tzvi Pesach Frank, *Har Tzvi*, (Jerusalem, 5737), OC chap.56

(82) Rabbi Shmuel Moholiver, *Chikrai Halacha Veshelot Vetshuvot*, (Mosad Harav Kook: Jerusalem, 5704), part 1, chap.1

(83) Information from Rabbi Eliahu Velet, Chief Rabbi of Sao Paulo, Brazil, Tevet 5752

(84) Information from Rabbi Mordechai Herbst, formerly Chief Rabbi of the Bet Haknesset Hagadol in Buenos Aires, Argentina, Tevet 5752

(85) Information from Rabbi Dr. Yehoshua Kemelman, formerly Av Bet Din in Sydney, Australia, Tevet 5752

(86) Information from Rabbi Moshe Shirkin, formerly from Cape Town, South Africa, Tevet 5752

(87) Talmud Bavli, Eruvin 56a

(88) Rabbi Avraham ibn Ezra, *Sefer Haibbur*, (Hevrat Mekitze Nirdamim: Lyck, 5634/1874), p.8; see also: *Peirush Harav ibn Ezra al haTorah*, Shemot chap.12 verse 2

(89) Rabbi Avraham bar Chyiah the Prince, *Sefer Haibbur*, (Longman: London, 5611), third paper, fourth gate, (p.87)

(90) Rabbenu Ovadia Peirush on Rambam Mishneh Torah, Hilchot Kiddush Hachodesh, chap.10, par.1

(91) Charles Elkin, "Birkath Hachamah: Blessing of the Sun", Proceedings of the Associations of Orthodox Jewish Scientists, vol. 6, (Feldheim Publishers: Jerusalem, 5741/1980), pp.108-09,

(92) Ibid., p.104

(93) Rabbenu Ovadia, Kiddush Hachodesh, op. cit., chap.9 par.3

(94) Rashi, Talmud Bavli, Rosh Hashanah 11a

(95) *Yeshuat Yaakov*, op. cit., p.80a

(96) e.g. Talmud Bavli, Sanhedrin 97a; see also Zohar, Bereshit p.128a

(97) Rabbi Avraham bar Chyiah, op. cit., p.86

(98) *Tashbatz*, op. cit.

(99) Rabbi Avraham ibn Ezra, op. cit., p.8b

(100) Rabbenu Yitzchak haYisraeli, *Yesod Olam*, (Berlin, 5537), second paper, chap.13, (p.33a)

(101) *Luach l'Eretz Yisrael*, arranged by Rabbi Yechiel Michel Tucazinsky, for example year 5710. Chodesh Tishrei

(102) for further details see: Chaim Simons, "hakesher bein hamolad uvein zeman kiddush levana", *Sinai*, (Mosad Harav Kook: Jerusalem), vol.117, pp.84-87

(103) Rabbi Azariah dei Rossi min ha'Adumim, *Mtzaref l'Kesef*, (Vilna, 5626), second paper, chap.5

(104) *Yesod Olam*, op. cit.

(105) Rabbi Mordecai Yoffe, *Levush*, OC chap.428 (end), first words: od yesh li; Rabbi Shimon ben Natan Neta Woltish, *Nvah Kodesh*, (Berlin, 5546), Rambam, Mishneh Torah, Hilchot Kiddush Hachodesh, chap.6 par.2

(106) Rabbenu Ovadia, Kiddush Hachodesh, op. cit., chap.9 par.3

(107) *Luach l'Sheshet Alafim Shanah,* arranged by Avraham Akavia, (Mosad Harav Kook: Jerusalem 5736), p.3, (date according to Julian calendar); Elkins, op. cit., p.99 (formula to convert dates on Julian calendar to that of Gregorian calendar)

(108) Rabbi Meir Posen, *Ohr Meir*, (London, 5733), tables

(109) Hugo Mandelbaum, "The Problem of Molad Tohu," *Proceedings of the Associations of Orthodox Jewish Scientists,* Volume 3 - 4 (Feldheim Publishers: Jerusalem, 5736/1976). pp.72-73,

(110) see: *Nvah Kodesh*, op. cit., par.6

(111) SA OC chap.426 par.3, Rama

(112) Rabbi Yechiel Avraham Zilber "Shigagat minhagim", *Otzrot Yerushalayim*, vol.72, (Jerusalem, 5731), p.1151; Yad L'Achim Wall Calendar, Adar 5750

(113) Harav Zilber, op. cit.

(114) SA Yoreh De'ah, chap 116, par.5, Rama

(115) Talmud Bavli, Chulin 10a

(116) Rabbi Menachem Mendel of Lubavitch, *Tzemach Tzedek*, she'elot uteshuvot, (Jerusalem, 5728), responsum 14

(117) *Luach l'Eretz Yisrael*, arranged by R' Avraham Moshe Lunz, (Ariel: Jerusalem), selection of papers, vol.2, pp.56-57

(118) Rabbi David ben Yosef Avudarham, Avudarham *Hashalem*, (Jerusalem, 5723), p.110

(119) Rabbi Nachman Kahana of Spinka, *Orchot Chaim*, (Sighet, 5658), chap.117 par.2

(120) Elkins, op. cit., p.104

(121) MS Budapest-Kaufmann 405/1 folio 51; MS London-British Museum, catalogue Margoliouth 1165; MS London British Museum, Or. 10727 folio 72b

(122) editions of Avudarham: Eshbona 5250, Kushta (Constantinople) 5274, Fez, 5277, Venice, 5306, Venice 5326, Prague 5544

(123) Rabbi Yosef Halevi Zweig, *Porat Yosef*, (Bilgoraj, 5693), OC chap.3

(124) *Tashbatz*, op. cit.

(125) *Sefer Maharil – Minhagim*, (Machon Yerushalayim, 5749), Hilchot Shabbat Bereshit, p.401

(126) editions of Sefer Maharil: Sabbioneta 5316, Cremona 5318, Cremona 5326, Lublin 5350, Hanuva 5388, Amsterdam 5490

(127) MS Bet Hamidrash laRabbanim New York, Rab.532; MS Vienna 77 folio 79a

(128) MS Frankfurt am Main 94/1 8^0 folio 123b and tables at the end; MS Parma 1421/1 folio 118

(129) MS Bet Hamidrash laRabbanim New York, Rab. 532

(130) Beit Yosef, Tur, OC chap.117 first words: v'katav R' D. Avudarham

(131) Ibid., e.g.: Even Ha'ezer chap.122, first word: v'katav; Even Ha'ezer chap.134, first word: b'inyan; Even Ha'ezer chap.143, first word: hamgaresh

(132) Rabbi Shmuel ben Natan Neta Halevi Kelin, *Machazit haShekel,* SA OC chap.117 par.1

(133) *Luach l'Sheshet Alafim Shanah*, op. cit., p.608

(134) *Kitzur Shulchan Aruch*, chap.19 par.5

(135) *Kitzur Shulchan Aruch* version edited by Rabbi David the son of Rabbi Sraya Davlitzki, (Bnei Brak, 5738), chap.19 par.5 footnote

(136) *Kitzur Shulchan Aruch* version edited by Rabbi David Feldman, (Manchester, 5711), p.164

(137) Ibid., chap.19 par.5, footnote – (two stars)

(138) *Orchot Chaim*, (Rabbi Kahana), op. cit.

(139) What is the situation when a person who is in doubt whether or not he has asked for rain when he says the amidah? The answer is that if he has already asked for it ninety times, he can assume that he has indeed not erred, and there is no need to repeat the amidah. There is also a rule that can one can say ninety times in a row the words "v'et kol minei tevuata letova, v'ten tal umatar livrocho" (Mishnah Berurah chap.114 par.40) and after that when in doubt one does not have to repeat the amidah

This rule also applies to other changes in the order of service. However, with regard to the "Hamelech Hakadosh" which is said during the "aseret yemai teshuvah," there is a problem, namely in saying the words "Baruch Ata Ad-noi Hamelech Hakadosh" ninety times in a row, since one would be saying the Divine name in vain which is strictly forbidden, and so this solution is not viable in this case. (Mishnah Berurah chap.582 par.3). However, there is room to investigate if in the following situations one would have to repeat the amidah if one was doubtful if one had said "Hamelch Hakodesh":

a) One says the amidah ninety times as a "tefillah n'dava" (voluntary prayer) on the fast of Gedalia!

b) A person who has not been observant and has never said the amidah, and he repented at the end of Elul and then started to pray each day! Such a person was, unlike other people, not accustomed to say "Ha–l Hakadosh."

c)   The "Tiferet Shmuel" writes (on the Rosh, Berachot, chap.1, par.40) that some say "Hamelech Hakadosh" on Hoshana Rabba even though it is not mentioned in the Talmud. We can learn from this that if throughout the year one says "Hamelech Hakadosh" one does not have to pray again. (This is also brought by the Ba'er Heteiv OC chap.118 par.1 and by the Mishnah Berurah chap.118 par.1). However, the "Kitzur Shulchan Aruch" (chap.129 par.3) adds "by mistake," and one needs to investigate from his words if he intentionally says "Hamelech Hakadosh" does he have to say the amidah again? If he does not have to say the amidah again, then if a person were to start saying "Hamelech Hakadosh" from Rosh Chodesh Elul, by the time he has reached Rosh Hashanah he has already said these words more than ninety times!

(140) SA OC chap.117 par.4

(141) SA OC chap.117 par.3

(142) Rabbi David ben Shlomo ibn Zimra ("Radbaz"), *She'elot uTeshuvot haRadbaz*, (New Jersey, date of printing not stated, maybe 5710?), part 6, responsum 2055,

(143) *Torat Chaim*, op. cit.

(144) *Pri Chadash*, op.cit., SA OC chap.117 par.2

(145) *Mishnah Berurah* SA OC chap.117 par.13

(146) Rabbi Yaakov Chaim Sofer, *Kaf Hachaim* SA OC chap.117 par.8

(147) In order not to embarrass the publishers of Siddurim and calendars where incorrect or unclear information on the dates to begin asking for rain is given, their names have not been published.

(148) Rabbi Shmuel David Halevi Yungreiss, *Shut Maharshda*, (New York, 5718), OC responsum 9

DIFFERENCES IN THE READING OF THE PARASHIOT BETWEEN ERETZ YISRAEL AND THE DIASPORA

(Parashiot = Shabbat readings from the Torah)

(In almost every case in this paper, the word "Parashah" is omitted before giving the name of the Parashah. An exception is the writing of the word Parashah before Bamidbar (and likewise Vayikra and Devarim), in order not to cause confusion between Parashat Bamidbar and Sefer Bamidbar.)

General Rules for the Reading of the Parashiot

In both Eretz Yisrael and the Diaspora, the Reading of the Torah is currently completed in the course of one year, and this occurs at the end of the Festival of Sukkot (Simchat Torah). However, on Yom Tov and Chol Hamoed which occur on Shabbat, one does not read the Parashah for that Shabbat, but instead a reading appertaining to the Festival.

The Gemara in Masechet (tractate) Megillah,[1] gives a number of rules according to which the order of reading of certain of the Parashiot is determined: "Ezra made a regulation for the Jewish people that they should read the curses in the book of Vayikra [Bechukotai] before Shavuot and those in the book of Devarim [Ki-Tavo] before Rosh Hashanah ... in order that the year may end together with its curses." On this, the Tosafot[2] brings a question which was asked of Rabbenu Nissim Gaon (11th century): "Why does one divide Nitzavim and Vayelech into two when there are two Shabbatot between Rosh Hashanah and Sukkot, namely one between Rosh Hashanah and Yom Kippur, and one between Yom Kippur and Sukkot, rather than not dividing up Matot and Massey which are of a greater length?"

After the Tosafot rejects the answer of Rabbenu Nissim, the Tosafot replies: "It seems to me that the reason that we divide them up is that we want to make a break and read one Shabbat before Rosh Hashanah with a Parashah which does not speak of curses in order not to mention the curses close to Rosh Hashanah, and similarly for this reason we read Parashat Bamidbar before Shavuot in order not to read the curses in Bechukotai near to Shavuot." On this Rabbi Mordechai Yoffe[3] the "Levush" (16th-17th centuries), explains that the reason is not to give the Satan the power of speech to accuse [the Jewish people]."

On this subject, the Maharit[4] (Rabbi Yosef ben Moshe Matrani, one of the Sages of Turkey in the sixteenth and seventeenth centuries) emphasizes that there should be *only one Shabbat* between Bechukotai and Shavuot, because if there were two Shabbatot it would "not be recognisable that the curses were finished" near to Shavuot, namely one must take care to read Parashat Bamidbar, and not Naso, on the Shabbat before Shavuot. The Sages[5] gave this principle the mnemonic "m'no v'itzru" ["m'no" is Hebrew for counting numbers, namely Parashat Bamidbar whose content is the numbering the Jewish people; "itzru" is Atzeret which is Shavuot] indicating that these two Parashiot must read in close proximity. However, according to our custom today, in the event that Rosh Hashanah occurs on a Thursday in a leap year, we read Naso on the Shabbat before Shavuot. However, Rabbi Menachem Meiri,[6] Rabbi Yissachar ben Mordechai Susan[7] author of the book "Tikkun Yissachar," and also a manuscript[8] from the thirteenth century state that there were communities that in those years divided Ki-Tissa into two, instead of dividing Matot and Massey, in order to observe the rule of "m'no v'itzru."[9]

Another rule established by the Sages[10] was that on the first Shabbat which occurs after the fast of Tisha b'Av, one reads Vaetchanan, and here the mnemonic is "tzumu v'tzalu" ["tzumu" is the fast of Tisha b'Av and "tzalu" is supplication namely Vaetchanan"; the Parashah read after Tisha b'Av is Vaetchanan].

It seems that until about the thirteenth century,[11] the reading of the Torah in Eretz Yisrael was completed after three years. The transition to an annual cycle was caused by the immigrants who arrived in Eretz Yisrael at that period. With this transition arose the problem which occurred on the occasions when the last day (eighth day) of Pesach and the second day of Shavuot occurred on Shabbat in the Diaspora. These days in Eretz Yisrael are "isru chag" (the day after the Festival) and therefore there one reads on them the Parashot of that week. In contrast, in the Diaspora they are Yom Tov and the Reading of the Torah is the Festival reading. Thus after the Festival, it will be necessary to equalise the Parashiot of the Diaspora with the Parashiot of Eretz Yisrael, and to accomplish this one must divide two of the usually joined Parashiot in Eretz Yisrael into two, whilst in the Diaspora they remain joined, or, alternatively to join two Parashiot in the Diaspora together when in Eretz Yisrael they are separate.

Eighth Day of Pesach which occurs on Shabbat in a non-leap year

In a non-leap year, Tazria is always joined to Metzora. Likewise, Acharei-Mot is joined to Kedoshim, and Behar is joined to Bechukotai. In the event that the eighth day of Pesach occurs in the Diaspora on Shabbat, that day will be isru chag (day after the Festival) in Eretz Yisrael and the Torah reading on it is Shemini. It would therefore be necessary to separate one of the above pairs of Parashiot. On which of them to separate has been the subject of controversy for generations.

The earliest source on this subject is the "Kaftor vaFerach,"[12] written in the year 5082 (1322), by Rabbi Ishtori Haparchi, after having done seven years of research in Eretz Yisrael. On this he writes: "In a non-leap year whose "simanei kviut" (designation) is hey-kaf-zayin [Rosh Hashanah on Thursday; 29 days Marcheshvan, 30 days Kislev; Pesach on Shabbat], namely the eighth day of Pesach will occur on Shabbat in Eretz Yisrael and they separate Tazria and Metzora." From this we can see that in the fourteenth century they utilised the first opportunity in Eretz Yisrael to equalise the Parashiot that were then between Eretz Yisrael and the Diaspora.

However, about two hundred years later, the "Tikkun Yissachar"[13] states that in such a year, in Eretz Yisrael they would separate Behar and Bechukotai. According to him, the reason that in Eretz Yisrael they did not want to separate Tazria and Metzora was "that they did not want to extend the reading of afflictions [Tazria and Metzora] to two Shabbatot, whilst in the Diaspora these two Parashiot are joined together on one Shabbat." However, there is a difficulty with this view of the "Tikkun Yissachar." The Sages fixed that in every leap year Tazria and Metzora would be separated. Were it preferable to read these two Parashiot together on the same Shabbat, why instead of separating them, did not the Sages divide Emor into two?

Furthermore, this reason regarding "afflictions" suggested by the "Tikkun Yissachar" is not relevant to Acharei-Mot and Kedoshim which were not separated in Eretz Yisrael in such a year. The "Tikkun Yissachar" gives another reason for specifically separating Behar and Bechukotai: "In those days the second day of a Festival was classed as a "custom of our ancestors in your hands." According to the "Tikkun Yissachar." Eretz Yisrael is the primary location, and there they observed just one day of Yom Tov. If one therefore separates two Parashiot in Eretz Yisrael in order to equalise it with the Diaspora, "then the primary [Eretz Yisrael] will then become secondary to the Diaspora, and it is not right to turn a primary into a secondary." From his words, if in Eretz Yisrael one would separate the Parashiot "it would give the appearance that the Diaspora is superior to Eretz Yisrael in that the people of Eretz Yisrael are secondary to the people of the Diaspora."[14]

From here the question arises: Why in *Eretz Yisrael* do they make a change and separate Behar and Bechukotai? On this the "Tikkun Yissachar" explains that it is necessary to read Parashat Bamidbar on the Shabbat before Shavuot and for this reason it is necessary to separate in Eretz Yisrael Behar from Bechukotai.

The above facts indicate that in the sixteenth century there was no uniform custom in Eretz Yisrael, and to understand this one has to look at the composition of the population and the state of immigration to Eretz Yisrael at that period.

In the year 1517, the Turks conquered Eretz Yisrael. From the government lists of taxpayers from 1525-26 one can see that in the city of Safed there were about sixty percent "Musta'arabi Jews" (original residents). As a result of the expulsion from Spain and Portugal there was a large influx of immigrants to Eretz Yisrael at that period, and in the year 1535 one of the immigrants from Italy wrote in his letter:[15] "... and he who saw Safed ten years ago and sees it now, is a wonder to his eyes, because all the time many Jews come ..." Also Rabbi Yosef Karo[16] who lived in Safed, mentioned in one of his responsa: "We see the number of the original inhabitants of Eretz Yisrael are very small compared with those who are coming from Spain, and they are the large majority compared to the small number of the original inhabitants."

At the beginning of the sixteenth century, there were two customs in the city of Safed insofar far as the subject hereby under consideration,[17] namely, the custom of the Musta'arabi Jews who followed the custom of Eretz Yisrael and they separated Behar and Bechukotai. On the other hand, some of the Spanish communities would separate Tazria and Metzora. On their custom the "Tikkun Yissachar" stated: "Their only reason was that they wanted to equalise with the Diaspora."

In the year 5305 (1545), a gathering of all the Sages of Safed was held, in which they decided on a uniform custom in Safed, in accordance with the custom of Eretz Yisrael, namely to separate Behar and Bechukotai. It seems that the Sefaradim, whose numbers and influence increased from day to day were not satisfied with this decision, and indeed three years later in the year 5308 (1548), all the Sefaradic sages gathered in Safed, and decided to separate Tazria and Metzora, and this they did. They did not find this to be sufficient and they even requested that the Musta'arabi Jews also do this. The latter did not agree to this and claimed that the separation between Behar and Bechukotai is "the custom of our fathers and ancestors from time immemorial."

From all of the above, an unclear picture emerges. During the period of the "Kaftor vaFerach," the vast majority of the country's inhabitants were Musta'arabi Jews. Their custom from "ancient times" was to separate Behar from Bechukotai. If so: a) Why does the "Kaftor vaFerach" *only* mention the separation between Tazria and Metzora? b) Why did the Spanish immigrants not follow the custom of most of the inhabitants of Eretz Yisrael, and what is more, they even pressured the others to separate Tazria from Metzora? The matter is difficult to understand, especially in light of the fact that there are two days of Yom Tov in the Diaspora, and the problem only arose with *their arrival* of the Spanish immigrants to Eretz Yisrael. However, an answer to the first question requires further study. As for the second question, it is possible that the immigrants wanted to prove their independence, and therefore chose a custom that only existed among the minority of residents. At the same time, one can see[18] the influence of the Sefaradim on the community in Safed regarding the laws of inheritance, namely in the year 5315 (1555), the Rabbis of Safed agreed to arrange the details of the custom of inheritance according to the customs of the Sefaradim who lived in the city.

Even a hundred years later, before the subject had reached a conclusion, and during the period of the "Magen Avraham"[19] (Rabbi Avraham Abele Gombiner, late 17th century) there were still two customs in the Eretz Yisrael. He wrote: "There are those who separate Tazria and Metzora whilst there are others who separate Behar and Bechukotai."

One hundred years later, we find in the book "Mizbach Adamah"[20] written by Rabbi Rafoel ben Shmuel Meuchas, who was regarded as one of the Sages of Eretz Yisrael and the Head of the Rabbis of Jerusalem in the eighteenth century, that the custom in Jerusalem was to connect Tazria and Metzora, Acharei-Mot and Kedoshim, and to separate Behar and Bechukotai. On this he explained: "It is more correct than joining them [Behar and Bechukotai] together, is to separate Acharei-Mot and Kedoshim, since in this way there is more recognition for the Shabbat before Shavuot by separating Behar and Bechukotai in Eretz Yisrael."

Rabbi Refoel Aharon ben Shimon, author of "Nehar Pakod"[21] was surprised at this reasoning of the "Mizbach Adamah," and he answered that even in the case that the people of Eretz Yisrael separated Acharei-Mot and Kedoshim, and connected Behar and Bechukotai, one would still read Parashat Bamidbar on the Shabbat before Shavuot. He regarded it as a strange thing to say that because one is reading Behar together with Bechukotai on the same Shabbat, it ceases to be a recognition of finishing a year with its curses before Shavuot.

The "Nehar Pakod" then continued that the reason to separate Behar and Bechukotai and not Tazria and Metzora is because anything which would equalise the same division of the Parashiot for Eretz Yisrael with that of the Diaspora [although not necessarily on the same Shabbat] is a good thing. Namely on non-leap years Tazria and Metzora and also Acharei-Mot and Kedoshim are joined, the reason being that in order that there should not be an additional difference between the readings in Eretz Yisrael and the Diaspora, these Parashiot are joined also in Eretz Yisrael, but when one arrives at Behar and Bechukotai, it will be necessary to separate them in order to observe the principle "m'no v'itzru."

In the "Sefer Hatakanot"[22] on the customs of Jerusalem, printed in 1883, it is written: "It is our custom to join Tazria and Metzora, and also Acharei-Mot and Kedoshim, and to separate Behar and Bechukotai." The Mishnah Berurah[23] which was published at about the same period states: "There are those who separate even in a non-leap year Tazria and Metzora and there are those who separate Behar and Bechukotai." Indeed the Mishnah Berurah brings the words of the "Magen Avraham" and it seems that he is just quoting his words, and not that it was his intention to say that at the end of the nineteenth century the *practical* custom was to separate Tazria and Metzora. This is also evident from the language of the book "Shoneh Halachot"[24] by Rabbi Chaim Kanievsky (son of the "Steipler") and Rabbi Elazar Turchin, (which was printed only in 5735). There it states in the present tense: "There are those who separate Tazria and Metzora and those who separate Behar and Bechukotai" even though there are no communities today in Eretz Yisrael that separate Tazria and Metzora. In the "Sefer Eretz Yisrael"[25] by Rabbi Yechiel Michel Tucazinsky, only the separation of Behar and Bechukotai is mentioned.

Eighth Day of Pesach which occurs on Shabbat in a leap year

In a leap year, there are no joined Parashiot in the Book of Vayikra. Therefore, when the eighth day of Pesach occurs on Shabbat, it is not possible to separate Behar from Bechukotai, in order to equalise the Parashiot between Eretz Yisrael and the Diaspora

The custom in Eretz Yisrael at the beginning of the fourteenth century was, as stated in the "Kaftor vaFerach":[26] "They divided up the Parashiot Matot and Massey," namely in Eretz Yisrael they divided Matot from Massey (whilst in the Diaspora they are connected).

The "Tikkun Yissachar"[27] also brings the same solution. However, he adds: "And I found it written that the Sage Rabbi Saadia Dayan of Tzova [the city of Aleppo in Syria] had the custom in such a year to separate these two Parashiot (Matot and Massey) and join in their place Korach and Chukat and according to this custom, the Diaspora will be equalised with Eretz Yisrael from Shabbat Korach onwards, since in Eretz Yisrael these two Parashiot are always separate."

The advantage of the method brought by Rabbi Saadia, is the advancement by four weeks of the equalising between Eretz Yisrael and the Diaspora. However, our custom today is not to advance even by a few weeks the equalising between Eretz Yisrael and the Diaspora. To the question, why during such a year "do we will not connect together two Parashiot immediately on the Shabbat after Pesach and read Acharei-Mot and Kedoshim in the same way as one connects them in a non-leap year?" This is answered by the Maharit:[28] "The Rishonim (great Rabbis who lived approximately between the eleventh and fifteenth centuries) took the trouble to arrange the Parashiot in the book of Vayikra in order to specifically read Parashat Bamidbar on the Shabbat before Shavuot. If instead on that year they had joined Acharei-Mot with Kedoshim, they would then have read Naso (and not Parashat Bamidbar) on the Shabbat before Shavuot. In Eretz Yisrael it is impossible to avoid this, but why in the Diaspora *enter* into the situation where one will read Naso on the Shabbat before Shavuot?

All of the above explains why one does not join two Parashiot in the Diaspora *before* Shavuot but it does not explain why one waits until about *two months* after Shavuot in order to equalise Eretz Yisrael with the Diaspora. Why not join Chukat and Balak as is done when the second day of Shavuot occurs on Shabbat?

It is true, the Maharit writes, that in the book "Tikkun Yissachar" it is written that in Syria it is customary to join Chukat and Balak in such a year. [In fact, we saw above that the "Tikkun Yissachar" writes that it was the custom in Syria to join Korach and Chukat.] However, according to the opinion of the Maharit that since one was not able to join them before Shavuot, one waits until Matot and Massey. Even though it is possible to join Chukat and Balak and to separate Matot and Massey, the Sages did not want to "change the yearly order that when Pesach did not occur on Shabbat they joined Matot and Massey."

In addition, the Maharit[29] states that one should equalise the Parashiot before Tisha b'Av in order to observe the rule "tzumu v'tzalu," and he brings a proof from the mitzvah of arava (willow) on Sukkot, to prove why one postpones the equalising until the latest date.

The Second Day of Shavuot which occurs on Shabbat

As we saw above, when the eighth day of Pesach occurs on Shabbat, the change in the reading of the Parashiot is made in *Eretz Yisrael*, namely one divides Behar from Bechukotai or Matot from Massey in *Eretz Yisrael*, whilst they are joined in the *Diaspora*. On the other hand, when the second day of Shavuot occurs on a Shabbat whether in a non-leap year or in a leap year, the change is made in the *Diaspora*, namely, one joins together Chukat and Balak in the *Diaspora*, and this already appears in manuscripts from the twelfth century.

The first manuscript[30] in our possession, in which the division of the Parashiot is mentioned, was written in the year 4939-4940 (1179-1180), which was before the transition to an annual cycle of reading the Torah in Eretz Yisrael. In the case when the second day of Shavuot occurs on Shabbat, one finds in a manuscript calendar as follows: "Zot Chukat with Vayar Balak on the twelfth [of Tammuz]." About one hundred years later (the year 1291) the same solution was found in another manuscript[31] and there it states: "When Shavuot occurs on a Friday one joins Chukat and Balak because of [the second day of] Shavuot occurring on Shabbat."

Rabbi Menachem Meiri[32] who lived at that period, brings the same answer in relation to a non-leap year. With regards to a leap year, he writes: "In every leap year whose "simanei kviut" is bet-chet-hey [Rosh Hashanah on Monday; 29 days both Marcheshvan and Kislev; Pesach on Thursday] or zayin-shin-aleph [Rosh Hashanah on Shabbat; 30 days both Marcheshvan and Kislev; Pesach on Sunday] (namely, the second day of Shavuot on Shabbat), one has two joinings of Parashiot, namely Chukat with Balak and Matot with Massey, and some say Behar with Bechukotai and not Chukat with Balak." [It is clear that there is a scribal error here because there is no such year with a "simanei kviut" zayin-shin-aleph and one needs to write zayin-shin-hey.]

The question that arises is why, according to the opinion of the "some say" one joins Behar with Bechukotai *only* in a year when the "simanei kviut" is bet-chet-hey and *not* in a year whose "simanei kviut" is zayin-shin-hey? In general, the explanation of the "some say" is puzzling because the joining of Behar with Bechukotai will result in reading Naso instead of Parashat Bamidbar on the Shabbat prior to Shavuot, and the matter thus requires further investigation.

A different division of Parashiot is to be found in the book "Kaftor vaFerach."[33] According to his opinion one divides Matot from Massey when Shavuot occurs on Friday.

It is true that according to the "Kaftor vaFerach," one arrives at equalising Eretz Yisrael with the Diaspora. However, according to his opinion one would read Parashat Devarim *after* Tisha b'Av (namely one does not observe the rule "tzumu v'tzalu"), and one would also finish reading the Torah a week *after* Simchat Torah! One must accordingly investigate his words.

There is a manuscript[34] from the period of the "Kaftor vaFerach" confirming the earlier manuscripts, (and not the words of the "Kaftor vaFerach") and this is its language: "Chukat and Balak are also read separately whether in a non-leap year and a leap year except when (first day of) Shavuot occurs on Friday, because then there would be lacking in the month of Sivan one Parashah and therefore one joins them together even in a leap year." He concludes: "This is what I found in the Siddur of the "Anshei Knesset Hagedolah" (Men of the Great Assembly).

Although the "Tikkun Yissachar"[35] brings this custom, he adds two other customs (instead of joining Chukat and Balak). One custom: is joining Shelach-Lecha and Korach, and the second is to join Korach and Chukat. As for the second custom he states that he found it written in "Ibur Shana" from the Sage Rabbi Saadia, the Dayan of Tzova which states: "it is always customary to separate these [Chukat and Balak] and these [Shelach-Lecaha and Korach] and in such a year join Korach with Chukat." However, a number of calendars[36] from the fifteenth and sixteenth centuries, and also from the "Levush,"[37] which was written in the middle of the sixteenth century, brings only the joining together of Chukat and Balak in such a year.

The "Leket haKemach"[38] (Rabbi Moshe ben Yaakov Chagiz who was one of the Sages of Eretz Yisrael and a Rabbi in Jerusalem in the eighteenth century) offers a different opinion, namely: "It would be proper to join together at the end of the Book of Bamidbar (Matot and Massey), but one is reaching the three (Haftarot) of Rebuke which are (i) Pinchas (ii) Matot-Massey (iii) Devarim, and it is proper for everyone [Eretz Yisrael and the Diaspora] to read them at the same time in order to exempt all the Jews from rebuke at the same time and not a different periods ... therefore before one reaches these three [Haftarot] of Rebuke on which one replaces the normal weekly Shabbat Haftarah, and thus since it would be necessary to uproot the Haftarah for them, they did want to make a change by joining two Parashiot and thus uprooting one of their haftarot."

However, it is difficult to understand the words of the "Leket haKemach." On the three Shabbatot between the fasts of the seventeenth of Tammuz and the ninth of Av, *only* the Haftarah is special for those Shabbatot, and not the Parashah from the Torah. And if so, why is it a problem if they read in the Diaspora a) Pinchas, ii) Matot-Massey, iii) Devarim, whilst in Eretz Yisrael they read i) Matot, ii) Massey, iii) Devarim; the Haftarot would be the same in Eretz Yisrael and the Diaspora and accordingly "all the Jews would be exempted from rebuke together."

The "Leket haKemach" himself agrees that this reasoning is "forced" and brings a different reason for specifically joining Chukat with Balak. All the Parashiot from Behalotecha begin with an "open paragraph" (the words begin on a new line). The *first* Parashah after Shavuot which begins with a "closed paragraph" (the words are on the same line but after a certain space) is Balak. Accordingly, Chukat and Balak are considered as one Parashah and it is thus proper to join them. This reasoning is also brought in the name of the "Leket haKemach" by the "Emet l'Yaakov"[39] (Rabbi Yisrael Yaakov Algazi) who was one of the Sages of Turkey in the eighteenth century.

Summary

- For several centuries, a uniform change has been made in order to equalise the Parashiot between Eretz Yisrael and the Diaspora as follows:
- Eighth day of Pesach on Shabbat in a non-leap year: in Eretz Yisrael one separates Behar from Bechukotai.
- Eighth day of Pesach on Shabbat in a leap year: in Eretz Yisrael one separates Matot from Massey.
- Second day of Shavuot on Shabbat in either a non-leap year or a leap year: in the Diaspora one joins Chukat with Balak.

References

Abbreviations
SA = Shulchan Aruch
OC = Orach Chaim
(1)   Talmud Bavli Megillah 31b
(2)   Tosafot, Megillah 31b, first word: klalot
(3)   Rabbi Mordechai Yoffe, *Levush*, OC chap.428 par.4

(4)   Rabbi Yosef Matrani (Maharit), *She'elot uTeshuvot Maharit*, (5621), vol.2 responsum 4

(5)   SA OC chap.428 par.4

(6)   Rabbi Menachem Meiri, *Kiryat Sefer*, (Jerusalem, 5716), fifth paper, b'Inyanei Kriyat Sefer Torah

(7)   Rabbi Yissachar ben Mordechai Susan, *Sefer Ibur Shanim – Tikkun Yissachar*, (Venice, 5339), p.42b

(8)   MS London British Museum, catalogue Margoliouth 664 folio 86b

(9)   Over the course of the generations there have been changes in the arrangement of the Parashiot on the Shabbatot of the year. The Gemara (Megilla 29b-30a) discusses the reading for Shekalim on a Shabbat when the Torah Parashah of the week is Tetzaveh or Ki-Tissa, and it follows that this was possible during the period of the Gemara. However, a number of Rishonim, namely the Ran (Megilla, on the Rif p.10a. first word Rosh Chodesh), Hagahot Maimoniot (Rambam, Hilchot Tefillah chap.13 par.100) and the Meiri (Megilla p.106) write that the situation today is that the reading for Shabbat Shekalim is not read when one reads Tetzaveh or Ki-Tissa. We could note that the Meiri writes that in a non-leap year Shabbat Shekalim occurs on either Mishpatim or Terumah, and in a leap year on either Pekudey or Vayikra. All of conforms to the arrangements for today, *except* that today one never reads Vayikra on Shabbat Shekalim, and this subject needs investigation on how was the arrangement of the Parashiot was made to enable one to say the reading for Shekelim on Parashat Vayikra? The Meiri also mentions that in the Provence region in southern France they sometimes read Tetzaveh or Ki-Tissa on Shabbat Shekalim.

Additional changes in the arrangement of the Parashiot are mentioned by the Meiri and the *"Tikkun Yissachar"* (ibid., p.33a), and both write about the division of Vaera into two.

The *"Tikkun Yissachar"* also writes that in another place in the West, there are years that one divides Miketz into two, and in this place both in non-leap year or a leap year they join Behar with Bechukotai, Chukat with Balak and Matot with Massey. Also, the *"Tikkun Yissachar"* mentions in the name of Rabbenu Behaye (Bahya ben Asher ibn Halawa) the dividing of Mishpatim into two.

(10)   SA OC op. cit.

(11)   see: Chaim Simons, "The Reading of the Torah on Shemini Atzeret (Simchat Torah) in Eretz Yisrael," Journal *Sinai* (Mossad Harav Kook: Jerusalem), 1989, vol.103, p.239

(12)   Rabbi Ishtori Haparchi, *Kaftor vaFerach*, (Hirsch Edelman: Berlin, 5611), chap.14, p.55

(13)   *Tikkun Yissachar*, op. cit., pp.32a-32b

(14)   One could note that there are cases where the inhabitants of Eretz Yisrael follow the inhabitants of the Diaspora. For example:

    a)   The Hoshana piyyut "Eeroch Shui" is specifically said on the *first day of Chol Hamoed Sukkot* (*Machazit haShekel* (Rabbi Shmuel Löw) SA OC chap.663). However, in Eretz Yisrael it is not said on the first day of Chol Hamoed Sukkot in Eretz Yisrael, but on the day that is the first day of Chol Hamoed in the Diaspora, (Rabbi Eliyahu Weissfish, *Sefer Arba'at Haminim Hashalem*, (Jerusalem, 5735), p.375, note 10)

b) There are those who are accustomed (according to the Ashkenazi rite) not to say "Lamnatzeach ...Yaancha Hashem" on "isru chag" the reason being because the majority of the Jews of the world live in the Diaspora, and this day is Yom Tov for them (*Luach Davar b'Ito*, year 5752, (Achiezer: Bnei Brak), p.671)

(15) R' Avraham Yaari, *Igrot Eretz Yisrael*, (Masada: Ramat Gan, 1971), p.184

(16) Rabbi Yosef Karo, *She'elot uTeshuvot v'Shitotov l'Rav Yosef Karo*, (Mantova, 5490), dinei Nashim, p.16b

(17) *Tikkun Yissachar*, op. cit., p.32b

(18) Rabbi Moshe Alshich, *She'elot uTeshuvot l'Rav Moshe Alshich*, (Lvov (Lemberg), 5649), responsum 27

(19) Rabbi Avraham Abele Gombiner, *Magen Avraham*, SA OC chap.428 par.6

(20) Rabbi Refoel ben Shmuel Meuchas, *Mizbach Adama*, (Salonika (Thessaloniki), 5537), OC chap.428

(21) Rabbi Refoel Aharon ben Shimon, *Nehar Pakod*, (Na Amon, Alexandria, 5668), Sha'ar haMefaked, hilchot Kriyat Sefer Torah, chap.12, (p.27)

(22) *Sefer haTakanot*, (Jerusalem, 5643), dinei Kriyat Sefer Torah, p.53a

(23) Mishneh Berura, SA OC chap.428 par.10

(24) Rabbi Chaim Kanievsky and Rabbi Elazar Tzadok Turchin, *Shoneh Halachot*, (Bnei Brak, 5735), vol.2, chap.428 par.13

(25) Rabbi Yechiel Michel Tucazinsky, *Sefer Eretz Yisrael*, (Lewin Epstein: Jerusalem, 5726), chap.17 par.15, chap.20

(26) *Kaftor vaFerach*, op. cit.

(27) *Tikkun Yissachar*, op. cit., pp.38a-38b

(28) *She'elot uTeshuvot Maharit*, op. cit.

(29) Talmud Bavli Sukkah 43b

(30) MS London British Museum, catalogue Margoliouth 677 folios 179a, 183a, 184a

(31) MS London British Museum, catalogue Margoliouth 664 folio 83b

(32) Meiri, op. cit.

(33) *Kaftor vaFerach*, op. cit

(34) MS Bodleian, catalogue Neubauer 2275/5 folio 49a

(35) *Tikkun Yissachar*, op. cit., p.24a

(36) e.g. MS Hamburg 293/1; MS Vatican, Vat Ebr, folio 117b; MS Budapest Kaufmann A 516/1; MS Bodleian, catalogue Neubauer 1483 folios 183a, 183b, 186a, 188b; MS Leiden, Warn 66/2, folios 14b, 16a, 20b

(37) *Levush*, OC chap.428 par.8

(38) Rabbi Moshe ben Yaakov Chagiz, *Leket haKemach*, (Amsterdam, 5467), OC hilchot Shabbat, p.38

(39) Rabbi Yisrael Yaakov Algazi, *Emet l'Yaakov*, (Kushta (Constantinople), 5524), Mishpat Seder Kriyat haParashiot b'Shanim Pshutot u'Meubarot, chap.1, (pp.63b-64a)

THE CONNECTION BETWEEN THE MOLAD AND THE TIME FOR RECITING KIDDUSH LEVANA

(molad = mean conjunction – the "birth of the New Moon")

(kiddush levana = monthly blessing for sanctification of the moon)

It is stated in Masechet (tractate) Sanhedrin:[1] "Until what day of the month may the berachah (blessing) over the new moon be recited? Until the concavity is filled up. And how long is that? Rav Yaakov ben Idi said in the name of Rav Yehudah, seven days. The Nehardeans said sixteen days."

This Gemara gives only the latest time for the mitzvah (commandment) of kiddush levana, and does not explicitly write the earliest time to perform this mitzvah. However, this can be found in the Talmud Yerushalmi[2] which states that the moon does not reach a full moon until fourteen days have passed. Therefore, for the first fourteen days one may recite the berachah, namely one can recite the berachah from day one.

This also appears in the Siddur of Rav Saadia Gaon,[3] (9th–10th centuries) which states that one can say the berachah "from the first night." [In the version of his Siddur which we have before us,[4] it states "from the fourth night," but Rabbi Mordechai Hakohen who researched the subject and examined the original Arabic text, and in addition various manuscripts, found that there has been a tampering with the text and instead of "fourth" there should be written "first."[5]] The early Rishonim (great Rabbis who lived approximately between the eleventh and fifteenth centuries), such as Rabbi Moshe miKotzi[6] (13th century) who is known as the SMaG, the Rambam (Maimonides),[7] and the Rabbi Menachem ben Shlomo Meiri,[8] (13th-14th centuries), also ruled

111

in this way. In addition, Rabbi Avraham Gombiner[9] known as the "Magen Avraham" (17th century) states in the name of the "Sefer Hakanah" (Rabbi Elkanah ben Yerucham, 15th century), that the ideal way to fulfil this mitzvah is on the first day, and Rabbi Yaakov ben Yosef Reischer,[10] who was known as the "Shvut Yaakov" (17th-18th centuries) quotes the Talmudic principle that "the conscientious perform a mitzvah as early as possible."[11]

The question arises, (not only in the matter of the earliest time of kiddush levana, but also in connection with the latest time), that are these days to be counted from Rosh Chodesh, or from the time of the molad? Rabbi Shlomo Kagan Hakohen,[12] (who was one of the scholars of Poland-Ashkenaz in the nineteenth century) who discusses this subject in connection with the latest time for kiddush levana, in his book "Atzei Beroshim" is convinced that it depends on the growth and decrease in size of the moon and therefore he writes: "It is obvious that the Gemara is speaking about sixteen days from the time of the molad, and hence it was not necessary for the Gemara to explain this since it is obvious, and there is no Posek who has doubted this." As we shall see below, this rule is correct for the beginning of the time of kiddush levana according to all the different opinions. However, there is an almost solitary opinion that states that one starts to count from Rosh Chodesh, and this is the opinion of Rabbi Zvi Elimelech from Dinov.[13] He was one of the scholars from Poland-Ashkenaz and was the author of the book "Bnei Yissaschar," and was a Rebbe of the third generation of Hasidism at the beginning of the nineteenth century, but even he writes in connection with his opinion that he was apprehensive to rule like this since the Rishonim and Acharonim (great Rabbis who lived from about the 16th century) had not ruled in this way.

Although from the language of the Rambam:[14] "If a person did not recite the berachah on the first night, he may recite the berachah until the sixteenth day of the month until there is a full moon," one might think[15] that the Rambam's intention on this subject is unclear since the word "month" indicates days from Rosh Chodesh. But the phrase "until there is a full moon" means from the time of the molad. However, there is an answer to this from the "Aruch Hashalchan"[16] who interprets the words of the Rambam: "On the first day from the molad he is able to recite the berachah," namely also the Rambam counts from the molad.

Another opinion regarding the day that one can start reciting the berachah for kiddush levana is to be found in the commentary of Rabbenu Yonah ben Avraham Gerondi[17] (13th century) on Masechet Berachot. In the version of Masechet Soferim[18] that was before him (and also before the "Kol Bo"[19]) it states: "One says the berachah for kiddush levana when one is 'mevusam' (ostensibly 'perfumed')." He rejects the interpretation in which one says the time for this mitzvah is on the termination of Shabbat; apparently he did not see the version which appears in our books today which states "and one does not say the berachah for kiddush levana except on the termination of Shabbat when a person is 'mevusam'," [20] and he interprets the word "mevusam" as "from the hour when its light is sweet and a person benefits from it which is after two or three days." Also Rabbi David Avudarham[21] author of the book "Avudarham" gives the time as two or three days. From the words of Rabbenu Yonah, most of the Acharonim[22] rule that it is possible to say kiddush levana after three days. Although the "Beit Yosef"[23] mentions "three days" in his commentary on the Tur, he does not mention this opinion in the Shulchan Aruch. The "Rema" (Rabbi Moshe Isserlis, 16th century) also does not bring this opinion. It should be noted that even though the Acharonim rule according to the opinion of Rabbenu Yonah, they do not say "after two days or three days," but they omit the words "two days" and only write "after three days"; [The reason for this is given below.]

The question is, does one count these three days "me'et le'at" [periods of twenty-four hours] from the hour of the molad, or from the day of the molad? The meaning of this question is that according to the first option, the time for kiddush levana will always be seventy-two hours after the time of the molad. However, according to the second option, if the time of the molad is at the end of the day, it would be possible to say the berachah a little more than forty-eight hours from the time of the molad. However, if the molad occurred at the beginning of the night, the earliest time for kiddush levana will be close to seventy-two hours after the time of the molad.

We shall start by answering the opinion of Rabbenu Yonah who holds: "After two days or three days." It seems that his intention is "me'et la'et," because if his intention is "from the day of the molad," two days from the molad could occur a little more than twenty-four hours from the time of the molad, and Rabbenu Yonah[24] writes explicitly that a moon of just one day, because of its small size, is not sufficient for a man to benefit from its light. [The routine use of the phrase "ben yomo" is *throughout the entire* first day. However, in the matter before us, (as we shall see below), that the moon is invisible until an average of twenty-four hours after the time of the molad, we must say that Rabbi Yonah's intention in his expression "ben yomo" is at the *end* of an entire day from the time of the molad.]

Is there a difference in this answer between the opinion of Rabbenu Yonah and the Poskim (Rabbinical arbiters) who write that the time of the beginning of kiddush levana is "after three days" counted from the time of the molad? Although most of them do not *explicitly* write that these days are "me'et le'at" from the time of the molad, the "Pri Megadim"[25] (Rabbi Yosef ben Meir Teomim, 18th century) and the Mishnah Berurah[26] write that it is so.

According to these Acharonim, is it possible to wait less than these seventy-two hours? On this, Rabbi Shmuel Halevi Kolin,[27] the "Machazit haShekel" (18th century), (based on the opinion of Rabbenu Yonah who writes: "After two days or three days"), allows one on the termination of Shabbat to recite kiddush levana, after two days and eighteen hours.

We are left with the question, why do the Acharonim write "after three days" while Rabbenu Yonah has written "after two days or three days." The "Bnei Tzion" [28] offers an answer based on the fact that the true molad (astronomical time of a new moon) can occur fourteen hours after the average molad. It follows, that two days after the average molad can only occur one and a third days after the true molad, and the moon will not yet be ready for a berachah because of its smallness. Therefore the Acharonim wrote "after three days," and it will always happen at least after two and a third days after the true molad, and the moon will then be of sufficient size for the berachah. Because we are not proficient in the calculation of the true molad, we always wait until after three days. However, the question remains, why does Rabbenu Yonah give two options, namely two days and three days? As an answer, it can be suggested that he did not worry about the non-proficiency of knowing the time of the true molad, and therefore when the true molad occurs at the same time as the average molad, or even before it, one can recite the berachah only two days after the average molad. However, if the true molad occurs several hours after the average molad, then it is necessary to wait for three days.

There is another opinion about the time of the beginning of the berachah for kiddush levana, which was brought by the Beit Yosef in his commentary on the Tur,[29] and also in the Shulchan Aruch,[30] which is based on the Kabbalah, and this is the opinion of Rabbi Yosef ben Avraham Gikatilla, who lived in the eleventh century and was author of the book "Sha'are Ora." According to this opinion, one can only recite this berachah after seven days from the molad. There are a number of Poskim[31] who are punctilious that *seven full days* have passed from the time of the molad.

On the other hand, there are a number of Poskim[32] who allow one to recite the berachah of kiddush levana if several hours are missing from seven full days, namely, they are counted from the day of the molad, and not from the hour of the molad. In Egypt there was a "compromise" on this issue, and only if on the termination of Shabbat "some hours are missing to complete seven full days 'me'et l'et'" can one recite the berachah.[33] There are also other Poskim[34] who allow one to recite the berachah on the termination of Shabbat that occurred between three and seven days after the time of the molad. One can note that the Poskim are more lenient to permit before seven days, as compared with those who permit before three days. On this matter, Rabbi Ovadia Yosef[35] in his responsa "Yechave Daat" explains: "Since the ruling on this matter as found in the Talmud and the Poskim is that one can recite the berachah immediately after three days, one need not be strict on it."

It should be noted that not all Kabbalists followed the opinion which states "after seven days." For example, Rabbi Chaim Vital,[36] (16th-17th centuries), the main disciple of the Ari, writes in his book "Pri Etz Chaim": "We say the berachah of kiddush levana on Rosh Chodesh."

So far we have stated that the source of seven days is only from the Kabbalah. However, Rabbi Menachem Azaria miPano[37] (the Rama miPano, 16th-17th centuries), learns this opinion from the Gemara in Masechet Sanhedrin,[38] and one needs to investigate his words. [One could note that Rabbenu Yerucham ben Meshullam[39] (14th century) learns this subject from the Gemara which states that "from the outset, one should recite this berachah before seven days have passed," and this is the opposite of the opinion of the Rama miPano!]

There are also different opinions regarding the end of the time of kiddush levana, and it is possible to ask a number of questions about the Gemara in Sanhedrin that was mentioned above:

1) Does "until sixteen" include the sixteenth day? On this, the Tur[40] and the Shulchan Aruch[41] write that it does not include the sixteenth day.

2) Does one count specifically from the day of the molad or "me'et la'et" namely from the time of the molad? The Tur[42] explicitly writes that one counts these fifteen days "from the hour of the molad." However, in contrast, the language of the Shulchan Aruch[43] is "from the day of the molad." The "Magen Avraham"[44] mentions this contradiction, but Rabbi Shmuel haLevi Loew[45] (18th century) author of "Machazit haShekel," and the Mishnah Berurah[46] explain that the intention of the Shulchan Aruch is in fact "from the time of the molad."

The length of time until there is a full moon is half a month. Its length [according to the average molad] is fifteen days less than six hours, so whoever says the berachah on kiddush levana in the last hours of the fifteenth day says the blessing already while the moon begins to decrease in size. From this it is possible to understand the gloss of the Rema[47] who quotes the "Maharil"[48] (Rabbi Yaakov Levi Moelin, (14th-15th centuries), and writes: "And one does not recite kiddush levana except until half of 29 days 12 hours 793 parts from the molad." Also, Rabbi Yaacov Weil[49] (Mahari Weil), who was a disciple of the Maharil who lived in Germany in the fifteenth century and was one of the great Rabbis of the Ashkenazim of his generation, quotes this ruling of his Rabbi, adding "and there is no doubt about it."

According to astronomy, however, this is inaccurate, because the Rema speaks of the average molad and the time for a full moon depends on the true molad. Therefore, it is very possible that if someone would recite the berachah during the last hours according to the opinion of the Rema, he would be saying the berachah at the time when the moon was beginning to get smaller. This problem can be resolved according to a responsum of Rabbi Moshe Sofer[50] known as the "Chatam Sofer" (18th-19th centuries). He proves from the Gemara,[51] that at least in the first eighteen hours after a full moon, the contraction of the moon would not be noticeable, and it is possible to recite the berachah of kiddush levana during these hours.[52] The difference between the true molad and the average molad is always less than eighteen hours and then even if one were to recite the berachah of kiddush levana right at the end of the time set by the Rema, he would always be saying it within the first eighteen hours of the contraction of the moon. [However, it should be noted that the "Chatam Sofer" allows the recitation of the berachah of kiddush levana up to eighteen hours *after the end of time fixed by the Rema*. For example, if in a particular month the true molad was fourteen hours before the average molad, then by saying the berachah of kiddush levana at the end of the time according to the opinion of the "Chatam Sofer" would be at a time when the contraction of the moon would already be noticeable. Rabbi Yechiel Tucazinsky[53] raises this issue in his "Luach l'Eretz Yisrael," and writes "it is difficult to move from the ruling of the Rema"; However, the "Bnei Tzion"[54] does not accept the opinion

117

RABBI DR. CHAIM SIMONS

of Rabbi Tucazinsky, and says that in this one relies on the majority, adding that in times of hardship one can rely on the "Chatam Sofer."

The question is whether one should be concerned about the accuracy of astronomy regarding the berachah of kiddush levana, because in the case of "birchat hachamah" (a berachah recited every twenty-eight years regarding the sun) we do not? But there really is no comparison between them. The Rabbis[55] decided that "birchat hachamah" goes according to the tekufah of Shmuel, which is *less accurate* than the tekufah of Rav Adda, and it follows that today this berachah is recited, instead of on the day of the *astronomical* tekufah of Nissan, but eighteen days later.[56] This is not so regarding the berachah for kiddush levana, which depends on the time of the molad, and on this there are no different opinions about its time. If one questions why the Rabbis ruled to recite the berachah *not* according to the real molad which is astronomical, but according to the average molad which is only fictitious, one can answer that the time difference between the real molad and the average molad is always less than fourteen hours, and during this time period the contraction of the moon will not be noticeable.

We have seen above, that the Talmud Bavli gives the latest time for kiddush levana "up to sixteen [days]," and the Talmud Yerushalmi "up to fourteen [days]" However in both of them it is written until there is a full moon. It is possible that this contradiction can be solved as follows: the Talmud Bavli speaks of "until and not including," namely until the end of fifteen days from the molad. Although it is about six hours later than the opinion of the Rema, it is possible to use the opinion of the "Chatam Sofer" that was mentioned above in the deliberation of the opinion of the Rema; [In fact, it is possible to reach a situation in extreme cases where the end of fifteen days will be about two hours above those eighteen hours. However, it is possible that the Rabbis made their ruling according to the majority of the cases.] On the other hand, the Talmud Yerushalmi says is "until and including," is based on the fact that the true molad can precede the average molad by fourteen hours, and therefore in every case "until the end of fourteen days" from the time of the average molad will be before a full moon is reached.

On this contradiction between the Babylonians and the Jerusalemites, the "Bnei Tzion"[57] writes another answer, which is based on the words of the "Marei Panim"[58] (Rabbi Moshe Margolit, 18th century). He says that both the Talmud Bavli and the Talmud Yerushalmi agree that the latest time for kiddush levana is after half of 29 days 12 hours and 793 parts, "and the intention of the Talmud Yerushalmi is more than fourteen days, whilst the intention of the Talmud Bavli is less than fifteen days."

In addition, the "Bnei Tzion"[59] gives another explanation for the solving of this contradiction, which is based on the opinion that the time of the molad was established for Jerusalem (see below). In the case that half of the 29 days 19 hours and 793 parts of an hour occur at about midnight on the fifteenth of the month, in the "far East", the time will be the end of the fifteenth of the month, and in the "far west", (in the opinion of "Baal haMaor"[60] (Rabbi Zerachiah Halevi of Girona, 12th century) and the "Kuzari"[61] (Rabbi Yehudah Halevi, 11th-12th centuries), the time will be the end of the fourteenth day of the month. According to the "Bnei Tzion," the Talmud Bavli refers to the far east, and the Talmud Yerushalmi to the far west.

There are Poskim who give a later time for the end of kiddush levana. We have already mentioned the "Chatam Sofer" who gives a time which is eighteen hours later than the opinion of the Rema. The pupils of Rabbenu Peretz and the "Meiri" learn[62] that "until the sixteenth means "until and including" which is an entire day above that of the Shulchan Aruch and about twelve hours above that of the opinion of the "Chatam Sofer." There are also a small number of Acharonim[63] who hold that one may recite the berachah for kiddush levana with "Shem uMalchut" (the Divine Name and the Divine Kingship) until the end of sixteen days.

What will be the size of the moon at the end of sixteen days? It roughly depends on the time difference between the true molad and the average molad. In the case of the true molad occurring about twelve hours after the average molad, the contraction of the moon will not be noticeable at the end of sixteen days. But when this difference is less than twelve hours, and all the more so when the real molad occurs before the average molad, the contraction of the moon will be noticeable at the end of sixteen days. It can be deduced from this, that according to these Poskim, the Rabbis were not particular about this *small* contraction which occurred after the full moon.

In conclusion, we see from all the discussions above, that all opinions for the beginning and end of the time of kiddush levana depend on the molad. Therefore, one needs to examine carefully in order to understand the meaning of the times of the molad that appear in our calendars today

In which Location in the World was the Molad Established?

The true molad is the moment when the moon is exactly between the earth and the sun, and it occurs at that same moment throughout the whole world. It therefore follows that even the average molad occurs everywhere in the world at that exact moment. However, no two places that are not on the same longitude have the same time on the clock. Thus the question arises, at what longitude is the time of the average molad that appears in the yearly calendars?

There are several opinions in the Rabbinical literature, and the following are the suggested locations where the molad was established:

Opinion 1: "Sof hamizrach" – the Extreme East. In the responsa of Rabbi Shimon ben Zemach Duran[64] known as the "Tashbatz," (14th-15th centuries), it is stated that the molad was established in the "extreme east," and it is located about 130 degrees east of Jerusalem. He writes that Rav Saadia Gaon who lived in Baghdad which is 120 degrees west of the extreme east, saw an eclipse of the sun eight hours before the time of the average molad that appears on the calendar; (Solar eclipses occur exactly at the time of true molad). According to Rav Saadia Gaon (or, it is possible that this is just the commentary of the 'Tashbatz'[65]), the place in the world that the eclipse occurred corresponds to the time of the molad that appears in the calendar, and this is the place where the molad was established. For every degree to the east, the clock is four minutes ahead, and therefore an advancement of eight hours is 120 degrees east. It follows that according to Rav Saadia Gaon (or the Tashbatz), the molad was established 120 degrees east of Baghdad, namely at the "extreme east."

However, Rabbi Menachem Kasher[66] in his book "Kaf haTa'arich haYisraeli" (The Jewish Date Line) says that "this method is difficult in itself" because "it is known that the eclipse [of the sun] occurs during true molad, and the molad that we calculate is average molad and there is a difference between them."

Opinion 2: "Ketze hamizrach" (the Edge of the East): The elderly Dayan Rav Hassan ben Marachasan, who lived in Cordova before the period of Rav Hai Gaon, [67] held that the molad was established at the "edge of the east." The distance between the "edge of the east" and the "edge of the west" was considered to be 180 degrees. The question is, where is the location of the middle? Some say that it is Jerusalem,[68] whilst others say that it is 24 degrees east of Jerusalem[69] (and this place is called "the centre of settlement" or "the navel of the land"). Therefore the "edge of the east" is 90 or 114 degrees east of Jerusalem.

Rav Hassan based his view on the Gemara in Masechet Rosh Hashanah:[70] "If it [the moon] is born before midday, then certainly it will have been seen shortly before sunset. If it was not born before midday, certainly it will not have been seen shortly before sunset." It is a known fact in astronomy that the moon distances itself from the sun by half a degree every hour, and only between eighteen and thirty hours after the true molad is it possible to see the moon, and the Rabbis agreed to an average time of twenty-four hours[71] as they wrote "the moon is invisible for twenty-four hours."[72]

Rav Hassan[73] explained that if the molad occurs at one "part" (a few seconds) before midnight, *only* at the edge of the east, it will be possible to see the moon on that *very day* (at the time of sunset), but it will only be seen at the edge of the west, namely eighteen hours later, the minimum time that one can see the moon after the molad.

Also, R' Yitzchak Rakofiel is of the same opinion.[74] However, Rabbi Yitzchak Baruch rejected this view,[75] and proves in a logical way that "the moon which is visible at the edge of the west is not beneficial to the people in the East."

Opinion 3: "Tabor Haaretz" - The Navel of the Land – The Centre of Settlement: Rabbenu Yitzchak haYisraeli,[76] author of the book "Yesod Olam" and a student of the Rosh, who lived in the city of Tulaytulah (Toledo in Spain) in the fourteenth century, believed that the molad was established at the "Navel of the Land," which is twenty-four degrees east of Jerusalem. He came to his conclusion in the above way:

a) An eclipse of the moon occurs while there is a conjunction between the moon, the sun and the earth, and this occurs exactly at moment of a full moon. He ascertained the times of the peak of a number of lunar eclipses in his city of Toledo, namely the times of "the molad."
b) From times of the molad, he calculated the average molad times in the city of Toledo, and thus the average time between one molad and the next.
c) The city of Toledo is located 62 degrees west of the "navel of the land," and from this he calculated the time of the average molad at the "navel of the land," that is, 4 hours and 8 minutes earlier (because there is a change of four minutes per degree).
d) He calculated from molad "baharad" [this is explained below] the average molad for that month and added 14 days 18 hours and 396 parts (average length of half a month), to get the time of the average molad.
e) In every case he found that the times stated in the above c) and d) were almost identical, and therefore he came to the conclusion that the place where the molad was established is the "navel of the land."

Although Rabbi Avraham Yeshaya Karelitz,[77] known as the "Chazon Ish" (19th-20th centuries) rejects the "centre of settlement" brought by the Rambam (which is the "navel of the land" of Rabbenu Yitzchak haYisraeli) and writes that although he is one of the great Rabbis of the world, it is not fit for this opinion to be brought into the Beit haMidrash, (Study Hall) whilst we are discussing halacha." It should be noted that the calculations regarding the calculation of the location of the molad by the author of the book "Yesod Olam" are only appropriate for this location.

Opinion 4: Close to Nusaybin (then) in Bavel (Babylon): Rav Yitzchak son of Rabbi Yitzchak ben Shlomo Chadav, author of "Orach Selulah" who lived in the fifteenth century, believed that the molad was established in Nusaybin in Babylon. [The location of Nusaybin is today in South Turkey on the border with Syria.] He writes[78] that the place on which 'baharad' was established is to be found in Babylon close to Nusaybin, and perhaps this was the location of Shmuel who was an expert in matters regarding. the Heavenly bodies. According to this opinion, one can see that the determination of the location of the establishment of the molad is not "Halacha l'Moshe miSinai" (a law given to Moshe on Mount Sinai), but was established *according to the location of the observatory* of the Amoraim Shmuel and Rav Adda bar Ahavah. (Amoraim are great Rabbis of the period of the Gemara about 200-500 C.E.)

Rabbi Azariah dei Rossi ha-Adumim[79] known as the "Meor Einayim," who lived in the fifteenth century, brings a number of other great Rabbis who agreed to the opinion of the "Orach Selula." They are: Rabbi Yehuda son of Asher, Rabbi Yehuda ben Virga, Rabbi Mordechai Penzi, and Rabbi Emanuel ba'al haKenafayim.

Opinion 5: Eretz Yisrael: A number of Rabbis stated that the molad was established in Eretz Yisrael. However not everyone agreed on a uniform location in Eretz Yisrael.

The "Baal haMaor"[80] writes: "The calculation of the molad that we say that it is on a certain day on this or that hour of the day or night, refers to Jerusalem." He came to his conclusion from an interpretation (different from the one brought by Rav Hassan) to the Gemara in Masechet Rosh Hashanah that was mentioned above. According to the Baal haMaor the "Jewish date line" is 90 degrees east of Jerusalem, namely, the clock at the "edge of the east" does not precede the clocks of Eretz Yisrael by six hours (as is customary with the nations of the world) but is eighteen hours later. Therefore, if for example on Shabbat the molad occurs just before midnight according to Jerusalem time, in the edge of the east it will occur prior to the commencement of Shabbat. At the time of the sunset on Shabbat at the "edge of the east," namely twenty-four hours after the molad, the moon will be visible in the "edge of the east," and so one can proclaim that Rosh Chodesh all over the world occurred on Shabbat. Furthermore the "Kuzari"[81] and also Rav Avraham ibn Ezra[82] in his "Igeret haShabbat" and his "Sefer haIbur"[83] holds that the molad was established in Jerusalem.

Rabbi Yitzchak ben Moshe Halevi Profiat Duran known as the "Efodi," (14th-15th centuries) stated in his book "Choshev haEifod"[84] that the molad was established in Eretz Yisrael. He wrote that from the first molad "baharad" (the mnemonic for the second day of the week the fifth hour and 204 parts of an hour) the various intricate details of "sod haibur," the principle of the intercalation of the Jewish calendar were made. He was of the opinion that the establishment of the molad was in the holy city of Jerusalem or the "eastern border of Eretz Yisrael." One needs to investigate what is the meaning of "the eastern border of Eretz Yisrael" – is it the area of Jericho or the other side of the Jordan.

Also, the astrologer Rav Avraham Zakut (15th-16th centuries), held that the molad was established in Eretz Yisrael.[85] He wrote: "The location of determining whether to declare a leap year was Jerusalem or Yavneh, since these were the places where the Rabbinical courts sat in connection with moon sightings." In a similar language, Rabbi Shimon ben Natan Neta Woltish[86] in his book "Navah Kodesh" wrote that it was established in Jerusalem since it was there that the accepting of sightings of the moon and the question of leap years were determined.

For the same reason as written above regarding Nusaybin, the fixing the place of the molad in Jerusalem or Yavneh *because it was the place of the Rabbinical courts,* shows that the molad "baharad" is not "Halacha l'Moshe miSinai."

In modern times, Rabbi Yechiel Michel Tucazinsky also believes that the place where the molad was established was in Jerusalem, and on this issue there was a heated debate[87] between him and Rabbi Menachem Kasher, who wrote that Rabbi Tucazinsky ignores the opinions of the Rishonim who bring different places in the world. Every year it is written in the "Luach l'Eretz Yisrael" of Rabbi Tucazinsky:[88] The average molad... calculated from early times from the time of 'baharad' until today is from the location of Jerusalem." Indeed, in the year 5753, perhaps because of the criticisms against his late father on the subject, his son Rabbi Nissan added a page on this calendar to explain (and defend!) his father's opinion.[89]

In addition, the "Chazon Ish"[90] holds (but without citing a source) that the Rambam's opinion is that "our accepted time of the molad refers to Jerusalem." However on this, the "Meor Einayim"[91] wrote that the Rambam "did not mention any place." Therefore one needs to study further the words of the "Chazon Ish."

In fact, there has been no decision on the question on "which place the molad was established" There is a difference of 130 degrees between the opinions, namely 520 minutes - 8 hours and 40 minutes. This length of time is the doubt which enters the question of fixing the earliest and the latest times for kiddush levana.

However, Rabbi Yonah Merzbach[92] believes (but does not give it as a practical ruling) that no calculations need to be made regarding the place where the molad was established. Jews living in Paris should think that the molad was established *specifically in Paris,* and Jews living in California should think that molad was established *specifically in California,* etc! He argues that this was the method taken by the Great Rabbis of Europe. He concludes that even if by this method, the time of the full moon is passed, because one is speaking only of limited differences, it would not be noticeable in the decrease in the size of the full moon

On the other hand, the "Bnei Tzion"[93] (who believes that the location of the molad was established in Jerusalem) writes that for every place in the world one needs to calculate the latest time for kiddush levana according to the time of the molad in Jerusalem. [He learned this rule from the commentary of the Rambam on the Mishnah in connection with the laws of "chadash" (new produce). One learns from the Mishnah[94] that the offering of the Omer in Jerusalem on the sixteenth of Nissan was always before noon, and therefore places which are situated a long distance from Jerusalem, "chadash" is permitted for them from noon of that day. On this the Rambam[95] commented: "And therefore it is necessary for people to go according to the time of midday in Jerusalem, and not at midday of their location which could possibly earlier or later."] From this we see the importance of knowing the place where the molad was established and every place in the world needs to calculate the time of kiddush levana in relation to the place where the molad was established.

The Length of a Day

An additional problem in determining the exact time to recite kiddush levana arises from the change in the length of the day throughout the year.

Although the Rambam[96] writes that the length of each day is twenty-four hours, accurate measurements show us that it is only the average, and during the year the length of the day is a few seconds more or less than twenty-four hours.) One concludes from this that the time for midday (the time when the sun is located in the highest part of the sky) is not exactly at twelve o'clock (according to "local time"), and midnight can therefore occur from 11.44 to 12.14, and only on four days a year is it exactly at twelve o'clock,[97] (and they are not on the days on which the tekufot of Tishrei and Nissan occur!)

In the laws of Kiddush Hachodesh, the Rambam[98] writes that one calculates the molad from the month of Tishrei of the "first year of Creation." This molad is known as "molad baharad." The meaning of "baharad" is that this molad occurred on the second day of the week and 5 hours and 204 parts ("part" is a number of minutes). Regarding the molad, the day begins throughout the year at six o'clock in the evening, namely, *six hours after midday*,[99] and therefore "baharad" is 11 hours and 204 parts after midnight of the first day of the week.

One can ask how does one calculate these six hours? There are the following possibilities on how to calculate these six hours after midday. They are:

(a) Using the average time for midday, namely i.e. at 11.59 (but not exactly 12.00!)
(b) Using the days that the tekufah of Tishrei or the tekufah of Nissan occurs
(c) Six hours after midday on the very day that the molad was baharad

According to the first option, you only need to allow for the one minute between 11.59 and 12.00. According to the second option, the method proposed by Rabbi Yonah Merzbach,[100] a problem arises, namely the time of midday today for the tekufah of Tishrei is not the same time as for midday for the tekufah of Nissan! Therefore which one will one use!? (Tekufot go according to a solar calendar). According to the third option, one needs to calculate what was the date on the solar calendar of molad baharad and after that to calculate what was the time of midday of that day.

In conclusion, there are opinions regarding changing the length of the day during the year. These opinions can cause a doubt of up to a quarter of an hour for the beginning and the end times of kiddush levana. Are these minutes significant? Rabbi Yechiel Zilber[101] discusses this issue and concludes: "A ruling of the changes in the time of midnight regarding kiddush levana have not been decided upon, but one should be strict regarding these changes for both the starting and finishing times."

Lunar and Solar Eclipses

The issue of lunar eclipses and solar eclipses is found in the laws of kiddush levana, and to understand these laws, one needs to look at the details of these eclipses and the differences between them.

An eclipse of the sun occurs when the moon is between the sun and the earth. This event is called "conjunction," and takes place exactly at the moment of the true molad. Solar eclipses do not occur at every Rosh Chodesh, and the reason is that during its journey around the earth the moon will pass sometimes "above" the sun and sometimes "below" it. A total eclipse is seen only in a small area of the earth, and that is in places that are completely in the shade of the moon. In the nearby areas, only a partial eclipse is seen. Due to the rotation of the earth, the area in the world where the eclipse is seen varies during the eclipse, and at a certain place in the orbit of the eclipse, the maximum time of the total eclipse anywhere, does not exceed seven and a half minutes.

A lunar eclipse occurs when the earth is to be found between the moon and the sun. This is called a "conjunction" and occurs close to the time of the full moon. In the case when the sun, earth and moon are not exactly in one line, there will be a partial eclipse of the moon. A full lunar eclipse is seen all over the world when the moon is above the horizon, namely in that hemisphere. It starts and ends everywhere at the same time, and can continue for several hours.

The first to associate an eclipse with the times for kiddush levana was the "Maharil"[102] who writes: "If there is an eclipse of the moon ... one cannot recite the berachah for kiddush levana after that time." In addition, Rabbi Mordechai Yoffe,[103] the "Levush" rules like this.

The question arises that since one utilises a lunar eclipse, namely, the true moment of a full moon to determine the end of a kiddush levana, can one likewise use a solar eclipse, namely, the moment of a true molad to determine the time for the beginning of kiddush levana? In this case, the Maharil[104] answers in the negative, basing his answer on the Rambam in the laws of kiddush hachodesh. There the Rambam[105] explains the postponement "ad'u" in order that the first day of Tishrei will occur on the day of the true molad. Even though Rabbi Avraham ben David[106] the Ra'avad, (who hardly writes any of his comments on kiddish hachodesh!) vehemently rejects the words of the Rambam, the Maharil uses this Rambam to explain why a solar eclipse is not used to determine when it will be the time of the beginning of kiddush levana. In addition, the words of the Maharil were quoted by the Beit Yosef[107] in his commentary on the Tur.

A number of questions can be asked on the subject of eclipses and kiddush levana:

1) If one knows from astronomical tables that there will be a lunar eclipse *after* the end of the time of kiddush levana according to the opinion of the Rema, is it possible to use this fact as a leniency?

2) If one knows from the astronomical tables when the real conjunction will be, is it possible to use this fact as a leniency or a strictness to determine the end of the time of kiddush levana? In one of his responsa, the Maharil[108] has doubts on this subject, but however in another responsum[109] writes that "when there is no lunar eclipse, one relies on the calculations that our Rabbis have passed down to us on the molad." However, there is also a partial answer to this question from the grandson of the "Yeshuat Yaakov"[110] who came to a different conclusion. During the discussion on a case that was asked whether it is possible to say kiddush levana until the end of fifteen days after the molad in a particular month, he checked the real time of the true molad, and found that it occurred between the end of time according to the Rema, and the end of fifteen days after the average molad, and he utilised this fact as a *secondary reason* to permit the saying of kiddush levana.

3) In a case where there is only a partial eclipse of the moon, will this determine the end of the time of kiddush levana, and if so what percentage - 90 percent - 50 percent - 20 percent?

According to the opinion of the Maharil who forbids saying kiddush levana after a lunar eclipse, Rabbi Chaim Elazar Spira, the Gaon of Munkatch[111] writes: "And I was surprised, and I did not know their reason." According to his reasoning there should be equality between covering the moon during an eclipse, and the covering of the moon by clouds. Although it is forbidden to recite the berachah while the clouds cover the moon, it is permissible to recite it after the clouds have dissipated. Therefore even in the case of a lunar eclipse, the berachah should be allowed after the eclipse is over.

Summary

- There are differing opinions among the Poskim for the beginning and end of the time of kiddush levana, but almost everyone agrees that these times are from the time of the average molad and not from the day of Rosh Chodesh.
- The indecision to this very day regarding the place in the world where the molad was established causes a doubt of more than eight hours at the beginning and the end of the time for kiddush levana.
- A change in the length of the day during the year causes a doubt of up to about a quarter of an hour at the beginning and end of the time for kiddush levana.
- In the case where there is a lunar eclipse, the time of the eclipse (if it occurs before the average molad) will be the end of the time for the recital of kiddush levana.

Announcing the Molad

The *only* occasion that we use the time of the average molad each month is to fix the times for the beginning and end of kiddush levana. However, even though this is not *explicitly* written in the Rabbinical literature, there is room to suggest[112] that one reason for announcing the molad on the Shabbat before Rosh Chodesh is to inform the public about the times for the beginning and the end of kiddush levana.

Although the prayer for the new moon recited in the synagogues is already mentioned in the Siddur of Rav Amram Gaon,[113] the announcing of the molad during this prayer appears only in books of recent times.

In the book "Sha'arei Ephraim" written by Rabbi Efraim Zalman Margoliot[114] it states on this subject: "There are those who wrote that it is 'good to know' at the time when the prayer for the new moon is recited when the molad will occur since the main thing of the name 'month' is the renewal of the moon which is the molad." However, for this reason, it would be more accurate to announce the time of the "true molad" instead of the "average molad" because the true molad time is the time of the "renewal of the moon."[115]

In the book "Kranot Tzadik" by Rabbi Eliyahu Saliman Mani,[116] the Head of the Rabbinical Court in Hebron, there is a difference in his language regarding the instructions for the announcement of the molad: "On the Shabbat before Rosh Chodesh at the time when one says the prayer for the new month ... he should remember the time of the molad, namely, which day and which hour and which portion of the hour." We can see that unlike the "Sha'arei Ephraim" which says "it is good to know," Rabbi Mani writes that he will "remember" *exactly* the time of the molad.

Why does Rabbi Mani use the word "remember"? It is possible to answer this question from an experiment that took place about ten years ago. For half a year, immediately after the announcing of the molad, about a hundred people were asked to say what was the time of the molad that was announced? None of them could repeat the time of the molad exactly![117] From this experiment we learn that it is not enough to hear the molad passively, but it is necessary to make an effort to *remember it!*

Rabbi Mani[118] gives a source for this remark: "So I heard in the name of the Kabbalists." As is well known, the Kabbalists are most particular that seven days pass from the time of the molad before saying the kiddush levana. Therefore, it is important for a person to *remember* the time of the molad.

The question arises, if this is a reason for announcing the molad, why not announce the times for the beginning and end for kiddush levana? As an answer, it can be suggested that there are no disagreements of the time of the molad, but there are differences of opinion of when is the beginning and end of the time for reciting kiddush levana, and it is impossible to announce all the opinions!

There are also different opinions according to which time to announce the molad, namely according to the time that appears on the calendar, or according to the "local time." [For every 15 degrees longitude around the world, the clock changes by one hour. For convenience, the world is divided into only 24 time zones, with the difference of an hour between two neighbouring zones. The clock in a particular area is called the "official clock", and the real time is called the "local clock". In Jerusalem, the difference between these two clocks is about 21 minutes.]

Therefore, those (in Eretz Yisrael) who hold[119] that the molad should be announced according to the "local time," should subtract twenty-one minutes from the time of the molad that appears in the calendar. [Actually, one can be more precise. The molad is fixed to the nearest "part," namely, three and a third seconds. Twenty-one minutes is the closest minute, and the difference between "local time" and "official time" varies from place to place within the country. Therefore it is necessary to make the calculation for every place in the country to the accuracy of a "part," and declare the molad accordingly!]

Also some want to make a correction of one hour while there is a "summer time" in the country.

Had there not been doubts, which as we have seen above, exceed more than eight hours regarding the fixing of the location of the place where the molad was established, it would be good to announce the molad with the above amendments, to help the public with the mitzvah of kiddush levana. However, because there are doubts, it seems that it is preferable to announce the molad according to what appears in the yearly calendars.

References

Abbreviations
SA = Shulchan Aruch
OC = Orach Chaim
(1)   Talmud Bavli, Sanhedrin 41b
(2)   Talmud Yerushalmi, Berachot chap.9 halacha 2

(3)   see: Rabbi Mordechai Hakohen, "kiddush hachodesh v'kiddush levana", *Machanayim*, Masechet l'Chayalei Tzava Hagana l'Yisrael l'Rosh Chodesh, (Tzava Hagana l'Yisrael, haRabanut Hatzvait Harashit, 5724), vol.90 p.28

(4)   *Siddur Rav Saadia Gaon*, (published by Yisrael Davidson, Simha Assaf and Yissachar Yoel, Hevrat "Mekitze Nirdamim": Jerusalem, 5730), p.90

(5)   Rabbi Mordechai Hakohen, op. cit.

(6)   Rabbi Moshe miKotzi, *Sefer Mitzvot Gadol*, (SMaG), (Venice, 5307), Mitzvat Aseh, Birchot Levana, (p.114)

(7)   Maimonides, Rambam Mishneh Torah, hilchot berachot chap.10 halacha 17

(8)   Rabbi Menachem Meiri, Sanhedrin 41b

(9)   Rabbi Avraham Gombiner, *Magen Avraham*, SA OC chap.426 par.13

(10)   Rabbi Yaakov ben Yosef Reischer, *Shevus Yaakov*, (Metz, 5549), responsa, vol.3, question 39

(11)   Talmud Bavli: Pesachim 4a, Rosh Hashanah 32b

(12)   Rabbi Shlomo Kagan Hakohen, *Atzei Beroshim*, (Vilna, 5663), includes she'elot uteshuvot, chidushim ubeurim on SA Yoreh De'ah, chidushim meinyanim shonim, chap.63, first word: veata

(13)   Rabbi Zvi Elimelech Spira of Dinov, *Bnei Yissachar*, (Lemberg, 5620), part 1, ma'amarei Rosh Chodesh, ma'amar one, kiddush hachodesh, par.13, (p.21)

(14)   Maimonides Rambam, Mishneh Torah, hilchot berachot chap 10 halacha 17

(15)   see: Rabbi David Shapira, *Shu't Bnei Tzion*, (Jerusalem, 5698), part 1, chap.42 par.5

(16)   Aruch Hashulchan OC chap.426 par.13

(17)   Rabbi Yonah ben Avraham Gerondi, (Rabbenu Yonah), commentary on Rav Alfas l'masechet Berachot, chap.4, 21a, first word: Nahardei

(18)   The version quoted by Rabbenu Yonah does not appear in the different versions assembled by Dr. Michael Higger, Masechet Soferim, ("Devai-Rabbanon": New York, 5697), pp.337-38

(19)   *Sefer Kol Bo*, (Fürth, 5542), chap.43, (p.33). This version is also brought by Rabbi David Avudarham, *Avudarham*, (Venice, 5326), hilchot berachot, gate 8, but he brings this in the name of the Talmud Yerushalmi. However, in the Yerushalmi we have today it does not appear.

(20)   Masechet Soferim, chap.20 halacha 1

(21)   *Avudarham*, op. cit.

(22)   see: Mishnah Berurah, SA OC chap.426 par.20

(23)   Beit Yosef, Tur, OC chap.426, first word: garsinan

(24)   Rabbenu Yonah, op. cit.

(25)   Rabbi Yosef ben Meir Teomim, *Pri Megadim*, Eshel Avraham, OC chap.426 par.13

(26)   Mishnah Berurah, op. cit.

(27)   Rabbi Shmuel ben Nathan Halevi Kolin, *Machazit haShekel*, SA OC chap.426 par.13

(28)   *Bnei Tzion*, op. cit., par.11

(29)   Beit Yosef, op. cit., first word: katav

(30)   SA OC chap.426 par.4

(31)   e.g. Rabbi Chaim Yosef David Azulai ("Chida"), *Moreh b'Etzbah*, (Amsterdam, 5584), chap.6 par.182; Rabbi Kalfon Moshe Hakohen, *Shoel v'Nishal*, (Djerba, 5721), OC part 1, chap.47; Rabbi Yosef Chaim, *Ben Ish Chai*, second year, Parashat Vayikra, par.23

(32)   e.g. Rabbi Menachem Azaria MiPano, *Shu't haRama*, (Diharnport, 5548), responsum 78; Rabbi Tzvi Elimelech Shapiro from Dinov, *Igra Depirka*, (Sighet, 5750), part 2, remez (par.)247, (p.87)

(33)   Rabbi Refoel Menachem ben Shimon, *Nahar Mitzrayim*, (No-Amon, 5668), vol.1, hilchot Rosh Chodesh, chap.4

(34)   e.g. Rabbi Yoel ben Shmuel Sirkis ("Bach"), *Bach*, commentary on Tur OC chap.426, first word Ar'I; Rabbi David Halevi Segal ("Taz"), *Taz*, commentary on SA OC chap.426 par.3; Ba'er Heteiv, SA OC chap.426 par.10

(35)   Rabbi Ovadia Yosef, *Yechave Daat*, part 2, (Jerusalem, 5738), chap.24, (p.99)

(36)   Rabbi Chaim Vital, *Pri Etz Chaim*, (Korzec, 5542), sha'ar Rosh Chodesh, chap.3, (p.93)

(37)   *Shu't haRama*, Rama miPano, op. cit.

(38)   see for example: *Yechave Daat*, op. cit., (p.94); Rabbi Avraham Haputa, "zeman kiddush levana", *Noam*, annual to investigate problems in halacha, vol.14, (Machon Torah Shelemah: Jerusalem, 5731), p.136

(39)   Rabbi Yerucham ben Meshullam (Rabbenu Yerucham), *Toldot Adam v'Chavah*, (Venice, 5313), section 11, part 1, p.63

(40)   Tur, OC chap.426(

(41)   SA OC chap.426 par.32

(42)   Tur, op. cit.

(43)   SA op. cit.

(44)   Rabbi Avraham Abele Gombiner, *Magen Avraham*, SA OC chap.426 par.12

(45)   Rabbi Shmuel ben Nathan Halevi Loew (Kelin), *Machazit haShekel*, SA OC chap.426 par.12

(46)   Mishnah Berurah, SA OC chap.426 par.17

(47)   SA OC chap 426 par.3, Rema

(48)   Rabbi Yaakov ben Moshe Levi Moelin, ("Maharil"), *Shu't Maharil*, (Hanover, 5370), chaps.19, 166

(49)   Rabbi Yaakov Weil, *Shu't Mahari Weil*, (Kapost, 5595), chap.159

(50)   Rabbi Moshe Sofer, *Shu't Chatam Sofer*, (Vienna, 5655), vol.1, OC chap.102

(51)   Talmud Bavli Rosh Hashanah 20b

(52)   Some find difficulties with this opinion of the Chatam Sofer, for example: Rabbi Yona Merzbach, *Aleh Yonah*, (Jerusalem, 5749), zeman hamolad l'inyan kiddush levana, p.33 note 1

(53)   *Luach l'Eretz Yisrael*, arranged by Rabbi Yechiel Michel Tucazinsky, for example year 5746, chodesh Tishrei, note 3

(54) *Bnei Tzion,* op. cit., par.10

(55) Sha'arei Teshuvah, SA OC chap.229 par.3

(56) see: Charles Elkin, "Birkath Hachamah: Blessing of the Sun", *Proceedings of the Associations of Orthodox Jewish Scientists,* vol.6, (Feldheim Publishers, Jerusalem, 5741/1980), p.96,

(57) *Bnei Tzion,* op. cit., par.4

(58) Rabbi Moshe Margolit, *Mareh haPanim,* commentary on Talmud Yerushalmi, Berachot chap.9 halacha 2, first words: ad yud-daled yom

(59) *Bnei Tzion,* op. cit., par.9

(60) Rabbi Zerachiah Halevi of Girona (Baal haMaor), *haMaor haKatan,* commentary on Rav Alfas l'Masechet Rosh Hashanah 5a, first words: ki salik

(61) Rabbi Yehudah Halevi, *Sefer haKuzari,* (Vilna, 5665), second paper, par.20

(62) Rabbi Chaim ben Yisrael Benvenist, *Shiyurei Knesset haGedolah,* (Livorno, 5553), OC Tur chap.426, par.2

(63) e.g. Rabbi Moshe Teitelbaum, *Heshiv Moshe,* (Lemberg, 5626), OC, responsum 14; Rabbi Yosef Shaul Nathanson, *Shoel u'Meishiv,* (Brooklyn New York, 5714), vol.3 chap.151

(64) Rabbi Shimon ben Zemach Duran, *Sefer Tashbatz,* (Amsterdam, 5501), part 3, subject 215, (p.45b)

(65) Rabbi Menachem Mendel Kasher, *Kav haTa'arich haYisraeli,* (Bet Torah Sheleimah: Jerusalem, 5737), p.91 note 12

(66) Ibid., p.91

(67) Rabbi Avraham bar Chiyah, *Sefer Haibur,* (London, 5611), second paper, gate seven, (p.54)

(68) *Baal Hamaor,* op. cit.; *haKuzari,* op. cit., *Midrash Tanchuma,* (Lvov), parashat Kedoshim, par.10

(69) Maimonides, Rambam, Mishneh Torah, Hilchot Kiddush Hachodesh, chap.11 halacha 17

(70) Talmud Bavli, Rosh Hashanah 20b

(71) Rabbenu Yitzchak ben Yosef haYisraeli, *Yesod Olam,* (Berlin, 5608), third paper, chap.15; also the Rambam brings the range of the degrees when it is possible to see the moon; this begins from 9 degrees, namely from 18 hours after the new moon: (Rambam Mishneh Torah hilchot kiddush hachodesh chap.12 halachot 15-21)

(72) Talmud Bavli, Rosh Hashanah 20b

(73) *Sefer Haibur,* op. cit.

(74) *Yesod Olam,* op. cit., fourth paper, chap.7

(75) *Sefer Haibur,* op. cit.

(76) *Yesod Olam,* op. cit., fourth paper, chap 7

(77) Rabbi Avraham Yeshaya Karelitz ("Chazon Ish"), *Kuntrus 18 Shaot,* chap.64, par.14, first word: ken hevi

(78) MS Paris, heb 1078/4 folios 41b-42a; MS Vatican Ebr 379/1

(79) Rabbi Azariah dei Rossi min ha'Adumim, *Meor Einayim,* (Vilna, 5624), matsref lakesef, second paper, chap.5, p.51, first words: ach nashuv

(80) *Baal Hamaor*, op. cit.

(81) *Hakuzari*, op. cit.

(82) Rabbi Avraham ibn Ezra, *Igeret haShabbat*, (Livorno, 5590), second gate, p.63

(83) Rabbi Avraham ibn Ezra, *Sefer Haibur*, (Hevrat Mekitze Nirdamim: Lyck, 5634/1874), p.10b

(84) MS Bodleian catalogue Neubauer 2047/1 folios 33b-34a

(85) *Meor Einayim*, op. cit., p.51, first words: vekvar yadayti

(86) Rabbi Shimon ben Natan Neta Woltish, *Navah Kodesh*, (Berlin, 5546), Rambam Mishneh Torah, Hilchot Kiddush Hachodesh, chap.6 par.6

(87) see: *Kav haTa'arich haYisraeli*, op. cit., pp.14-16

(88) *Luach l'Eretz Yisrael*, op. cit., for example year 5752, chodesh Tishrei, note 2

(89) Ibid., year 5753, hashaot bizmanei hamoladot, (towards the end of the luach)

(90) Chazon Ish, op. cit., par.16, first word: heivi

(91) *Meor Einayim*, op. cit., p.52, first words: ukvar yadati

(92) *Aleh Yonah*, op. cit., p.35

(93) *Bnei Tzion*, op. cit., par.9. After that he writes: "Indeed one needs to investigate that they were not particular in the western part of the world on the longtitude of Jerusalem with regards to the end of the time of kiddush levana."

(94) Mishnah, Menachot chap.10 mishnah 5

(95) Peirush of Rambam on the Mishnah, Menachot chap.10 Mishnah 5; Rabbi Yom Tov Lipmann Heller, *Tosafot Yom Tov* on the Mishnah, Menachot, chap.10, Mishnah 5

(96) Maimonides Rambam Mishneh Torah hilchot kiddush hachodesh chap.6 halacha 2

(97) see for example: Rabbi Meir Posen, *Or Meir*, (London, 5733), tables of times

(98) Rambam Maimonides, Mishneh Torah hilchot kiddush hachodesh, chap.6 halacha 8

(99) *Navah Kodesh*, op. cit.

(100) *Aleh Yonah*, op. cit., p.36

(101) Rabbi Yechiel Avraham Zilber "shigagat minhagim", *Otzrot Yerushalayim*, vol.72, (Jerusalem, 5731), p.1151

(102) *Maharil*, op. cit., chap.19

(103) Rabbi Mordechai Yoffe, *Levush*, OC chap.426 par.4

(104) *Maharil*, op. cit., chap.19

(105) Rambam Maimonides, Mishneh Torah hilchot kiddush hachodesh, chap.7 halacha 7

(106) Rabbi Avraham ben David ("Ra'avad"), *Hasagat haRa'avad* on Rambam Mishneh Torah, hilchot kiddush hachodesh, chap.7 halacha 7

(107) Beit Yosef, Tur, OC chap426, first words: uma shekatav Rabbenu

(108) Maharil, op. cit., chap.166

(109) Ibid., chap.19

(110) Rabbi Tzvi Hirsch Orenstein, *Yeshuat Yaakov*, (Lorje and Sperling: Lemberg, 5623), OC after chap.426

(111) Rabbi Chaim Elazar Spira, *Nimukei Orach Chaim*, (Slovakia, 5690), chap.426 par.3

(112) Charles Elkin and Martin Schechter, "The Molad will be --- when", *Intercom*, vol. 20, no. 1, 1983, (Association of Orthodox Jewish Scientists, New York), p.21,

(113) *Seder Rav Amram Gaon*, (Warsaw, 5625), section one, seder Rosh Chodesh, p.33. However, this prayer appears in his siddur amongst the prayers for Rosh Chodesh. There are those who say (note to the siddur of Rav Amram Gaon, op. cit.) that this prayer "was copied to here by mistake, and in fact it is part of the service for the Shabbat before Rosh Chodesh" and there are those who say (Rabbi Rachamim Sar-Shalom, Sha'arim l'Luach HaIvri, p.124, bircat hachodesh, (Tel-Aviv,5744), that originally they would recite bircat hachodesh on Rosh Chodesh itself, and only at a later date they transferred bircat hachodesh to the Shabbat prior to Rosh Chodesh."

(114) Rabbi Efraim Zalman Margoliot, *Sha'arei Efraim*, (Dubno, 5580), gate yud, par.37, (p.23),

(115) Professor Efraim Halevi Frankel, *Encyclopedia Hebraica* (Encyclopedia HaIvrit), (Company for publishing the Encyclopedia: Jerusalem, 5729), vol.21, p.343, "luach"

(116) Rabbi Eliyahu Saliman Mani, *Kranot Tzadik*, (Jerusalem, 5653), chap.3 par.93, (p.62b)

(117) Elkin and Schechter, op. cit., p.22

(118) *Kranot Tzadik,* op. cit.

(119) Rabbi Moshe Sternbuch, *Moadim Uzemanim Hashalem*, (Jerusalem, 5724), part one, chap.20, note 1, (p.48); Rabbi Shimon Wirtzberger, "zeman hamolad b'luchot hamitparsim b'Aretz", *Hama'ayan*, (Jerusalem), vol.24 issue 3, Nissan 5744, pp.26-28

Appendix

Follow on to publication of my paper.

The material below is taken from my unpublished autobiography.

Soon after my paper was published in "Sinai" I received a letter from Engineer Ya'acov Levinger from Tel-Aviv. (I asked the family of Rabbi Moshe Levinger whether he was a relative and the answer was in the negative.) After saying that he enjoyed my article "which summed up well the subject," he put forward a number of comments.

The first two Rabbinic authorities who had given an actual date in the secular calendar (Julian as it then was) to start asking for rain, were the Avudarham and the Tashbetz. I had shown in my article that whereas the Tashbetz was completely accurate, this was not so with the Avudarham. Levinger tried to forcefully argue, using some "acrobatics," that the Avudarham was also correct. However, Levinger's explanation falls down since the tenses used by the Avudarham are wrong.

Levinger also held that the place in the world where the Tekufot (and Moladot) were fixed was the "centre of settlement." In fact, this is only one of the many opinions. The Chazon Ish went as far as to say that this opinion was a non-Jewish one and had no place in the Bet Hamidrash. Two other points in which he disagreed with me was that one must regard a day as exactly 24 hours, irrespective of its variable length and that the time of midday today is not the same as it was thousands of years ago. These points are open to further study although it would seem that on the latter point Levinger is correct.

READING THE MEGILLOT
SHIR HASHIRIM, RUT, EICHAH
AND KOHELET

Introduction

Among the twenty-four books in the Tanach there are five which have the designation "Megillah." They are: Shir Hashirim (Song of Songs), Rut (Ruth), Eichah (Lamentations), Kohelet (Ecclesiastes) and Esther. However, only the reading of Esther is mentioned in the Talmud. The reading of the other four is not mentioned neither in the Babylonian Talmud nor in the Jerusalem Talmud, but in the Masechet (tractate) called Soferim, which is one of the "Minor Tractates."

In Masechet Soferim, chapter 14, it is written that one recites the berachah (blessing) "Al Mikra Megillah" (and has commanded us on the reading of the megillah) before the reading of Rut, Shir Hashirim. Eichah and Esther, and according to some versions also before Kohelet.

In that chapter of Masechet Soferim it is stated that there are different opinions for the time for reading Shir Hashirim and Rut, and in chapter 18, the time for reading Eichah is given. However, the time for reading Kohelet is not mentioned at all in Masechet Soferim, and this includes the versions which state that one says the berachah before reading Kohelet.

Megillat Shir Hashirim: The Optimal Date for Reading it according to the Various Communities

It is written in Masechet Soferim:[1] "Shir Hashirim is read in the Diaspora (where there are two days of Yom-Tov) on the evenings of the last two days of Pesach, namely, half on the first night and the other half on the second night. The first half of Rut is read at the termination of the first day of Shavuot and the remainder at the termination of the second day. However, some say that one begins to read both of these megillot on the termination of the Shabbat prior these Festivals. The custom is to act according to these opinions, since the halacha is not fixed until a custom is established.

One can conclude from this that the time for reading Shir Hashirim is either on the last nights of Pesach, or that one begins to read it on the termination of Shabbat prior to Pesach. From the language of Masechet Soferim, the custom was according to the second opinion, and on this the book "Mikra Soferim"[2] writes that one does not prevent a person from doing it this way since the reading was not an established halacha, and it is possible that a custom can change a halacha, and that the halacha of the time to do this reading was determined by a custom.

The reading of the megillot on the termination of Shabbat prior to the Festivals is not so strange! Masechet Soferim[3] also mentions the reading of Megillat Esther, namely half on the termination of the first Shabbat of the month of Adar, and half on the termination of the second Shabbat of Adar. Although it is not in accordance to the Babylonian Talmud, the Jerusalem Talmud[4] brings an opinion to the reading of Megillat Esther from the beginning of the month of Adar. In addition, the Shulchan Aruch[5] quotes this opinion. However, the comparison between the other megillot and Megillat Esther requires further investigation since the reading of Megillat Esther from the beginning of Adar can be learned from the verse "And the month which was turned for them from sorrow to gladness."[6]

The question arises whether the time of reading Shir Hashirim is determined today according to Masechet Soferim? Apparently, there is no evidence that in practice one reads Shir Hashirim on the last nights of Pesach or on the termination of Shabbat before Pesach.

Although there are a number of Sefaradi communities who do not read Shir Hashirim on Pesach, (except on the Seder nights[7]) there are communities that read it on the last days of Pesach. One should note that this is close to, but not identical, to the first opinion mentioned in Masechet Soferim. One community that practices this is the Yemenite Jewish community that reads Shir Hashirim (in the Diaspora) half on the afternoon on the seventh day of Pesach and the other half on the last day of the Festival,[8] and there are those who read the entire megillah on the last day of Pesach.[9] [In Eretz Yisrael the first half is read on Shabbat Chol Hamoed (the intermediate Shabbat of Pesach) and the second half on the seventh day of Pesach.[10]]. The custom is to read it together with the Aramaic translation.[11] Rabbi Yosef Kapach adds:[12] "In the synagogue of my late grandfather, and in several other synagogues, it was customary to add the translation by Rav Saadia Gaon into Arabic." On the other hand, Rabbi Yosef Tzubiri (Chief Rabbi of Yemenite Jewry in Tel Aviv) writes: [13] "We have never seen or heard of using the translation into Arabic by Rav Saadia Gaon in the synagogue reading ... nor in the reading in the other three Megillot."

A similar custom (that is brought by Rabbi Kapach) is found among the Jews of Libya,[14] who on the last two days of Pesach (in Eretz Yisrael on the seventh day) gather together in several houses, and read Shir Hashirim verse by verse in a pleasant tune, together with the Targum (translation by) Yonatan Ben Uziel, and also with a translation into Arabic. The hosts serve refreshments, drinks and a variety of dainties.

The Jews of Georgia also divide Shir Hashirim into two parts, namely on the seventh and eighth days of Pesach, and add a translation into Georgian. The content of the Rabbi's sermon before the minchah (afternoon service) on these two days is, the love between G-d and the Jewish people, which is the message of Shir Hashirim.[15]

According to the custom of the Jews of Algiers,[16] Shir Hashirim is recited after the minchah service of the last day of Pesach. There is a similar custom among the Jews of Rome.[17] It should be mentioned that although Italy is in the middle of Europe, the details of their prayers are not the same as the Ashkenazi order of prayers.

There are also other customs regarding the reading of Shir Hashirim on the days of Yom Tov of Pesach. One of them is to found in Mahzor Romania[18] from the year 5280 which mentions (among other things) the division of this book into four parts, namely for the first, second, seventh and eighth days of Pesach. The Romanian order of service was in use until approximately the end of the sixteenth century.

The custom of the Jews of Bukhara is to divide the reading of Shir Hashirim and its translation into Persian, into four or five days during Pesach. The children of the Talmud Torah there studied its commentary before Pesach so that it would be conversant with them on the Festival.[19] In addition, the Jews of Bukhara read Shir Hashirim together with the commentary on Shabbat Chol Hamoed Sukkot.[20]

Another custom is found in the Spanish-Portuguese community in North America,[21] who read Shir Hashirim on the first two days of Pesach. In the Sefaradi communities in England, the entire Shir Hashirim is read on one of the days of Pesach.[22]

An entirely different custom, which is practiced by the Ashkenazim, is to read Shir Hashirim on Shabbat Chol Hamoed Pesach. The earliest source seems to be found in the customs of the Maharam of Rothenburg[23] who writes that it is said after Hallel on Shabbat Chol Hamoed Pesach. This is also mentioned by a number of Rishonim (great Rabbis who lived approximately between the eleventh and fifteenth centuries) and Acharonim (great Rabbis who lived from about the 16th century).[24]

A question which arises is what is the source of reading Shir Hashirim specifically on Shabbat Chol Hamoed Pesach, and not on the last day(s) of Pesach? The reason given by the Vilna Gaon[25] is that "Shabbat is a day for people to gather together, and even on Yom-Tov people are in the fields."

The Mishnah Berurah[26] gives another reason for its reading on Shabbat Chol Hamoed: "Because the piyyutim (liturgical poems) of that Shabbat are related to Shir Hashirim." However, also the piyyutim of the first and second days of Pesach are also related to Shir Hashirim, so why according to his reasoning does one not read Shir Hashirim on the first or second days of Pesach? Let us examine the possibility that the piyyutim for Shabbat Chol Hamoed, relating to the Shir Hashirim were written prior to the piyyutim for the first and second days, and the custom of reading Shir Hashirim on Shabbat Chol Hamoed was established before the writing of the piyyutim for the first and second days. The piyyutim for the first day were written by Shelomo haBavli (or possibly by one of his friends or relatives)[27] who lived in the middle of the tenth century; for the second day they were written by Meshullam the son of Rabbi Kalonymos,[28] who was a pupil of Shelomo haBavli, and lived in the late tenth and early eleventh centuries; for Shabbat Chol Hamoed by Shimon ben Yitzchak (Shimon ibn Avun)[29] who lived in the second half of the tenth century. We see from this that apparently the chronological order of writing these piyyutim was: first day of Pesach, Shabbat Chol Hamoed, second day of Pesach. Therefore, the words of the Mishnah Berurah require further study.

However, not in every year is there is a Shabbat Chol Hamoed! The time of reading Shir Hashirim in a year which does not have a Shabbat Chol Hamoed is mentioned in the book of customs for Ashkenazim written by Rabbi Chaim Paltiel[30] at the end of the thirteenth century. He writes in connection with the last two days of Pesach: "If the seventh day of Pesach or the eighth day is on Shabbat one reads Shir Hashirim on the day which is Shabbat." It is of interest to note that in his customs for Shabbat Chol Hamoed, Rabbi Paltiel does not include reading Shir Hashirim! Glosses on the book of Rabbi Paltiel were written by Rabbi Avraham Klauzner who lived one hundred years later. We have a number of manuscripts of these glosses,[31] but not all of them are identical! In one of them[32] it is written in the margin by the customs of Shabbat Chol Hamoed Pesach: "Yozer ahuvecha ... Berach Dodi Shir Hashirim." It is not known if these glosses were written by Rabbi Klauzner or by someone else.

In the year in which the first day of Pesach occurs on Sunday, the seventh day of Pesach will fall on Shabbat, and everyone agrees that those who are accustomed to read Shir Hashirim on Shabbat Chol Hamoed, read it on that year on the seventh day of Pesach.[33] However, in a year when the first day of Pesach is on Shabbat, in the Diaspora there are two Shabbatot during Pesach, that is, on the first day and the eighth day and the question arises as to which of them does one read Shir Hashirim? The "Poskim" (Rabbinical arbiters) write that it is read on the eighth day.[34] Rabbi Yosef ben Meir Teomim[35] known as the "Pri Megadim," (18th century) explains that Shir Hashirim is not said on the first day which falls on Shabbat, because there are many piyyutim in the prayer for tal (dew). However in Eretz Yisrael, there is only one Shabbat during Pesach, namely on the first day, and therefore Shir Hashirim is read on that Shabbat.[36] One should mention that since the establishment of the Ashkenazi community in Jerusalem by the students of the Vilna Gaon, it is customary in Jerusalem (and apparently in the entire Eretz Yisrael) not to add in piyyutim during the service, with the exception of Rosh Hashana and Yom Kippur.[37] Therefore the reason given by the Pri Megadim is not relevant!

Megillat Rut: The Optimal Date for Reading it according to the Various Communities

We have seen above that it is written in Masechet Soferim that half of Rut is read on the termination of the first day of Shavuot, and half on the termination of the second day, or alternatively one begins reading it on the termination of Shabbat before the Festival. As in the case of Shir Hashirim, these two possibilities are not utilised today.

However, the division of the book into two, namely half on the first day and half on the second day, takes place in a number of Sefaradi communities that read Rut in the afternoon of Shavuot, after having read the "azharot, (a group of piyyutim summarising the 613 commandments in the Torah).[38] In some congregations, each verse is first read in the original "Lashon Kodesh" (Hebrew), and then the same verse is translated,[39] for example, into the Ladino language. As in the reading of Shir Hashirim, that took place in Rabbi Kapach's grandfather's synagogue, they also read Rav Saadia Gaon's translation into Arabic.[40]

Also in "Mahzor Romania"[41] the division of Rut into two parts is mentioned, but it does not mention that a translation should be added. On the other hand, the Algiers custom[42] is not to divide Megillat Rut into two parts, but to read the entire book on the second day after minchah (afternoon service).

In the case of the Jews of Afghanistan the head of the family reads the "azharot" at home after they have finished the Festival meal, and then they return to the synagogue to read Megillat Rut before minchah.[43] Among the Jews of Tunisia, the Synagogue arranges the reading of Rut and the "azharot," only for those who did not read them at home.[44]

The custom among Ashkenazim is to read Rut on the second day of Shavuot before the reading of the Torah.[45] The "Pri Megadim"[46] gives two reasons for reading Megillat Rut specifically on the second day of the Festival and not on the first day. The first reason is because the Torah was given on the seventh day of Sivan, namely the second day of Shavuot in the Diaspora. The second reason is that because on the first day, the men were awake all night, and during the Synagogue service there are many piyyutim.

The order of "Tikkun Leil Shavuot" includes, among other things, the beginning and end of each parashah in the Torah, and likewise in each book of the Nevi'im (Prophets) and Ketuvim (Writings). However, only Megillat Rut is read in its entirety. The question arises, whether in reading Rut during the "Tikkun Leil Shavuot" a person fulfils his obligation of reading Megillat Rut on Shavuot.

Megillat Eichah: The Optimal Date for Reading it according to the Various Communities

It is written in Masechet Soferim:[47] "Some read Kinot (Megillat Eichah) in the evening, whilst others delay the reading until the morning after the reading of the Torah." According to Masechet Soferim there are two options, namely the first which is in the evening, and the second which is in the morning. The "Tur"[48] (Rabbi Yaacov ben Asher, 13th-14th centuries) brings the words of Masechet Soferim. The "Beit Yosef" (Rabbi Yosef Karo, 16th century) in his commentary on the "Tur"[49] mentions both of these two opinions, but the Shulchan Aruch mentions only the reading at night! Rabbi Shem Tov Gagin (who was head of the Sefaradi Rabbinical Court in England, 20th century) was surprised at this omission! In general, one can say that there are the following two reasons for this. Sometimes the Beit Yosef finds authorities who disagree with what he has specifically written on the Tur, and therefore he retracts on what he has written in the past, and sometimes he omits writing it in the Shulchan Aruch, because in his opinion the matter is obvious.[51] One needs to investigate which of these two reasons (or even one of them) is relevant in this case!

The reading of Eichah on Tisha b'Av is also mentioned in the "Midrash Eichah Rabati":[52] "Since they are sitting and eating and drinking and getting drunk at the Tisha b'Av meal, they sit and recite kinot (dirges) and wailings and Eichah." According to this source, the reading is done at night. Rabbenu Yerucham[53] (14th century) cites the Midrash Eichah Rabbati as a source for reciting Eichah at night, but he does not mention Masechet Soferim. However, the "Hagahot Maimuniot"[54] (Rabbi Meir Hakohen, 13th century), brings both Masechet Soferim and Eichah Rabbati.

There are attempts to prove from the Gemara that one must recite Eichah on Tisha b'Av. The Rema (Rabbi Moshe Isserles, 16th century) writes:[55] Reading [Megillat Eichah] is mentioned in the Gemara... as it states one reads kinot [Eichah] ... but the reading of the Megillot [Shir Hashirim, Rut and Kohelet] are only a custom." However, one need to investigate these words of the Rema, since the Gemara is talking about things which are permitted to learn on Tisha b'Av. In that same Gemara,[56] it is also written that it is permitted to learn the book of Iyov on Tisha b'Av, but there is no obligation to read the book of Iyov on Tisha b'Av, even though there are some Sefaradi communities who read Iyov on the morning of Tisha b'Av.[57] Rabbi Shem Tov Gagin[58] also wants to bring proof from the Jerusalem Talmud[59] that one is obligated to read Eichah on Tisha b'Av. There it is written that a number of Rabbis were sitting and learning Eichah from minchah onwards on the day before Tisha b'Av, and they said that they would finish it on the next day, namely on Tisha b'Av. But for the same reason as before, this requires further investigation![60]

According to the customs of all the Jews throughout the world,[61] Eichah is read on the night of Tisha b'Av.

In the Jewish communities of Persia they read Megillat Eichah and the kinot that follow it in a translation into Persian.[62]

Regarding the reading of Eichah on the morning of Tisha b'Av, the "Shelah"[63] (Rabbi Yeshayahu ben Avraham Halevi Horowitz, 16th–17th centuries) basing himself on the Tur writes: "It is a good custom for every individual to read again Megillat Eichah for himself on the morning of Tisha b'Av. The Mishnah Berurah[64] and the "Ba'er Heteiv"[65] (Rabbi Yehudah ben Shimon Ashkenazi, 18th century) bring the words of the "Shelah."

In a synagogue where it was customary on the night of Tisha b'Av to read Eichah from a hand written parchment scroll with a berachah, and on one year the worshipers by force of circumstances could not do so, the question arose whether they could read it in the morning with a berachah?

According to Rabbi Chanoch Zundel Grossberg (20th century, Jerusalem) in his book "Zer HaTorah,"[66] the answer is positive, and he brings a support from Masechet Soferim. The "Rivevot Ephraim"[67] (Rabbi Ephraim Greenblatt, 20th –21st centuries), agrees with this opinion, and he also brings that this is the opinion of Rabbi Neta Zvi Greenblatt (20th century, USA) and Rabbi Shmuel Monk.

On the other hand, Rabbi Yechiel Tucazinsky[68] writes that in 5708 during the War of Independence on the night of Tisha b'Av there was a shelling of Jerusalem, and many Synagogues did not read the Megillah at night. In the morning they asked if it could be read with a berachah. He answered in the negative since the obligation to read Eichah in public is only at night, and we learn this from the verses: "And the people wept that night,"[69] and "She weeps in the night."[70]

The question also arises as to whether it is possible to recite Eichah from the time of "plag haminchah" (about an hour and a quarter before sunset) of the evening of Tisha b'Av. This question was submitted to Rabbi Yitzchak Weiss,[71] Head of the Rabbinical

Court in the city of Verbo in Slovakia during the Second World War. There was "an order from the non-Jews that the Jews were not allowed to go outside after eight o'clock," and in that year Tisha b'Av began at the termination of Shabbat. Rabbi Weiss answered this question in the negative, because reading Eichah before the termination of Shabbat was a "strange thing for the multitude." In addition, it is written in the Gemara: "One does not advance a calamity,"[72] and according to this, even if Tisha b'Av did not begin at the termination of Shabbat, one cannot advance the reading of the megillah.

There are different opinions among the Rabbinical authorities as to whether one can only read Eichah amongst a congregation. According to the "Chayei Adam" (Rabbi Avraham Danzig, 18th–19th centuries),[73] an individual can read Eichah at night, but the "Levush" (Rabbi Mordechai Yoffe, 16th-17th centuries)[74] writes that one reads it amongst a congregation. According to Rabbi Yitzchak Weiss,[75] (20th century) one fulfils the opinion of the "Levush" if one reads it at night individually, but on the following day reads it in a congregation after the kinot.

The question can be asked whether in the opinion of "Chayei Adam" an individual who reads Eichah at night in his home can also recite the berachah? Rabbi Tucazinsky's "Luach l'Eretz Yisrael"[76] which brings the ruling of "Chayei Adam," adds that an individual who reads Eichah from a kosher megillah does not recite the berachah.

Megillat Kohelet: The Optimal Date for Reading it according to the Various Communities

The time for reading Megillat Kohelet is not mentioned at all in Masechet Soferim. This even includes the versions that write about the berachot recited over Kohelet.

Probably the earliest sources for reading Kohelet on Sukkot are in Rashi's Siddur and Rabbenu Simchah's "Machzor Vitry." In one place[77] it is written that one reads Kohelet on Shabbat Chol Hamoed Sukkot, and in another place[78] it states that "if one has not yet read it," then it is read on Shemini Atzeret.

Namely, lechatchila (from the outset) the time for reading it is on Shabbat Chol Hamoed Sukkot, but b'diavad (in retrospect) it is read on Shemini Atzeret. These two sources learn the time for reading Kohelet from the verse in Kohelet: "Divide a portion into seven, and even into eight."[79] "Seven" is the seven days of the Festival of Sukkot, and "Eight" is the eighth day, Shemini Atzeret. However, according to this interpretation, it seems that it can be read on any of the seven days of the Festival, and not necessarily on Shabbat Chol Hamoed! However, the books "Yafe laLev"[80] (Rabbi Yitzchak Palagi (19th century, Izmir Turkey), and "Bechorei Yaacov"[81] (Rabbi Yaakov Ettlinger, 19th century), learn from the word "Seven" that it means specifically on Shabbat, namely the seventh day of the week.

Among the other Rishonim there are disagreements on the date of its reading, with some saying[82] Shabbat Chol Hamoed, whilst others[83] say Shemini Atzeret. It should be noted that whilst the "Orchot Chaim" (Rabbi Aharon Hakohen m'Lunel, 13th-14th centuries), writes that Shir Hashirim is read on Shabbat Chol Hamoed Pesach, for Kohelet he writes "The custom of France is to read it on Shemini Atzeret" and he makes no mention at all of Shabbat Chol Hamoed Sukkot!

It seems that today there are those who read the entire megillah on Shabbat Chol Hamoed. However those who read it on Shemini Atzeret in the Diaspora, divide the book between the two last days, namely Shemini Atzeret and Simchat Torah.

The vast majority of the Sefaradi communities do not read Kohelet. The exceptions are the Yemenite communities, namely both Baladi and Shami. Their custom in the Diaspora is to divide the megillah between Shemini Atzeret and Simchat Torah.[84] (In Eretz Yisrael, it is divided into three days, namely the first day of Sukkot, on Shabbat Chol Hamoed and Shemini Atzeret[85]) and it is read with an Aramaic translation, and there are some congregations, who in addition, read the translation in Arabic of Rav Saadia Gaon.[86]

According to Mahzor Romania,[87] there are congregations that divide the megillah into four parts, the first day of Sukkot, the second day of Sukkot, Shemini Atzeret, and Simchat Torah. The custom of the Bnei Roma[88] is to read Kohelet before minchah on Simchat Torah.

In the years when there is no Shabbat Chol Hamoed, the first day of Sukkot will occur on Shabbat and (the first day of) Shemini Atzeret will therefore also occur on Shabbat. The question arises, on which of them is Kohelet read? The "Pri Megadim"[89] brings the two possibilities and asks if on Pesach they do not recite Shir Hashirim when the first day of Pesach occurs on Shabbat, because of the lengthy piyyutim recited with "tefillat tal" (prayer for dew). In a similar way, why read Kohelet on Shemini Atzeret when there are the lenghty piyyutim for "tefillat geshem" (prayer for rain)?

However, the custom (in the Diaspora) is, as mentioned by Rabbi Yaakov ben Moshe Levi Moelin,[90] known as the "Maharil," (14th-15th centuries), and quoted by the Rema in his "Darchei Moshe"[91] and in his glosses on the Shulchan Aruch,[92] to recite Kohelet on Shemini Atzeret in the years when there is no Shabbat Chol Hamoed. However Rabbi Sraya Davlitzki[93] (20th-21st centuries, Eretz Yisrael) writes, that in Eretz Yisrael the followers of the Vilna Gaon were accustomed to read Kohelet on the first day of Sukkot. The reason for this is the extremely lengthy service is as a result of the "hakafot" (circuits made around the Synagogue with the Sifrei Torah) and the Reading of the Torah (where all those present get called up to the Torah). It should be noted that in Eretz Yisrael, Shemini Atzeret is also Simchat Torah. He does not mention the piyyutim recited in the prayer for rain as a reason for the extension of the service. It is possible that because reciting all the piyyutim is of relatively less time compared with the hakafot and Torah reading! In addition, as we have seen above, the custom in Eretz Yisrael is not to recite piyyutim during the prayers, except on Rosh Hashanah and Yom Kippur.

The "Bechorei Yaakov"[94] has a different opinion to read Kohelet specifically on Shemini Atzeret occurring on a Shabbat. On the verse "Divide a portion into seven, and even into eight." he writes that when Shabbat does not occur on Shemini Atzeret seven takes precedence over eight, but when Shabbat and Shemini Atzeret occur on the same day then Shemini Atzeret is the day for reading Kohelet.

In Eretz Yisrael, however, the practice on this subject, as it appears in the "Luach l'Eretz Yisrael" by Avraham Lunz[95] and by Rabbi Yechiel Tucazinsky,[96] is according to the custom of the students of the Vilna Gaon.

An Ashkenazi Rabbi from Switzerland wrote to Rabbi Sraya Davlitzki, saying that in his opinion (with the exception of the Ashkenazi Perushim) the Ashkenazi Jews in Eretz Yisrael should "act according to the Rema and read [Kohelet] on Shemini Atzeret [in the years in which it falls on Shabbat]." In his reply Rabbi Davlitzki wrote that the custom to read Kohelet on the first day of Sukkot "goes back to the time of the foundation of the Ashkenazi community in Eretz Yisrael." Regarding the leadership of the Hasidim (those whose custom is to read Kohelet) in this matter, he asked the Hasidim of Slonim and the answer he received was "that in Tiberias and Jerusalem they do not read it at all, but in Bnei Brak they do read it, and when the first day occurs on Shabbat, they read it on the first day [of Sukkot]." Rabbi Davlitzki also clarified the custom of the Great Synagogue in Safed, "from a Safed man who was well versed in their customs and today lives in Jerusalem" and the answer was "that in principle they read it on Shabbat Chol Hamoed, but when the first day [of Sukkot] falls on Shabbat, it is possible that they do not read it at all. However, he does not remember it ever being read on Shemini Atzeret that occurred on Shabbat." Furthermore, Rabbi Davlitzki also held that among the Hasidim who read Kohelet, "will be found that everyone reads it on the first day when it occurs on Shabbat."[97]

There is another custom mentioned by the "Yafe laLev"[98] that there are congregations that read Kohelet on the first day and the last day of Sukkot even when it occurs on a weekday.

The manner of reading as it appears in the Siddur of Rashi,[99] and in "Machzor Vitry"[100] and the customs of Rabbi Chaim Paltiel[101] is that the whole congregation reads Kohelet "bayeshivah." The question that can be asked is "What is the meaning of the word "bayeshivah"? Perhaps its intention is that every individual reads it, and that they can read it whilst sitting. However, if it is read by the "shaliach tzibur" (the person conducting the service), then the shaliach tzibur has to stand out of respect for the congregation.[102]

General subjects regarding the times of reading the Megillot

For communities that read Shir Hashirim and Kohelet on Shabbat Chol Hamoed, its location in the service is before the reading of the Torah at Shacharit. These same communities also read Rut before the reading of the Torah on Shavuot. This location was mentioned in the Siddur of Rashi,[103] and in "Machzor Vitry"[104] and by the Maharam (Meir) of Rothenburg (13th century).[105]

It should be noted that it is possible that these megillot can also be read later on the same day. It is written about the Vilna Gaon[106] that on one occasion whilst the megillah was being read to him, he felt suddenly very weak, and said that the megillah should be read to him together with the berachot at minchah.

To my recollection, in the 1950s at the United Synagogue in Edgware in North London they read Shir Hashirim and Kohelet before "aleinu leshabeach" at minchah on Shabbat Chol Hamoed, or on Shabbat in the years when it occurred at the end of the Festival, and Rut before "aleinu leshabeach" at minchah on the second day of Shavuot. However, at least since the 1980s, these megillot have been read in that Synagogue at shacharit before the reading of the Torah.[107]

The question arises, why read Megillat Esther after the reading of the Torah, but Shir Hashirim, Rut and Kohelet before the reading of the Torah. The book "Heorot Hashemesh" [108] written by Rabbi Moshe Shmuel Shapira (and also the book "Minchat David" written by Rabbi David Hakohen Rosenberg[109]) writes that he heard from a number of great Rabbis that the principle: "tadir v'she'eino tadir, tadir kodem" (that when one has something which is frequently said or done, and something which is not frequently said or done, the frequent one comes first.)[110] However this only applies if both of them are obligatory, but if only one of them is obligatory, this principle does not apply. The reading of Megillat Esther is obligatory, and therefore one reads it after the reading of the Torah which is also obligatory. In the case of the other megillot however, their reading is not obligatory and therefore one is not particular about the order. It seems that the Maharam from Rothenburg[111] was not pleased with reading these megillot before the reading of the Torah because he writes: "there are places where it is customary to say verses such as 'ein kamocha' and 'malchutcha' between reading these megillot and taking out the Sefer Torah from the Ark in order to make a break between 'divrei kabbalah' (the megillot) and reading the Torah, so it should not appear that one is putting 'divrei kabbalah' before words from the Torah."

The question arises: Why then read Eichah on the morning of Tisha b'Av after the reading of the Torah? This question is answered in the book "Heorot haShemesh" that there is no mention of an obligation for the congregation to read it in the morning.[112] However, there is a custom among the Jews of Ghardaia (Algeria) and the Jews of Bichan (Arabian Peninsula) to read the "Megillat Antiochus" on Shabbat Chanukah after the haftarah.[113] Because it is read specifically on Shabbat Chanukah, it is a public reading (see below), so there is room to investigate why not to read it before reading the Torah?

Ashkenazi congregations always recite Shir Hashirim and Kohelet on Shabbat, and in the Diaspora, Rut is read on Shabbat if the second day of Shavuot occurs on Shabbat.[114] If so, how can it be permitted to read Shir Hashirim and Kohelet on Shabbat, since when Purim occurs on Shabbat (this is only possible in walled cities) one advances the reading of Megillat Esther to Friday because of the "gezeira d'Raba" (the apprehension that one might carry the megillah in a public thoroughfare on Shabbat). If so, how it is possible that the Rabbis decreed to read Shir Hashirim and Kohelet specifically on Shabbat?! What is the difference between Megillat Esther and the other megillot? The commentary "Peulat Sachir"[115] to Maaseh Rav of the Vilna Gaon explains that the reading of Megillat Esther is an obligation on every individual, whilst reading the other megillot is only an obligation on the congregation in the same way as reading the Torah is, and one does not make a decree on something which is an obligation of the congregation since the people would remind each other not to carry.[116]

However, Sefaradi congregations (who have the custom to read these megillot) do not read Shir Hashirim, Rut and Eichah before the morning reading of the Torah, but instead during the day.[117] It should be noted that there are congregations in Cochin (southern India) that read Shir Hashirim on the seventh of Pesach before the reading of the Torah.[118]

There are a number of questions which still remain: 1) Do these megillot have to be read specifically during the daytime and not at night? On this question Rabbi Tucazinsky[119] writes that the regulation regarding the reading of Shir Hashirim, Rut and Kohelet is limited to the daytime. 2) Even though there are specific defined times during the Festivals to read these megillot, does a person fulfil his duty if he reads them at another time during the Festivals? 3) By the reading of Shir Hashirim following the Pesach Seder,[120] or at the beginning of Shabbat Chol Hamoed Pesach, does a person fulfil his duty of the reading of Shir Hashirim on Pesach? (See below the opinion of Rabbi Sraya Davlitzki on these questions.)

Does one recite a Berachah on Reading these Megillot?

It is written in Masechet Soferim:[121] "For [the Megillot] Rut, and Shir Hashirim and Eichah and Megillat Esther one recites the berachah "al mikra megillah" (who has commanded us to read the megillah),

We can immediately see that Kohelet is not mentioned. However, there are several versions of Masechet Soferim in which Kohelet is included in this list, and they are to be found in: "Machzor Vitry," "Sefer haAgudah" and Nuschaot haGra (Vilna Gaon) to Masechet Soferim.

1)   "Machzor Vitry" was written by Rabbenu Simcha who was a student of Rashi, and included in it is Masechet Soferim in which Kohelet is also mentioned.[122] It should be noted that this is not a "late addition," because we have a manuscript[123]

of "Machzor Vitry" from the middle of the thirteenth century, and Kohelet is mentioned there. In addition, this passage from the Masechet Soferim is also quoted in two other places[124] in Machzor Vitry.

2) "Sefer haAgudah" was written by Rabbi Aleksander haKohen Zuslin in the fourteenth century and includes commentaries and novella to Masechet Soferim. There it is written:[125] "Anyone who reads [Megillot] Rut, and Shir Hashirim has to recite the berachah 'al mikra megillah' and for Eichah, Esther and Kohelet."

3) In the version of the Vilna Gaon to Masechet Soferim, Kohelet appears in parentheses.[126]

In addition, there are Rishonim[127] who mention that it is written in Masechet Soferim that one should say the berachah on all five megillot, namely this includes Kohelet. However, there are other Rishonim[128] who had a version of Masechet Soferim in which Kohelet did not appear. From this we see that already in the period of the Rishonim there were different versions of Masechet Soferim.

In contrast to the versions who mention Kohelet, there is a manuscript in Moscow[129] in which the halacha of saying the berachah on reading these megillot does not appear at all! However, it is possible that the copyist accidentally missed out this halacha.

Masechet Soferim is considered one of the "Small Masechtot (tractates)." Rabbenu Asher ben Yechiel,[130] who was known as the "Rosh" (13th-14th centuries) writes about the period of the composition these Masechtot: "Their composition was in recent generations and they are not mentioned in the Talmud." On this, Rabbi Chaim Yosef David Azulai known as the "Chida," (18th century) writes: "There is a proof from the Rosh that they were written during the period of the Geonim (great Rabbis who lived between about the 6th to the 11th centuries)."[131] A similar language is found in the book "Meor Einayim"[132] written by Rabbi Azaria ben Moshe Del Rossi Min ha-Adumim, (16th century). However, Rabbi Malachi ben Yaakov Hakohen (18th century) in his book "Yad Malachi"[133] writes that it is not the intention of the Rosh to say that they were written after the Talmudic period. In any case, the berachah on reading these megillot is not mentioned in the Talmud, and the Rabbis did not readily accept that a berachah should be said over things that do not appear in the Talmud![134]

Therefore there are a lot of discussions by the Rishonim and the Acharonim on this subject. On the one hand, Rabbi David Halevi Segal, [135] known as the "Taz" (17th century) writes: "Everyone who recites a berachah on reading these megillot is saying a berachah in vain," and on the other hand, the Vilna Gaon[136] says that one recites a berachah when reading these megillot, and this includes Kohelet.

We have already seen that according to Masechet Soferim, the reading of Shir Hashirim and Rut is accomplished in two parts. The question arises: Does one recite a berachah before reading each part, or only before the first part? It may be possible to compare this to counting all the days of the Omer, which is in fact just one mitzvah, but one recites the berachah every night and not just on the first night.[137]

A question that can be asked: Is the reading of these megillot performed just by the person reading it as in the case of Megillat Esther, or does everyone read it and say the berachah by himself? On this the "Levush"[138] writes that with Shir Hashirim and Rut everyone reads the megillah and recites the berachah silently. Why should everyone recite the berachah? There is a similar discussion in the "Aruch Hashalchan"[139] regarding the berachah on hallel on Rosh Chodesh on which there are different opinions as to whether or not to recite the berachah. He writes: "Regarding an individual worshiper who is praying together with the congregation, I do not know why each one will say the berachah, since they are able to answer "amen' on the berachah at the beginning and end of hallel of the person leading the service, and answering amen is as if one has himself recited the berachah, and why should everyone say berachot in a situation where there are great Rabbis who hold that it is a berachah in vain? And yes, I am accustomed to act like this [and not say such a berachah]."

The "Levush"[140] writes that everyone should say the berachah silently. Saying the berachah in a whisper is mentioned by the Maharam of Rothenburg[141] and by the "Maharil"[142] in connection with Eichah. This was the custom in Vermaiza for all the megillot. This was according to the books of the customs of that city, which was compiled in the period from the end of the sixteenth century to the end of the first third of the seventeenth century by Rabbi Yuda Löw Kircheim,[143] and for about twenty-five years in the middle of the seventeenth century by Rabbi Yuspa Shammash.[144] In Vermaiza they also recited the berachah for Kohelet.[145] However, Rabbi David ben Shelomo ibn Zimra[146] known as the "Radbaz" (15th-16th centuries) is amazed by those who say this berachah silently or in a whisper. According to him: "I do not know what is this compromise, since if it is not a berachah recited in vain, why should one not say it in a loud voice, and if it is a berachah in vain, is it permitted to say the Heavenly name in vain in a whisper?"

In order to avoid the apprehension of saying the Heavenly name in vain, the "Elia Raba"[147] (Rabbi Eliyahu Spira, 17th-18th centuries), writes that one should say the berachah before reading Kohelet omitting the Divine name and kingship. Rabbi Aharon Varmish (18th-19th centuries) in his book "Oad Lamoed"[148] did not want to disagree with the Rabbis who said that one should say the berachah on the megillot, and therefore writes: "At least one should say the berachah with a cognomen and kingship" in place of the word G-d."

In the responsa of the Rema[149] on this subject, he makes a distinction on the recital of a berachah between the various megillot, namely by saying the berachah for Eichah on the one hand, and on the other hand not for the other megillot, the reason being that like Megillat Esther the person conducting the service reads Eichah aloud. Therefore the Rema holds in his responsum and in his "Darchei Moshe"[150] that one says the berachah before reading Eichah but he does not mention this in his glosses to the Shulchan Aruch!

Masechet Soferim does not mention saying the berachah "shehechyanu" when reading these megillot. However, according to the "Levush,"[151] one recites the berachah shehecheyanu on reading Shir Hashirim and apparently, also on Rut. His explanation is that this berachah is recited on mitzvot which are performed "from time to time." According to the custom in Vermaiza[152] one does not recite the berachah shehecheyanu, but in the glosses[153] to these customs it is written: "But I was amazed why they do not say this berachah."

According to "Maaseh Rav" of the Vilna Gaon,[154] it is said that one recites shehecheyanu on Shir Hashirim, Rut and Kohelet, but with regards to Eichah this berachah is not mentioned at all. This statement is not ridiculous since the Vilna Gaon[155] holds that it is not forbidden to say this berachah during the "Three Weeks" (period between fast of 17 Tammuz and fast of 9 Av). He proves this from the last chapter of Masechet Berachot:[156] If his father died and he was an only son, and he thus is the sole inheritor, he says the berachah shehecheyanu, on the day his father died; the "Three Weeks" it is no more severe than the day his father died. However, in the calendars for Eretz Yisrael of Avraham Lunz[157] and of Rabbi Tucazinshy,[158] reciting only the berachah "al mikra megillah" is mentioned in connection with Eichah, namely one does not recite the berachah shehecheyanu.

It is also written in Masechet Soferim:[:159] "And the person called up to reading the Torah (on Tisha b'Av) says 'Baruch Dayan Haemet' [blessed be the True Judge)." On this, Rabbi Moshe ben Nachman – Nachmanides[160] known as the "Ramban," (13th century), writes: "He recites the berachah on reading the megillah 'al mikra megillah' ... and he also says the berachah 'baruch dayan haemet' and thus he recites two berachot." The author of "Orchot Chaim"[161] writes likewise. However the "Beit Yosef"[162] is more precise regarding to the language of Masechet Soferim, and wrote that the simple meaning is that the berachah "baruch dayan haemet" is recited when one is called up to the Torah. The "Shelah"[163] holds that the Shulchan Aruch omitted reciting the berachah "baruch dayan haemet" because there is a doubt whether this berachah is recited when one is called up to the Torah, or when one reads Eichah.

The "Levush"[(164)] rules that one recites "the berachah 'baruch dayan haemet' in a whisper before he says the berachah on the Torah."

There are a number of Poskim who write that one recites the berachah "al mikra megillah" also before reading Kohelet. They include (in addition to those mentioned above) the "Maharil"[(165)] and the Vilna Gaon,[(166)] and this also appears in "Machzor Saloniki,"[(167)] according to the Ashkenazi rite, which was published in the year 5315.

The question arises: Does one say the berachah on Shir Hashirim and Kohelet only if one reads them on Shabbat Chol Hamoed? This question was asked to Rabbi Moshe Feinstein[(168)] (20th century), author of the responsa "Igrot Moshe," by a community that used to read these megillot on Shabbat Chol Hamoed from a kosher megillah with a berachah. It happened that on one year that they did not have such a megillah on Shabbat Chol Hamoed, but they did have such a megillah only on the last day of Pesach. The "Igrot Moshe" ruled that since on Yom Tov there was also a gathering together of people, and therefore one could say the berachah. On this subject, the "Rivevot Ephraim"[(169)] writes in the name of Rabbi Shmuel Monk, that if one has already recited the megillah from a printed text on Shabbat Chol Hamoed, one cannot say the berachah on the last day of the Festival when reading from a parchment hand written text, on the grounds that one has already fulfilled one's duty according to the ruling of the Rema.

In the above question there was a discussion on the possibility of reading the megillah on the last day of the Festival. However, the question arises that if the only option is to read a parchment hand written text on a weekday of Chol Hamoed, is it permitted to recite the berachah, namely is reading on a day denoted as "a day of gathering the people together" only lechatchila but not b'diavad? Possibly the words of Rabbi David Abudraham[(170)] (14th century) are relevant in this case. He writes without stating a specific day that the Shir Hashirim is recited on Pesach, Rut on Shavuot and Kohelet on Sukkot; (he also holds that one recites a berachah before reading these megillot, with the exception of

Kohelet). Does he therefore mean that you can choose the day during the Festival when you want to read and recite the berachah over these megillot? However, Rabbi Sraya Davlitzki[171] writes that one cannot even consider the option that if one cannot read the megillah on the Shabbat during the Festival, that one can read it with a berachah on the last night of the Festival, or to begin reading it on the termination of Shabbat before the Festival, and in this way rely in a time of need of the opinions found in Masechet Soferim.

In contrast to what is written above, the Sefaradi communities do not recite a berachah on these megillot.[172] The question arises: if a Sefaradi reads the megillah in an Ashkenazi synagogue where it is customary to recite the berachah, can he himself recite the berachah? On this Rabbi Ovadia Yosef[173] (20th-21st centuries) in a responsa in "Yabia Omer" rules that he should not recite the berachah, but instead someone else should do so.

There are also quite a number of Ashkenazim who do not recite the berachah on these megillot. In the Ashkenazi customs of Rabbi Chaim Paltiel,[174] Rabbi Avraham Klausner,[175] Rabbi Yitzchak Madura[176] and also in "Mahzor Romania"[177] there is no mention at all of saying the berachah on these four megillot. In the "Sefer Haminhagim l'Rabbenu Eizik Tyrnau,"[178] who was a Rabbi in Austria and wrote his book at the end of the fourteenth century, only in the case of Eichah it is mentioned that one recites the berachah. However, in one of the glosses[179] written in the margin of this book it is written that one recites the berachah over all the megillot with the exception of Kohelet. Who wrote all the glosses that appear in the margin of this book? Some hold that the author himself wrote some of them, and others added the remainder at a later date. In one of his responsa, the Rema writes:[180] "Amongst the customs of Rabbi Eizik Tyrnau ... he did not write to recite the berachah over Shir Hashirim, Rut and Kohelet." On the other hand, the "Levush,"[181] who lived at the same period as the Rema, mentions this gloss but in the name of Rabbi Eizik Tyrnau. Therefore one needs to investigate if these glosses were written by Rabbi Eizik Tyrnau, or by someone else at the same period.

Another question that is asked is: Is it possible to divide the reading of these megillot amongst a number of readers, with the first reader reciting the berachah? On this Rabbi Efraim Greenblatt, the "Rivevot Ephraim"[182] writes, that because this possibility is permitted for Megillat Esther, it is also permitted for these megillot.[183]

In the case of the Sefer Torah and Megillat Esther, one can only say the berachah if the megillah is handwritten in ink on parchment. What is the law regarding the other megillot? According to Rabbi Avrahan Abele Gombiner[184] known as the "Magen Avraham" (17th century), rules that today one can say the berachot on both the haftarahs and these megillot which are printed on paper. Similarly, Rabbi Shmuel Loew (18th century) in his book "Machatzit haShekel" writes:[185] "Even if they are printed and written on paper, one can recite the berachah over these megillot, since, with the exception of a Sefer Torah it is permitted." The "Levush"[186] permitted the reading of Eichah in a book which is printed "because we are waiting and expecting every day that this day [Tisha b'Av] will become a day of joy, happiness and a festival." In addition, when the "Levush"[187] wrote in connection with Shir Hashirim and Rut that everyone will read it for himself with a berachah, perhaps his intention was when there was no kosher megillah in front of everyone. However according to the majority of the Rabbinical authorities,[188] one can only say a berachah on a megillah handwritten with ink on parchment.

In a Sefer Torah even one letter that has not been written correctly can invalidate the entire Sefer Torah. On the other hand, a Megillat Esther in which up to half is missing, provided that it has a beginning and an end, is a kosher megillah, and one can recite a berachah over it.[189] What is the law regarding the other megillot? Rabbi Aryeh Pomransig[190] (20th century), author of "Emek Berachah" discusses the subject and writes that the reason to permit a Megillat Esther despite the fact that part of it is missing, is because it is referred to in the Megillat Esther as a "Letter," which is not the case with the other megillot. Therefore, these other megillot have the same laws as for a Sefer Torah.

Summary

- The time to read Shir Hashirim is on Pesach, Rut on Shavuot, Eichah on the night of Tisha b'Av, and Kohelet on Sukkot (including Shemini Atzeret).
- Many of the different communities in the world read Rut and Eichah; however, some of the Sefaradi communities do not read Shir Hashirim and Rut, and the vast majority of the Sefaradi communities do not read Kohelet.
- There are different customs of the various communities regarding the day of the Festival in which Shir Hashirim and Kohelet are read.
- There are various customs on the question of their location in the order of the Synagogue service.
- All the Sefaradi communities and a considerable part of the Ashkenazi communities do not recite a berachah when reciting these megillot. Today, only the Ashkenazim who follow the Vilna Gaon recite a berachah before reading these megillot (and this includes Kohelet), provided they read them from a megillah written according to all the laws appertaining to the writing of a Sefer Torah.

References

Abbreviations
SA = Shulchan Aruch
OC = Orach Chaim
(1)   Masechet Soferim, chap.14 halacha 18
(2)   *Mikra Soferim*, Masechet Soferim, (Suwalki, 5622), chap.14 par.15
(3)   Masechet Soferim, chap.14 halacha 18
(4)   Talmud Yerushalmi, Megillah chap.1 halacha 1
(5)   SA OC chap.688 par.7
(6)   Biblical Esther chap.9 verse 22
(7)   e.g. Rabbi Kalphon Moshe Hakohen, *Brit Kehuna*, (Djerba, 5701), OC vol.1 p.55b
(8)   *Siddur Knesset kaGedola, Tichlal*, vol.2, arranged by Rabbi Yosef Tsobari, (Tel-Aviv, 5746), p.327
(9)   Rabbi Yosef Kapach, *Halichot Taiman*, (Machon Ben Zvi: Jerusalem, 5728), p.27
(10)   *Knesset haGedola*, op. cit., p.327
(11)   Ibid.

READING THE MEGILLOT SHIR HASHIRIM, RUT, EICHAH AND KOHELET

(12) *Halichot Taiman*, op. cit., p.27

(13) *Knesset kaGedola*, op. cit., p.328

(14) *Yalkut Minhagim Miminhagei Shivte Yisrael*, arranged by Asher Wassarteil, (Misrad haChinuch vehaTarbut: Jerusalem, 5756), p.234

(15) Ibid., p.256

(16) Rabbi Eliyahu Gig (Guedj), *Zeh haShulchan*, (Algiers, 5649), vol.2, chap 52 (p.129)

(17) *Machzor kol Hashanah Keminhag Italiani*, (Livorno, 5616), vol.1 p.131

(18) *Mahzor Romania*, (Venice, 5280), p.130b

(19) *Yalkut Minhagim*, op. cit., p.216

(20) Ibid., p.207

(21) Herbert C. Dobrinsky, *A Treasury of Sephardic Laws and Customs*, (Ktav: New Jersey, 1988), p.283

(22) Letter from Rabbi Dr. Avraham Levi, Rabbi of Sefaradi Synagogue "Sha'ar Hashamayim" London, to Chaim Simons, Shevat 5755

(23) Rabbi Meir ben Baruch (Maharam m'Rothenburg), *Sefer Minhagim*, (published by Rabbi Yisrael Elfenbein, Bet Hamidrash Harabbanim b'America, New York, 5698) p.28

(24) Rabbi Aharon Hakohen m'Lunel, *Orchot Chaim*, (Florence, 5510), Hilchot Tefillat haMoadim, p.172; Rabbi Yaakov ben Moshe Levi Moelin, *Sefer Maharil*, (Frankfurt am Main, 5448), p.50b; Rabbi Mordechai Yoffe, *Levush*, OC chap.490 par.5; SA OC chap.490 par.9, Rama; Rabbi Shneur Zalman of Liadi, *Shulchan Aruch Harav*, OC chap.490 par.17; Aruch Hashulchan OC chap.490 par.5

(25) Vilna Gaon, *Beur haGra*, SA OC chap.490 first word: b'Shabbat

(26) Mishnah Berurah SA OC chap.490 par.18

(27) Yisrael Davidson, *Otzar Hashira VeHapiut*, (Thesaurus of Mediaeval Hebrew Poetry), (Ktav: New York, 1970), vol.1 no.1962, "Ohr Yesha Meusharim"

(28) Davidson, op. cit., no.7129 (p.324), "Afik Renen v'Shirim"

(29) Davidson, op. cit., no.1387 (p.65), "Ahuvecha Ahaivucha Maishorim"

(30) MS Bodleian, catalogue Neubauer 910/1 folio 63a; MS Israel National Library Jerusalem1282 Heb 8⁰ folio 91a

(31) e.g. MS Cambridge-Trinity College 73.1; MS Bodleian, catalogue Neubauer 2256/1

(32) MS Bodleian, catalogue Neubauer 910/1 folio 62b

(33) e.g. SA OC chap.490 par.9; Rabbi Yechiel Michel Tucazinsky, *Sefer Eretz Yisrael*, (Jerusalem, 5726), p.63; *Yalkut Minhagim*, op. cit., p.30

(34) SA OC chap.490 par.9 Rema; Aruch Hashulchan OC chap.490 par.5; *Yalkut Minhagim*, op. cit., p.30

(35) Rabbi Yosef ben Meir Teomim, *Pri Megadim*, Eshel Avraham, SA OC chap 490 par.8

(36) Rabbi Sraya Davlitzki, *Zer Hashulchan b'Inyanei Orach Chaim v'Yoreh De'ah*, (Bnei Brak, 5718), p.39

(37) Rabbi Tzvi Pesach Frank, *Shut Har Tzvi*, OC vol.1 chap.55 first word: umilvad

(38) e.g. *Moadei Hashem keMinhag Kehal Kadosh Sefaradim*, (Livorno), p.178 (378); *Moadim l'Simchah*, Tefilot l'Shalosh Regalim keMinhag Baghdad, (Baghdad, 5667), pp.191, 204; *Prayers for the Festivals according to the custom of the Spanish and Portugese Jews* (in America), ed. David de Sola Pool, pp.352, 357, New York, 5717,

(39) *Moadei Hashem*, op. cit., p.178 (378)

(40) *Halichot Taiman*, op. cit., p.31

(41) *Mahzor Romania*, op. cit., pp.182b, 189b

(42) *Zeh haShulchan,* op. cit., chap.58 (p.136)

(43) *Yalkut Minhagim*, op. cit., p.46

(44) Ibid., p.308

(45) e.g. *Maharil*, op. cit.; *Luach Davar b'Ito* (Every Thing in its Proper Time), year 5755, (Achiezer: Bnei Brak, 5755), pp.847-848; Yad L'Achim Wall Calendar, Jerusalem, Sivan 5755

(46) *Pri Megadim*, op. cit., SA OC chap.490 par.8

(47) Masechet Soferim chap.18 halacha 4

(48) Tur, OC chap.559

(49) Beit Yosef on Tur OC chap.559 first word: avol

(50) Rabbi Shem Tov Gagin, *Keter Shem Tob*, (5714), vols 4 and 5, p.40

(51) A detailed index to the responsa "Yehave Daat," compiled by Rabbi Yitzchak Yosef ben Rabbi Ovadia Yosef, (Jerusalem, date of publication not given) p.36

(52) *Eichah Rabba*, (Buber), Pesichta 17; *Eichah Rabba* (Vilna), Pesichta 17

(53) Rabbenu Yerucham, *Sefer Toldot Adam v'Chava*, (Venice, 5313), section 18 part 2 p.164b

(54) Rabbi Meir Hakohen, *Hagahot Maimuniot*, Rambam, Hilchot Ta'aniyot chap.5 par.2 and Minhagei Tisha b'Av (at end of chapter)

(55) Rabbi Moshe Isserles, *Shut haRema*, (Jerusalem, 5731), chap.35

(56) Talmud Bavli, Ta'anit 30a

(57) *Halichot Taiman*, op. cit., pp.45-46; *Yalkut Minhagim*, op. cit., pp.357, 448

(58) *Keter Shem Tob,* op. cit., vols.4 and 5, p.42

(59) Talmud Yerushalmi, Shabbat, chap.16 halacha 1

(60) In connection with the laying of tefillin at minchah on Tisha b'Av, the Mishnah Berurah writes (chap.555 par.5): "One should now be careful to now read the paragraphs of the Shema and also the paragraph which begins 'kadesh li' since this is like reading the Torah, and the study of Torah is forbidden all day long." However, before and after putting on one's tallit, one recites certain verses from Tehillim and likewise when laying one's tefillin verses from Hoshea. One therefore needs to investigate why the Mishnah Berurah does not write about saying these verses on Tisha b'Av.

(61) *Yalkut Minhagim*, op. cit., pp.87, 117, 357, 448, 465, 548

(62) Ibid., p.284

(63) Rabbi Yeshayahu ben Avraham Halevi Horowitz, *Shney Luchot Habrit*, (Warsaw, 5612), vol.2, Masechet Ta'anit p.204b

(64) Mishnah Berurah, SA OC chap.559 par.2

(65) Rabbi Yehudah ben Shimon Ashkenazi, *Ba'er Heteiv* SA OC chap.559 par.5

(66) Rabbi Chanoch Zundel Grossberg, *Zer Hatorah*, (Jerusalem, 5724), pp.90-91

(67) Rabbi Ephraim Greenblatt, *Rivevot Ephraim,* (Brooklyn New York, 5735) OC chap.385

(68) *Sefer Eretz Yisrael*, op. cit., chap.21 footnote 1 (p.67)

(69) Biblical Bamidbar chap.14 verse 1

(70) Biblical Eichah chap.1 verse 2

(71) Rabbi Tzvi Hirsh Meizlish, *She'elot uTeshuvot Mikdashei Hashem*, (Brooklyn New York, 5753), part 1 chap.70

(72) Talmud Bavli Megillah 5a

(73) Rabbi Avraham Danzig, *Chayei Adam,* section 135 par.19

(74) Levush, OC chap.559 par.1

(75) *Mikdashei Hashem*, op. cit., chap.70

(76) *Luach l'Eretz Yisrael*, arranged by Rabbi Yechiel Michel Tucazinsky, for example for year 5727, month of Av, footnote 1, p.53

(77) *Siddur Rashi*, brought out by R' Shelomo Buber, (Hevrat Mekize Nirdamim: Berlin, 5672), p.104; Rabbenu Simchah, *Machzor Vitry*, brought out by Rabbi Shimon Halevi Ish Horowitz, (Nuremberg, 5683), p.440

(78) *Siddur Rashi*, op. cit., p.147; Machzor *Vitry*, op. cit., p.446

(79) Biblical Kohelet chap.11 verse 2

(80) Rabbi Yitzchak Palagi, *Yafe laLev*, (Izmir, 5636), vol.2 chap.663 par.2

(81) Rabbi Yaakov Ettlinger, *Bechorei Yaakov al Hilchot Sukkah v'Lulav*, (Altona, 5618), chap.663 par.4

(82) Maharam m'Rothenburg, op. cit., p.68; Maharil, op. cit., p.139

(83) Rabbi Avraham ben Natan haYarchei, *Sefer haManhig*, (Berlin, 5615), p.71; *Orchot Chaim*, op. cit., chap.43 p.260; Rabbi Yitzcak Abohav, *Menorat haMaor*, (Mantova, 5323), p.55

(84) *Halichot Taiman*, op. cit., p.34; *Knesset haGedola*, op. cit., p.327

(85) *Knesset haGedola*, op. cit., p.327

(86) *Halichot Taiman*, op. cit., p.34

(87) *Mahzor Romania*, op. cit., pp.404b, 406b, 408b, 435b

(88) *Machzor Italiani*, op. cit., vol.2 p.196

(89) *Pri Megadim*, op. cit., SA OC chap.490 par.8

(90) Maharil, op. cit., p.139

(91) Darcei Moshe, Tur, OC chap.663 par.2

(92) SA OC chap.663 par 2, Rama

(93) Rabbi Sraya Davlitzki, *Zer Hashulchan*, (Bnei Brak, 5726), vol.3 p.32 footnote 201

(94) *Bechorei Yaakov*, op. cit., chap.663 par.4

(95) *Luach l'Eretz Yisrael*, arranged by Rabbi Avraham Lunz, for example year 5656.

(96) *Luach* Tucazinsky, op. cit., for example for year 5730, p.20,

(97) Letter from Rabbi Sraya Davlitzki, *Luach Davar b'Ito*, op. cit., year 5757, pp.71-72

(98) *Yafe laLev,* op. cit., vol.2 chap.663 par.2

(99) *Siddur Rashi*, op. cit., p.104

(100) *Machzor Vitry*, op. cit., p.44

(101) MS Bodleian, catalogue Neubauer 910/1 folio 32b; MS Israel National Library Jerusalem 1282 Heb 8^0 folio 83a

(102) see: SA OC chap.690. par.1

(103) *Siddur Rashi*, op. cit., pp.104, 147

(104) *Machzor Vitry*, op. cit., pp.440, 446

(105) Maharam m'Rothenburg, op. cit., pp.28, 30

(106) Vilna Gaon, *Maaseh Rav*, (Jerusalem, 5656), Tosefet Maaseh Rav, chap.171

(107) Letter from Rabbi Binyamin Rabinowitz, Rabbi of Edgware United Synagogue, London, to Chaim Simons, Kislev 5755

(108) Rabbi Moshe Shmuel Shapiro, *Haiorot Hashemesh*, (Odessa, 5649), chap 7 par.10

(109) Rabbi David Hacohen Rosenberg, *Minchat David,* (Brooklyn New York, 5739), vol.1, chap.52 (45), (p.92)

(110) e.g. Talmud Bavli: Berachot 51b, Pesachim 114a, Zevachim 91a

(111) Maharam m'Rothenburg, op. cit., p.28

(112) *Haorot Hashemesh*, op. cit., chap.7 par.10

(113) *Luach Davar b'Ito*, year 5755, op. cit., p.463

(114) Talmud Bavli, Megilla 4b

(115) Vilna Gaon, *Maaseh Rav*, op. cit., Peulat Sachir, chap.171

(116) Talmud Bavli, Eruvin 3a

(117) *Yalkut Minhagim*, op. cit., pp.179, 382, 384, 447, 483, 489, 515, 528, 545, 548

(118) Ibid., p.488

(119) *Sefer Eretz Yisrael*, op. cit., p.67 footnote

(120) When the Seder night occurs on Shabbat, the salt water must be prepared before Shabbat, and those who did not do so before Shabbat are allowed to prepare only a limited amount on Shabbat, (Mishnah Berurah chap.493 par.21) According to the "Chayei Adam" (summary of the seder par.1) and the "Kitzur Shulchan Aruch" (chap.118 par.2) one needs to prepare the salt water on the eve of Yom Tov, even if Yom Tov occurs on a weekday.

Some people are accustomed to eat an egg during the meal on the night of the Seder (SA OC chap.476 par.2, Rema), and some dip this egg into salt water. Rabbi Shlomo Zevin suggests that the reason for dipping it in salt water is: "When Seder night occurs on Shabbat one is limited to make salt water to just the quantity that one requires for that meal, and even if Yom Tov occurs on a weekday, one should be strict about this, and thus only make a quantity for that meal, and in such a case it is completely permissible without any doubt to put salt into the water on that night." ("Hamoadim b"Halacha" (Tel-Aviv, 5724) p.290). It is possible that a support for Rabbi Zevin's explanation is based on Chabad customs. In the Chabad Haggadah it is written: "The meal... at the beginning... eating an egg... dipping it in salt water." The Shulchan Aruch Harav (Rabbi Shneur Zalman, the first rabbi of Chabad) discusses the various laws of Yom Tov in connection with the salting of radishes on Shabbat and also on Yom Tov and concludes: "It is good to follow the strict opinions." (OC chap.510 par.15). It seems that from the language of the "Misgeret haShulchan" that according to this ruling, even when Yom Tov occurs on a weekday one should prepare the salt water prior to Yom Tov, even though preparing salt water for the Seder is only mentioned in the Shulchan Aruch Harav in connection with Shabbat. (Misgeret haShulchan par.4, Kitzur Shulchan Aruch chap.118). Therefore there is room to investigate if there is a connection between this ruling in the Shulchan Aruch Harav, and the Chabad custom to dip the egg in salt water on the Seder night.

(121) Masechet Soferim chap.14 halacha 3

(122) Machzor Vitry, op. cit., p.706

(123) MS British Museum London, Catalogue Margoliouth 655, folio 223b

(124) *Machzor Vitry*, op. cit., pp.229, 344

(125) Rabbi Aleksander haKohen Zuslin, *Sefer haAgudah,* (Krakow, 5331), p.227b

(126) Masechet Avot and also Masechtot Ketanot – (versions of the Vilna Gaon) (Shklov, 5564), p.57a

(127) *haManhig,* op. cit., p.50; Mordechi, Masechet Megillah, par.783; *Orchot Chaim*, op. cit., p.260

(128) Rabbi Yitzchak ben Moshe of Vienna, *Or Zarua*, (Zhytomyr, 5622), vol.2 hilchot Rosh Chodesh, chap.455, (p.182); Rabbi David Abudraham, *Abudraham Hashalem*, (Jerusalem, 5723), p.240; *Hagahot Maimunot*, op. cit., Rambam, Hilchot Ta'aniyot chap.5 par.2 and Minhagei Tisha b'Av (at end of chapter); Rabbi Zedekiah ben Avraham Anaw, *Shibbolei haLeket Hashalem*, (Vilna, 5647), Inyan Rosh Chodesh chap.174, (p.138); Rabbi Moshe ben Avraham of Przemysl, *Mateh Moshe*, (London, 5718), chap.967 (p.292)

(129) MS Moscow catalogue Günzburg 515/3 folio 172b

(130) Rabbenu Asher ben Yechiel ("Rosh"), Halachot Katanot, Hilchot Sefer Torah, (after 60 and 61)

(131) Rabbi Chaim Yosef David Azulai ("Chida"), *Yair Ozen*, (Livorno, 5553), Ayin Zocher, section samech, par.31 p.60

(132) Rabbi Azaria ben Moshe Del Rossi Min haAdumim, *Meor Einayim*, (Vilna, 5626), Imrei Binah, chap19 (p.250)

(133) Rabbi Malachi ben Yaakov Hakohen, *Yad Malachi,* (Berlin), klalai shnai hatalmudim, par.12 (p.121)

(134) see: *Talmudic Encyclopedia*, (Talmudic Encyclopedia Publishing: Jerusalem 5717), vol.4, p.313

(135) Taz SA OC chap.490 par.6

(136) *Maaseh Rav*, op. cit., chap171

(137) Rabbi Zvi Cohen, *Sefirat haOmer*, (Israel, 5746), pp.21-22, 109 note 2

(138) Levush OC chap.490 par.5

(139) Aruch Hashulchan OC chap.422 par.8

(140) Levush OC chap.490 par.5

(141) Maharam m'Rothenburg, op. cit., p.28

(142) Rabbi Yaakov ben Moshe Levi Moelin, *Sefer Maharil Minhagim*, brought out by R' Shlomo Spitzer, (Machon Yerushalayim, 5749), p.251, different versions

(143) Rabbi Yuda Löw Kirchheim, *Minhagot Vermaiza*, (Machon Yerushalayim, 5747), pp. 184, 240, 260, 268

(144) Rabbi Yuspa Shammash, *Minhagim Vermaiza*, (Machon Yerushalayim, 5748), pp.64, 93, 113, 123, 212

(145) Vermaiza Kirchheim, op. cit., p.184; Vermaiza Shammash, op. cit., pp.64, 212

(146) Rabbi David ben Shelomo ibn Zimra, *Shut haRadbaz*, (Warsaw, 5642), vol.6, responsum 2091

(147) Rabbi Eliyahu Spira, *Elia Raba*, (Sulzbach, 5517), OC chap.663 par.6

(148) Rabbi Aharon Varmish, *Meorei Ohr – Sefer Od Lamoed*, (Metz, 5582), Pesachim, p.32b

(149) *Shut haRema*, op. cit., chap.35

(150) Darchei Moshe, Tur OC chap.559 par.2

(151) Levush OC chap.490 par.5

(152) Vermaiza Kirchheim, op. cit., pp. 240, 260; Vermaiza Shammash, op. cit., pp.64, 93, 113, 212

(153) Vermaiza Kirchheim, op. cit., p.240 note 4

(154) *Shut haRema*, op. cit., chap.171

(155) Vilna Gaon, *Beur haGra*, SA OC chap.551 first words: m'yud zayin tov

(156) Talmud Bavli Berachot 59b

(157) Luach Lunz, op. cit., for example yeas 5656, chodesh Av

(158) Luach Tucazinsky, op. cit., for example year 5727 p.53

(159) Masechet Soferim chap.18 halacha 7

(160) Ramban (Nachmanides), *kitvei Rabbenu Moshe ben Nachman, Torat haAdam*, (Mossad Harav Kook: Jerusalem, 5724), p.258

(161) *Orchot Chaim*, op. cit., p.212

(162) Beit Yosef, Tur OC chap.559 first word: avol

(163) *Shney Luchot Habrit*, op. cit., Masechet Ta'anit, first words: haBeit Yosef (p.205)

(164) Levush, OC chap.559 par.4

(165) Maharil, op. cit., p.139

(166) *Maaseh Rav*, op. cit., chap.171

(167) *Machzor mikol Hashanah keMinhag haKehilot haKdoshot v'haTohorot Kehilot Ashkenazim*, (Saloniki (Thessaloniki), 5315?), no page numbers in book

(168) Rabbi Moshe Feinstein, *Igrot Moshe*, OC vol.4 chap.99

(169) Rivevot Ephraim, op. cit., OC chap. 321

(170) *Abudraham*, op. cit., p.240

(171) *Zer Hashulchan*, op. cit., vol.3 p.70

(172) Beit Yosef, Tur, OC chap.559 first words: mem-shin shin-aleph-aleph; Rabbi Ovadia Yosef, *Yabia Omer*, (Jerusalem, 5746), vol.1 chap.29 par.2

(173) *Yabia Omer*, op. cit., vol.1 chap.29 par.2; Rabbi Yitzchak Yosef (son of Rabbi Ovadia), *Yalkut Yosef*, (Tel-Aviv, 5731), p.63

(174) MS Bodleian, catalogue Neubauer 910/1; MS Israel National Library Jerusalem, Heb 8^0 1282

(175) e.g. MS Cambridge-Trinity College 73.1; MS Bodleian catalogue Neubauer 2256/1

(176) MS Montefiore London 136/4

(177) *Machzor Romania*, op. cit.

(178) Rabbi Eizik Tyrnau, *Sefer haMinhagim*, (Machon Yerushalayim, 5739), p.80

(179) Minhagim Tyrnau, op. cit., notes on the Minhagim, p.134 par.1

(180) *Shut haRama*, op. cit., chap.35

(181) Levush OC chap.490 par.5

(182) Rabbi Ephraim Greenblatt, *Rivevot Ephraim*, (Memphis Tennessee; 5745), OC vol.4 chap.132

(183) SA OC chap.692 par.1 Rama

(184) Magen Avraham, SA OC chap.284, introduction

(185) Rabbi Shmuel Loew, *Machatzit haShekel*, SA OC chap 559, introduction

(186) Levush, OC chap.559 par.1

(187) Levush, OC chap.490 par.5

(188) *Shut haRema*, op. cit., chap.35; Mishnah Berurah, SA OC chap.490 par.19 and Sha'ar Hatziyun par.14; Shulchan Aruch Harav, OC chap.190 par.17

(189) SA OC chap.690 par.3

(190) Rabbi Aryeh ben Zeev Pomransig, *Emek Berachah*, (5708), Kriyat 4 megillot, p.43
Luchot

PILGRIMAGES TO THE GRAVES OF THE RIGHTEOUS IN THE MONTH OF IYAR

Amongst the writings of the Rishonim (great Rabbis who lived approximately between the 11th and 15th centuries) and the Acharonim (great Rabbis who lived from about the 16th century) are three days in the month of Iyar, on which people would visit the graves of specific righteous people and they are: Pesach Sheni, Lag b'Omer, and the day of the death of the Prophet Shmuel. There is no connection between these three events, and the only common factor (in this paper) is the visitation of the graves of righteous people during the month of Iyar.

Pilgrimage to Meron on Pesach Sheni

Some of the travelers to Eretz Yisrael in the twelfth and thirteenth centuries wrote about the miracle of the water which took place near the tombs of Hillel and Shammai in the village of Meron, but they make no mention of "Pesach Sheni."

The first book which mentions Pesach Sheni is called "Eile haMassaot" ("These are the Journeys"). There it states: "In Meron there is the Cave of Shammai and Hillel and their students, which can hold a total of thirty-two people. There the Jews and Moslems gather together on Pesach Sheni, and the Jews pray there and sing psalms and when they find water in the Cave everyone is happy since this is a sign that the year will be blessed."[1]

Who wrote this book and when? Rabbi Yehuda David Eisenstein (19th-20th centuries) in "Otzarot haMa'asot"[2] holds, (and it is also to be found in "Luach Davar b'Ito,"[3]) that the author was Rabbi Yaakov, the disciple of Rabbi Yechiel of Paris, and he wrote this book in the middle of the thirteenth century. Rabbi Yechiel had sent Rabbi Yaakov to the Galil in Eretz Yisrael to collect donations for his large yeshiva in Paris, and it was there that he wrote the book "Eile haMassaot." Rabbi Eisenstein bases his conclusion on a manuscript[4] which is now in Paris.

However, a study of this manuscript shows us that in fact it is only a list of cemetery sites in the country ("Simanei Hakvarot") written by the disciple Rabbi Yaakov, whilst the content of "Eile haMassaot" was not even included in this manuscript. If so, what caused Rabbi Eisenstein to make an error and think that the author was the disciple, Rabbi Yaakov? Perhaps the reason is that Rabbi Elyakim Carmoli (19th century), who in the year 1841 was the first to print "Eile haMassaot" and he printed it together with "Simanei Hakvarot" which was written by Rabbi Yaakov.

Another version of "Eile haMassaot" appears in more detail in the Leningrad manuscript (at present not available), and this version includes the words: "And my teacher Rabbi Moshe ben Rabbi Nachman [the Ramban]."[5] From these words it is clear that that the author of this version was a student of the Ramban, who lived about fifty years after the disciple Rabbi Yaakov.

The first person to print the version from the Leningrad manuscript was Rabbi Simcha Assaf[6] (20th century), and he argues that the student of the Ramban used various books which contained details of journeys and lists which were made at an earlier date, and in particular "Eile haMassaot" of the disciple Rabbi Yaakov. However, he did not accept that the author was indeed the disciple Rabbi Yaakov but "Ploni Almoni" (someone anonymous). [It seems that Rabbi Assaf is doubtful regarding this conclusion, because he writes in a footnote,[7] that it is necessary to study whether the student of the Ramban copied from "Ploni Almoni" or vice versa.]

This account of the pilgrimage to Meron is not limited only to Jewish sources. The Arab geographer el-Dimichqui,[8] who lived in Safed in the fourteenth century, writes: "In Meron there is a Cave and in it are cupboards, and it is dry throughout the year, and there is not even a drop of water there. And when one arrives at the period of the year, that Jews gather there from far and wide, and peasants etc., and they are there all that day, both inside and outside the Cave, and it is dry, and before they know anything, one suddenly sees water coming out of the cupboards and within an hour or two there are pools of water which later stop accumulating, and that day is a Festival for the Jews." However, el-Dimichqui does not specifically mention "Pesach Sheni," and it is not clear if his intention is that that day was already a Festival (even if there was no water miracle), or that it was a Festival because of the miracle of the water.

There are also other Arab sources for this event, and one of them is a manuscript found in the library of the University of Istanbul,[9] the contents of which are the history of Safed. In their university catalogue, the manuscript is named "Ta'rikh Safad," the author of which is al-Uthmani, and it was probably written between the years 1372-76. On the miracle of the water he writes:[10] "And in this district there is a village called Meron. And there are niches and pits inside a cave, where the water seeps in slowly. But on a certain day of the year, and it is the middle of the Jewish month of Iyar, many Jews from near and far gather and they spend the whole day in the area of this place. And the water seeps in on this day more than usual, and the Jews take this water to distant lands." In his composition, al-Uthmani makes great use of el-Dimichqui's book, and in particular he adds details about the miracles that took place in Safed. In the subject before us, al-Uthmani adds an important detail that does not appear in el-Dimichqui's book and this is the date on which the Jews gathered in Meron: "The middle of the Jewish month of Iyar," namely Pesach Sheni. This event also appears in the writings of the Arab historian Qalqachandi,[11] but he writes: "one day in the year" without specifying which day.

In the year of 5282, "Ploni Almoni"[12] (some say he was R' Moshe Basula) writes: "On the fifteenth of Iyar which is called Pesach Sheni a large convoy was formed at Meron ... this is a village in which there is a Cave in which Rabbi Shimon Bar Yochai (Rashbi) and his son were closeted for thirteen years .... and the spring is still there ... and we spent two days and two nights celebrating and rejoicing, and we prayed by the graves of the righteous people that are buried there." In this description there are a number of differences from the description given by earlier travelers. They are that the graves of Hillel and Shammai are not specifically mentioned, but the grave of Rashbi is mentioned, and secondly the miracle of the water is not specifically mentioned, although the word "spring" is mentioned.

Date of Pesach Sheni

In Jewish calendars, the date of "Pesach Sheni" is invariably given as the fourteenth of Iyar, but "Ploni Almoni" from the year 5282 states that "Pesach Sheni" occurs on the fifteenth of Iyar. [The other travelers and the Arab historians did not state the date of Pesach Sheni.] What indeed is the date of Pesach Sheni?

It is written in the Torah[13] that on the afternoon of the fourteenth of Iyar the Korban Pesach (Pascal Lamb) is slaughtered, and eaten with matzah and maror on the night of the fifteenth of Iyar. Thus Pesach Sheni is on the fourteenth and fifteenth of Iyar.

However, in Megillat Ta'anit[14] it is written: "on fourteenth of Iyar is the 'little Pesach' (the Tosafot Chadoshim explains that the Korban Pesach is then slaughtered), and it is forbidden to mourn. When one studies the language carefully, one can see that in Megillat Ta'anit the date of Pesach Sheni is not mentioned, but only the day of the slaughter of the korban Pesach, which is the fourteenth of Iyar. [It is of interest to note that even though Megillat Ta'anit has been nullified (namely one is permitted to fast on the days which this megillah forbade),[15] some say that even today it is forbidden to eulogise and fast on the fourteenth of Iyar.[16]]

It should be noted that in the matter of eating matzah, reciting the tachanun prayer, and eating roast meat, the laws and customs of Pesach Sheni spread over to the fifteenth of Iyar:

1) Eating matzah: In the "Luach Davar b'Ito"[17] it is written: "There are places where it is customary today [on fourteenth of Iyar] to eat matzah in memory of the slaughter of the korban Pesach Sheni which took place in the afternoon." Some are amazed at this statement, since the korban Pesach together with matzah and maror (bitter herbs) was eaten only on the night of the fifteenth. Therefore there are those who eat matzah on the night of the fifteenth. Some also include an egg and "chazeret" (bitter herbs), and there are those who drink

wine. It should be noted that there are those who suggest a reason to eat them during the day of the fourteenth. Some Hassidishe Rebbes have a "tish" (a gathering around their Rebbe with speeches, singing and refreshments) on the fourteenth of Iyar in honour of that day. The custom of the Nadvorna Hasidim is to have on the night of the fifteenth, matzah, maror and the four cups of wine, accompanied by singing and reciting praises. It was also the custom of the Rebbi of Munkatch[18] to eat matzah on the night of the fifteenth of Iyar (in addition to the fourteenth). Furthermore it is written in the book "Zichron Yehuda" authored by Rabbi Menachem Eisenstadt[19] (17th-18th centuries): "On the night of the fifteenth it was the custom of Rabbi David Deutsch and the father of Rabbi Eisenstadt to eat matzah and an egg and learn the Mishnayot and the laws regarding Pesach Sheni.

2) The prayer tachanun: The Shulchan Aruch does not bring "Pesach Sheni" as one of the days when "tachanun" is not said. However in the second edition of the book "Tikkun Yissachar" which was written by Rabbi Yissachar ibn Mordechai Susan[20] (16th century), and published at about the same time as the Shulchan Aruch it states: "the fourteenth (of Iyar) ... the date of Pesach Sheni, which is regarded as a festival and therefore it is customary not to recite tachanun." (In the first edition of the book, which was published in the year 5324, Pesach Sheni is not mentioned at all. However, this edition was published without the consent of the author, who had found a number of mistakes in it, and he himself brought out the second edition with corrections and additions.)[21] Also a number of books of the Acharonim state that one does not say tachanun on the fourteenth of Iyar.[22]

   Regarding the reciting of tachanun on the fifteenth, Rabbi Yisrael Elgazi[23] (17th-18th centuries) writes in his book "Shalmei Tzibur": "There are places where it is customary not to say it." There are two reasons for this custom for this which are given in the Rabbinical literature:

   a) In the "Machberet Hakodesh l'haArizal" (written by his disciples) it states: "In another place it is written not to recite tachanun also on the fifteenth of Iyar, because on that night they would eat the Korban Pesach and therefore there is a small residue of holiness also on the following morning."[24]

b) "Because of the doubt of the correct date of Yom Tov."[25] (However, the decree of the Rabbis of Yom Tov Sheni in the Diaspora is limited to "Yom-Tov,"[26] and does not include Pesach Sheni. Therefore, this reason requires further investigation.)

In the city of Salonici (Thessaloniki) there were different customs on this subject. The author Rabbi Yosef David of the "Bet David"[27] (17th-18th centuries) who lived in Salonici wrote that his custom was to go according to the person conducting the service. In the year 5484, however, an incident occurred and everyone said tachanun and vidui on the fifteenth of Iyar, and "from then on it is proper to protest against those who do not say tachanun on fifteenth of Iyar."[28] According to "Luach Davar b'Ito," [29] even today there are communities who do not say tachanun on the fifteenth of Iyar. On the other hand, there are those, including Rabbi Yeshaya Karelitz[30] (known as the "Chazon Ish") who say tachanun even on the fourteenth of Iyar.

We can see from the book "Itim leBina," written by Rabbi Yosef Ginsberg[31] that the law of not saying tachanun on the fourteenth of Iyar is most importantly at minchah (afternoon service). In the case of the fourteenth of Iyar occurring on the second Monday fast (some fast on the Monday, Thursday and the following Monday during the months following Pesach and Sukkot), according to "Itim leBina" at shacharit (morning service) one recites the selichot for that day, Ovinu Malkenu and tachanun, "but at minchah one does not say tachanun because that is the time of the sacrifice of the Korban Pesach. It is possible that from this custom one can understand why, according to most opinions, tachanun is said on the fifteenth of Iyar, the reason being that the time of observance of Pesach Sheni is in the afternoon the fourteenth (the time of the minchah prayer) and on the night of the fifteenth, and only in these times (and on shacharit of the fourteenth) one does not say tachanun. Because one does not say tachanun at night, there is no need to say that one does not say it on the fifteenth of Iyar.

The same argument can be used regarding the second Monday fast that occurred on Pesach Sheni. In the case that it occurred on the fourteenth of Iyar, it was ruled that one should not fast at all, or at least that one should not complete the fast.[32] However, when it falls on the fifteenth of Iyar, no such ruling is found. Why? The laws and customs of Pesach Sheni which apply on the fifteenth, are limited to the nighttime, whilst the time for fasting on the second Monday fast begins only at dawn.

According to the "Eshel Avraham"[33] (Rabbi Avraham David of Buczacz, 18th-19th centuries), fasting is also allowed in the event that this fast occurs on the fourteenth of Iyar. One of the reasons for this is: "And eating the korban Pesach on Pesach Sheni was only at night."

3) Eating Roast Meat: There have been discussions amongst the Acharonim as to whether it is permissible to eat roast meat on the night of the fifteenth of Iyar. There are those who completely forbid it, whilst others only forbid a lamb which has been wholly roasted (as was the Korban Pesach).[34]

From all of the above, it is possible to understand why the traveler from the year of 5282 describes the fifteenth of Iyar as Pesach Sheni. (Most of the opinions brought above from the words of the Acharonim were of a later date than 5282, but it is obvious that the source of these writings was from the words of the Rishonim.)

Hilula to Rashbi in Meron on Lag b'Omer

The first source for the hilula (celebration) of Rabbi Shimon Bar Yochai (Rashbi) and his son Eliezer on the eighteenth of Iyar, which is Lag b'Omer, at the place of their burial in Meron is found in the writings of the disciples of Rabbi Yitzchak Luria (the Ari), especially in the writings of Rabbi Chaim Vital.

To understand these writings, one needs to know the history of the life of the Ari.

The Ari was born in Jerusalem in the year 5293, and while still a boy his father died, and he went with his mother to Egypt. He learned "nigla" (the parts of the Torah which are revealed to everybody), and when he reached the age of fifteen he got married. Two years later he began studying the Zohar. For the first six years, his understandings were not the correct ones, nor were they what the Rashbi had intended. Therefore he continued to study the Zohar for two years in seclusion next to the River Nile, and then his head was filled with Jewish mysticism. The Ari returned from Egypt to Eretz Yisrael at the end of the year 5330. A year later, Rabbi Chaim Vital began studying with the Ari. On the fifth of Menachem Av 5332 the Ari died.[35]

We shall now try to understand the words of the disciples of the Ari regarding his pilgrimage to Meron on Lag b'Omer:

Rabbi Chaim Vital[36] (16th-17th centuries) writes in the "Shaar haKavonot": "There is the custom where the Jews go in groups on Lag b'Omer to the graves of Rashbi and Eliezer his son who are buried in the city of Meron, and there they eat and drink and rejoice there. I saw that the Ari went there on one occasion on this Lag b'Omer together with all the members of his household, and they remained there for the first three days of that week, and this was on the one occasion that he came from Egypt, but I do not know if by then he was proficient and had this exceptional wisdom which he afterwards obtained."

On which year is Rabbi Chaim Vital speaking about? He gives us four indications:

a) "He [the Ari] and all the members of his household":
    This means that the Ari already had a family. He married at the age of fifteen and at the time of his visit with "all the members of his household", it can be assumed that he was at least sixteen years old, that is, namely not before the year 5309.

b) "And he dwelt there for the first three days of that week":
    The commentators find it difficult to understand the meaning of these words. The Rebbi of Munkatch[37] writes: "I did not know what he meant when he wrote the 'first three days of that week'."

178

He understood that one should stay near the grave for three days, and one should also arrive the day before Lag b'Omer and leave on the day after Lag b'Omer. Therefore, he concluded that staying in Meron means the "first three days of that week." However, this is only possible if Lag b'Omer occurs on a Monday, but according to the Hebrew calendar this is impossible.

There is a source in the "Pesikta d'Rav Kahana"[38] about a three-day stay at the grave of Rashbi: "Rabbi Yehudah ben Giori, and Rabbi Yitzchak and the Rabbis went to hear 'Parashat Nesachim' from Rabbi Shimon Bar Yochai and they were there for three days." But where does the Rebbi from Munkatch learn that it is necessary to arrive precisely on the day before Lag b'Omer and depart precisely on the day after Lag b'Omer? However, when Lag b'Omer occurs on a Sunday, it is not possible to arrive on the day before Lag b'Omer, since it occurs on Shabbat. It is also written in the "Shaar haKavonot": "I saw my teacher [the Ari] who went there on one occasion on the day of Lag b'Omer," that is, the Ari went there on Lag b'Omer itself.

However, it is possible to explain the words of Rabbi Chaim Vital, namely, that in the same year that the Ari went up to Meron, Lag b'Omer fell on the Sunday of that week. This means that the Ari and his family arrived in Meron on Lag b'Omer, which was on the Sunday of that week, and stayed there for three days as was the custom, that is: "the first three days of the week."

During the life of the Ari, from the age of Bar-Mitzvah, Lag b'Omer occurred on a Sunday in the years 5307, 5310, 5317, 5321, 5324, 5327, 5328, 5330, and 5331.[39]

The "Bigdei Yesha"[40] (Rabbi Aharon Pereira, 19th century) suggested that there is a scribal error and instead of "the first three days of that week" it should read "the last three days of that week." However, he brought no evidence to support this, and therefore it requires further study.

c) "This was on the one occasion that he came from Egypt": From this we see that the Ari came to visit Eretz Yisrael for the hilula of Lag b'Omer, before he came to live permanently in Eretz Yisrael in the year 5330.

d) "But I do not know if by then he was proficient and had this exceptional wisdom which he afterwards obtained."

The Ari began studying Kabbalah in the year of 5310, and only after eight years, that is, from the year of 5318 was knowledgeable in this wisdom.

From all of the above, it seems that the Ari went from Egypt to Meron on Lag b'Omer in the year of 5317. The question that can be asked is whether it was the Ari who established the custom to go to Meron on Lag b'Omer to the graves of Rashbi and his son? From the words of Rabbi Chaim Vital: "A custom practiced by Israel," means that there was a custom before that. However, from the words of the traveler from 5282,[41] who went to Meron on the fifteenth of Iyar and left on the seventeenth of Iyar (namely he left before Lag b'Omer!), shows that the visit on Lag b'Omer was not yet known in the year 5282. However, it seems that remaining in Meron for three days by the grave side was already customary. Maybe one can learn this from the words of the traveler: "And we spent two days and two nights there," and according to the rule "part of the day is as a whole day,"[42] one can conclude that he stayed in Meron for three days

The Ari in his "Shaar haKavanot" continues:[43] "Rabbi Yonatan Sagish testified before me that in the previous year, I went to him to learn with the Ari who took his little son there together with his household and there they shaved his head according to the known custom and made it a day of feasting and rejoicing."

The Ari arrived in Eretz Yisrael at the end of the year 5330, apparently after Lag b'Omer. Therefore, the year mentioned in the words of Rabbi Yonatan Sagish is the year 5331. From this we can conclude thar Rabbi Chaim Vital began to learn with the Ari at the end of 5731.

In his book "Pri Etz Chaim,"[44] Rabbi Chaim Vital tells about his pilgrimage to Meron, but with some changes: "At this period, regarding going to the graves of Rashbi and his son Rabbi Eliezer in Meron on Lag b'Omer, I saw my teacher [the Ari] that for eight years he went with his wife and his household and was there for these three days." This also appears in the book "Nogid u'Mitzvah"[45] by Rabbi Yaakov Tzemach (17th century).

"For eight years": On which year did Rabbi Chaim Vital see the Ari in Meron on Lag b'Omer? From the words "my teacher, may he be remembered for blessing" one can understand that the Ari died in the period following these eight years. Therefore he is not speaking about the year 5317 as stated in "Shaar haKavanot." However, according to the information that the Ari arrived in Eretz Yisrael at the end of the year 5330 and died in Menachem Av 5332, one is compelled to say that Rabbi Chaim Vital is speaking about the year 5331 or 5332.

However, there remains the differences of opinion regarding the years of the Ari coming to Eretz Yisrael and of his death, and there are a number of possibilities.

On the question of the Ari coming to Eretz Yisrael, it is written in the book "Shivchei R' Chaim Vital":[46] "In the year 5326, on the night of Shabbat on the fifth day of Chanukah I recited kiddush and sat down at the table to eat ... after three months on the night of Shabbat Hagadol; and I slept and beheld a dream.... and I woke up and after two years, my teacher the Ashkenazi came to Safed and I learned with him." From this one can learn that the Ari arrived in Eretz Yisrael around Pesach 5328.

[One has to arrive at a conclusion that in the course of the generations there have been distortions in the writings of Rabbi Chaim Vital: in his book "Shivchei R' Chaim Vital" he writes in one place that the Ari came to Eretz Yisrael at the end of the year 5330,[47] and elsewhere in the same book, he writes on Pesach 5328.[48] The distortions are especially related to the dates. In the above quote it is written "on the night of Shabbat on the fifth day of Chanukah." However, in the year 5326, Shabbat Chanukah occurred on the seventh day of Chanukah,[49] (also Shabbat Chanukah can never fall on the fifth day of Chanukah!) In addition, the year 5326 was a leap year, and there were four months between Chanukah and Shabbat Hagadol, and not three months as written in the book. Elsewhere in the book he writes "on Shabbat twenty-fifth of Iyar,"[50] but in the Hebrew calendar twenty-fifth of Iyar cannot occur on Shabbat!]

There are also differences of opinion regarding the year of the death of the Ari. Yosef Sambari[51] (17th century) writes about this: "The Ari died in the month of Menachem [Av] in the year 5334. Also in the letter sent from Eretz Yisrael by the traveler Shimshon Bak[52] on the nineteenth of Tammuz 5342 it is written: "It is eight years since he died ... Rabbi Yitzchak Luria [the Ari]." Namely he died at the end of the year 5334.

From the above, it can be assumed that Rabbi Chaim Vital speaks of one of the seven years between the year 5328 and the year 5334. But the question arises: Why did the leading disciple Rabbi Chaim Vital go only once to the hilula in Meron on Lag b'Omer in order to see his Rabbi? In answer, we have to relate to the fact that Rabbi Chaim Vital was the disciple of the Ari from the end of 5331 until the death of the Ari in Menachem Av 5332, namely he was the disciple of the Ari for only one year. On Lag b'Omer of the year 5332 he went to see his Rabbi, the Ari. He wrote in his book "Pri Etz Chaim" in the year 5340, that he saw the Ari in Meron on Lag b'Omer "eight years ago."

In the books "Pri Etz Chaim" and "Nogid u'Mitzvah" it is written about Lag b'Omer of "eight years ago," and the language is: "And they were there for these three days," but not "the first three days of that week". The reason is simple: In the year 5332, Lag b'Omer fell on a Thursday[53] (and not on a Sunday as in the year 5317). Therefore the phrase "first three days of that week" is not appropriate for the year 5332 and therefore it does not appear there.

Also the Siddur with the "kavanot" (hidden meanings) of the Ari written by Rabbi Shabtai miRashkov[54] (17th-18th centuries), mentions the custom of the Ari and his family to go up to Meron on Lag b'Omer. In his Siddur, he talks about a custom for all the years, and therefore he writes: "and they remained there for three days" without specifying that it was the "first three days of that week," because in some years Lag b'Omer did not occur on Sunday.

The Origin of Lag b'Omer

It was stated above that the hilula in Meron on Lag b'Omer was not known until the sixteenth century. However, there are earlier sources in the Rabbinical literature which point on the singularity of the thirty-third day of the Omer, but they are no connected with the hilula in Meron.

The earliest source for Lag b'Omer is mentioned by Rabbi Avraham haYarchei (end of the 12th century) in his book "haManhig." [55] He quotes the Gemara in Masechet Yevamot:[56] "Rabbi Akiva had twelve thousand pairs of students ... all of them died between Pesach and Shavuot." He then writes: "It is a custom in France and Provence to get married from Lag b'Omer onwards." About a hundred years later, Rabbi Menachem ben Shelomo, known as the "Meiri"[57] (13th century), interprets this Gemara and he writes: "It is a tradition from the Geonim (great Rabbis who lived between about the 6th to the 11th centuries) that on Lag b'Omer the dying ceased ... from this it is a custom not to marry a woman from Pesach until that time." From the above we learn that Lag b'Omer was a special day in the Hebrew calendar, because the students of Rabbi Akiva ceased dying, and as a result, the mourning customary in the days during the counting of Omer ends.

It is interesting to note that the Rambam (Maimonides) makes no mention whatsoever in his Mishneh Torah of the laws of mourning during the days of the counting of the Omer. One would therefore ask why? Was he not aware of this regulation? On this Rabbi Yosef Kapach[58] writes: "There was no such regulation and it is not known who decreed it, and the Rambam only rules on things that appear in the Talmud, or could be learned from the Talmud. Also, when there were regulations from the Geonim, he would write accordingly that it was a regulation from the Geonim. The customs of mourning in the days of Omer are shrouded in mystery. In any case, in Yemen they did not practice these mourning customs such as not getting a haircut, but the reason it was customary for them not to marry a woman was not because of mourning but as the 'Rokeach' wrote 'that the days are likely [to cause something unpleasant]'."

According to the "Meiri," the tradition that the dying stopped on Lag b'Omer is from the period of the Geonim, that is, several centuries before the "Meiri." In addition, Rabbi Avraham haYarchei ("haManhig")[59] wrote that Rabbi Zerachiah of Gerondi, known as the "Baal haMaor" (12th century) found this "in an old book from Spain." However, it is not clear (from the language of those quoted above), if the custom of not getting married during these days of the Omer was also old. But in the responsa of the Geonim is found that Rav Natronai Gaon[60] and Rav Hai Gaon[61] wrote that not getting married between Pesach and Shavuot is because of the plague which killed the students of Rabbi Akiva; however, Lag b'Omer is not mentioned in the writings of the Geonim.

On the other hand, in the book "Tur Bareket"[62] by Rabbi Chaim Hakohen, (16th–17th centuries) who was the chief disciple of Rabbi Chaim Vital, it is explicitly mentioned that the custom of not marrying these days is "new," and is a result of the Kabbalistic custom of not getting a haircut in the days of the Omer: "The custom of not marrying a woman between Pesach and Shavuot until Lag b'Omer, is a new custom amongst the Jews, and arose because on seeing that one does not have haircuts during these days, they thought that the reason was because of mourning since in this period twenty-four thousand pupils of Rabbi Akiva died, but those who know the real reason why not to cut one's hair during these days, know it is not because of mourning but it is a mystical reason." Indeed, it is difficult to understand the meaning of the word "new" in the authorship of "Tur Bareket," since Rav Natronai and Rav Hai Gaon lived about six hundred years before Rabbi Chaim Hakohen, and therefore how is it possible to class a custom dating back at least six hundred years as "new"? It is possible that he did not have before him the writings of the Geonim and the Rishonim on this subject.

Furthermore, another student of Rabbi Chaim Vital[63] writes: "The reason we do not practice mourning on Omer but do not shave is "sod alfin" (a kabbalistic reason).

In addition, the book "Seder haYom,"[64] written by Rabbi Moshe ben Machir (16th century) in Safed, gives a Kabbalistic reason for not cutting one's hair: "And it is proper that one should not cut one's hair but let it grow long during the entire period that the spiritual uncleanliness is to be found in him." His book does not mention mourning or the prohibition to marry, and there is not even a mention of the day of Lag b'Omer. Even in "Tikkun

Yissachar" written at the same time and place, there is no mention of Lag b'Omer.

However, in contrast, the Shulchan Aruch, which was written in Safed at that same period, mentions Lag b'Omer, and writes the reason for it in accordance with the words of the Rishonim, who state that these prohibitions during the days of Omer were because of mourning. According to the Shulchan Aruch:[65] "It is customary not to marry a woman between Pesach and Atzeret (Shavuot) until Lag b'Omer because at that period the pupils of Rabbi Akiva died ... it is customary not to cut one's hair until Lag b'Omer, because then they stopped dying and one should not cut one's hair until the day after Lag b'Omer." We also see from the words of the Shulchan Aruch, that the laws of mourning also apply to Lag b'Omer, and only the next day the rules regarding mourning cease. In contrast to this, according to the Rema:[66] "From Lag b'Omer onwards all is permitted ... one can cut one's hair on Lag b'Omer and increases a bit in rejoicing and one does not recite tachanun."[67] [The discussions on whether the mourning restrictions finish on Lag b'Omer itself, or the next day have continued until the present time.[68]] The reason for the Rema saying "increasing a bit in rejoicing" explain the Acharonim[69] "that it was on Lag b'Omer that they completely ceased dying," but not because of the hilula for Rashbi.

From all the commentaries on the Shulchan Aruch (which today appear on the printed page of the Shulchan Aruch), only the "Ateret Zekenim"[70] mentions the hilula for Rashbi: "The custom in Eretz Yisrael is that people have the custom to go to the graves of Rashbi and his son Rabbi Eliezer on that day."

Rabbi Chaim Azulai[71] known as the "Chida," (18th century), joins together the death of the students of Rabbi Akiva, Rashbi, and Lag b'Omer, namely that the reason we rejoice on Lag b'Omer was that Rabbi Akiva had taught twenty-four thousand students but they all then died and as a result this caused a serious loss to Torah in the world, and to remedy this, on Lag b'Omer he began to teach Rashbi, Rabi Meir, Rabi Yosi and others, and this caused the Torah to return and therefore one rejoices on Lag b'Omer.

Pilgrimage to the Grave of the Prophet Shmuel on the Date of his Death

The original source, which mentions the pilgrimage to the grave of the prophet Shmuel Haramati on the date of his death, is Rabbi Yitzchak ben Alfara from the city of Malka, who sent a letter in the year 5201 about his sightings in Eretz Yisrael:[72] "To the surroundings of Jerusalem they come from Syria and Egypt and from No Amon and from the Land of Babylon together with the congregation of Jerusalem to prostrate themselves on the grave of the prophet Shmuel in Rama which is close to Jerusalem ... and the time of this is on the night of the twenty-eighth of Iyar, the anniversary of his death."

Forty years later, Meshulam writes from Volterra about the same incident.[73]

Rabbi Ovadia of Bartinura (15th-16th centuries) arrived in Eretz Yisrael from Italy in the year 5248, and whilst he was in Eretz Yisrael he wrote three letters that are still extant. The second one was written from Jerusalem to his brother in the year 5249.

During his visit, he visited the tomb of the prophet Shmuel and wrote:[74] "The grave of our master Shmuel Haramati is still in the hands of the Jews and they come from all the surroundings to prostrate themselves there every year on the twenty-eighth (and according to another version 'on the twenty-first') of Iyar, the day of his death, and large torches are lit ... "

The Sefaradim made this into a big event, and Rabbi Gedaliah of Simiatitz describes it in detail:[75] "The routine there is that every year on the twenty-eighth day of the month of Iyar, the president of the Sefaradim goes to Kfar Rama to the grave of Shmuel Haramati and remains there with other scholars, and any volunteers may also go and remain there throughout the night. And I also came once with the president and when the president came it was before the time for minchah (afternoon service). And we recited the prayer which had been compiled to be said at the grave of Shmuel Haramati, and we said this prayer copiously weeping until everyone was weeping, unless a person had a heart of stone. After that they then prayed minchah and ma'ariv (evening service), and everyone would then sit and study the Book of Shmuel and other things which are connected with the prophet Shmuel, and they then studied the Zohar until midnight. At midnight they extinguished all the lights which were in the cave, and they sat in the dark and said in a weeping voice 'tikkun chatzot' (prayers recited at midnight in memory of the destruction of the Temple). And after they finished tikkun chatzot, they learned a small amount of the Zohar, and they then brought something to drink which was hot coffee and they gave everyone to eat "shakatin" which the Sefaradim are accustomed to eat... and they also ate sweet things such as fruit, and after that sang songs and 'pizmonim' (liturgical poems) which were related to the night of the death of Shmuel, which they sang in a very pleasant and happy voice until the morning. And first thing at the light of day they prayed shacharit (morning service) and then went home."

This event is not limited to the burial site of the prophet Samuel, and has spread even to the Diaspora. Rabbi Chaim Palaji,[76] (19th century) who lived in Turkey, wrote: "On 28 Iyar, the prophet Shmuel died. In Kushta (Constantinople) we make a celebration like on Lag b'Omer and the day becomes like a Yom Tov."

Distortions in Rabbi Ovadia Bartinura's letter

We saw above that in the year of 5249, Rabbi Ovadia Bartinura sent a letter from Eretz Yisrael to his brother. The original letter is not extant today, but we do know of the existence of five manuscript copies:

1. A manuscript currently in the British Museum in London.[77] In the opinion of Dr. Menachem Hartum,[78] who researched the subject, this copy is the most accurate one that we have.
2. Moscow manuscript.[79] Until a few years ago this manuscript was not within reach. In 1863, however, Senior Sachs[80] published its contents.
3. Sinigalia manuscript. It was lost in recent years, and we know little about its content.

4. Warsaw manuscript.[81]
5. Manuscript at the New York Bet Hamidrash laRabbanim.[82]
    Manuscripts 4 and 5 are copied from manuscript 1.

Since manuscript 3 is lost, and manuscripts 4 and 5 are copies of manuscript 1, then in fact we have the text of only two manuscripts, namely manuscripts 1 and 2, and there are many differences between them. On the subject of this visit there are two differences between these two manuscripts.

The first difference is the date of the death of the prophet Shmuel. According to manuscript 1, he died on the twenty-first of Iyar, and according to manuscript 2, it was on the twenty-eighth of Iyar. (According to Dr. Hartum, even on manuscript 3, the date is twenty-first of Iyar.[83])

The second difference is that in manuscript 2 at least one line is missing, and this is how it appears in the manuscript: "On the twenty-eighth of Iyar the day he died people come from all the surrounding areas and lit a big fire."[84] The missing line is the line that mentions Shmuel Harmati.

The Moscow manuscript was the first that was printed, and the question thus arose: On the day of whose death? In answer, support can come from the notes of the publisher of this letter in the year 5682 from the city of Berlin,[85] who wrote "of the Sefaradi above (?)" – the Sefaradi is mentioned several lines earlier.

But about thirty years earlier in a publication called "Darchei Tzion"[86] there is a very interesting and strange remark: "On the twenty-eighth of Iyar on the day of his death (said the copyist from the plain text there appears to be a scribal error and one should say that it is the day of the death of the Tanna Rashbi, the day of his hilula) they come from the surrounding areas." In later editions[87] of "Darchei Tzion" this amendment was made.

It is very difficult to understand this correction, because immediately afterwards Rabbi Ovadia writes about the elders of Jerusalem, namely one is speaking about a location in Jerusalem and not in the Galilee.

Even though in the year 5701, in his book "Hilula d'Rashbi,"[88] Rabbi Asher Zelig Margoliot included this correction to a quote from the letter of Rabbi Ovadia that the first person who wrote about the travelling and the large gathering at Meron on Lag b'Omer (Rabbeno Ovadia miBertinura in his letter to his brother in the year 5249 was printed in the book "Darchei Tzion") on the eighteenth of Iyar came from all the surroundings and lit enormous torches.

From all of the above we see how distortion after distortion over the course of the years completely changed the content of Rabbi Ovadia's letter. He wrote about a pilgrimage to the tomb of the prophet Shmuel in Jerusalem on the twenty-eighth (or twenty-first) of Iyar, and the content got completely changed to a pilgrimage to the grave of Rashbi in Meron on Lag b'Omer, the eighteenth of Iyar!!

Date of the Death of the Prophet Shmuel

In the Shulchan Aruch[89] there is a list of "days on which one fasts", and amongst them: "On the twenty-eighth of [Iyar] the prophet Shmuel died." The date of this occurrence appears in almost all the lists brought in the Rabbinical literature.

At the end of the Megillat Ta'anit[90] there is also a list of days which one fasts, and there the date of the death of the prophet Shmuel is given as twenty-ninth of Iyar. Even though there are a number of differences between Megillat Ta'anit and other sources, and in the Amsterdam printing of Megillat Ta'anit from the year 5419 the variant versions are given, but it does not state by the date of the death of the prophet Shmuel that there also exists a version which states twenty-eighth of Iyar. Because Megillat Ta'anit was composed by the sages of the Talmud,[91] Rabbi Avraham Gombiner known as the "Magen Avraham" (17th century), writes: [92] "It seems to me one must follow them [the dates which appear in Megillat Ta'anit] because this is the source for them." Although according to the evidence from many sources (which were not in the possession of the Magen Avraham), the list of fasts was added as an appendix to Megillat Ta'anit only from the sixteenth century, and it does not form part of the original Megillat Ta'anit.[93] The date twenty-ninth of Iyar also appears in the siddur "Shaarei

Shamayim,"[94] in the siddur "Nehora haShalem,"[95] and in the book "Itim leBina."[96]

The doubt on the date of the death of the prophet Shmuel has not been decided to this day. In the "Luach l'Eretz Yisrael"[97] from 5666 onwards, this is explicitly stated: "On twenty-eighth [of Iyar] a large number of people from Jerusalem go to prostrate themselves on the grave of the prophet Shmuel which is situated in the village Nevi Samuel and they remain there all the night and on the following day they pray for the second time, since there is a doubt of the date of his death, namely whether it is the twenty-eighth or twenty-ninth of that month."

It seems that saying prayers near his burial place on these two days is not a new phenomenon. From the words of R' Gedaliah of Simiatitz (quoted above), it can be seen that the president arrived at the grave of the prophet Shmuel on the twenty-eighth day, and held the prayer service on the night of the twenty-ninth. On the other hand, in his letter, Rabbi Yitzchak ben Alfara[98] explicitly quotes the night of twenty-eighth of Iyar. Also in "Haadrot haKedoshot"[99] it states "one goes up to prostrate oneself at the grave of our master the prophet Shmuel on the night of his death which is twenty-eighth of Iyar." There is a manuscript[100] from the seventeenth century in which it is written: "Order of the prayers for the twenty-eighth day of Iyar, the death of Shmuel the prophet"

The book "Tzeda laDerech"[101] was written in the fourteenth century by Rabbi Menachem ibn Zerach. In it appears, as well as in all the manuscripts of "Tzeda laDerech"[102] that are in our possession, the date twenty-first of Iyar, as the date of the death of the prophet Shmuel. Rabbi Yechiel Zilber[103] thinks that in fact according to the opinion of the "Tzeda laDerech," twenty-eighth of Iyar is the date, but the quoted opinion of the "Tzeda laDerech" in his book was distorted. Although this date does not appear only in the "Tzeda laDerech," we have already seen above that in the most accurate manuscript of the letter of Rabbi Ovadia Bertinura appears the date of the twenty-first of Iyar.

The date the twenty-first of Iyar also appears in the letter of the traveler from the year 5282:[104] "On Sunday the twenty-first of Iyar in the year 5282, they went together, men and women to pray at the graves of the righteous which was in Tiberias." Maybe it was a onetime occasion that they went together to the graves of the righteous on the twenty-first of Iyar. However, it is impossible to rule out the possibility that they chose the twenty-first day of Iyar in order to mark and commemorate the day of the death of the prophet Shmuel, and if in the Diaspora they celebrated such an event (as was stated above), why not commemorate it in the Galilee as well?

Hilula to Rabbi Meir Baal Haness on Pesach Shani

Above we mentioned above, the pilgrimage to the graves of Hillel and Shammai on Pesach Sheni began (at least) from the thirteenth century. About one hundred and fifty years ago they began to have a hilula at the grave of Rabbi Meir Baal Haness in Tiberias on that same day.

Rabbi Shraga Weiss writes on this subject in the Torah journal called "Orita"[105] dated the year 5743: "Every year, on the fourteenth day of Iyar they celebrate in Tiberias the 'hilula of Rabbi Meir'. On that day all the Sefaradi communities gather together from Tiberias and many of the cities of Eretz Yisrael by the grave of the Holy Tanna [Rabbi Meir], they study his teachings, sing and rejoice, and kindle bonfires and lights in honour of "Rabbi Meir." This hilula was first established in the year 5627. With the Sefaradim, this hilula grew in numbers from year to year. The Rabbis of Tiberias even printed piyyutim and songs to say at this hilula.

The Sefaradim accepted the hilula with great enthusiasm, and even the Sefaradic communities abroad, in Morocco[106] and Tunisia[107] observed it. However, there was opposition among the Ashkenazim. Rabbi Mordechai Chaim who was the Admor of Slonim even published a proclamation against the hilula.[108] In his explanations he included the mixing of women and men when they came to the grave and he concluded the proclamation with the words: "But from then until now they have had to remember and abolish this custom altogether."

There were also other reasons which disturbed the holding of the hilula. In the year 5636 Rabbi Rahamim Yosef Chaim Ofelatka (h'Yerach) made a visitation there on Pesach Sheni and writes about his visit[109] and said that on that year an insufficient number of pilgrims came and there was thus a financial loss due to not selling all the candles.

Every year (from 5656 to 5676) in the "Luach l'Eretz Yisrael," [110] and also in the "Luach Yerushalayim"[111] from the years 5701 and 5702 it is stated: "On the night of the fourteenth (Pesach Sheni) they made a bonfire and a hilula by the grave of Rabbi Meir Baal Haness." In addition, the calendars of that period[112] mention the hilula of Rabbi Meir Baal Haness. However, these calendars do not mention the pilgrimage to the graves of Hillel and Shammai in Meron on Pesach Sheni. This means that at that time this pilgrimage no longer took place. It can be assumed that the reason for this lies in the fact that until the time of the Ari, the only hilula in Meron took place on Pesach Sheni. Participation in the hilula for Rashbi on Lag b'Omer continually increased in numbers from year to year. The public did not want to go up to Meron or stay there for Pesach Sheni and also a few days later for Lag b'Omer, so they stopped going to Meron for Pesach Sheni.

One can suggest that a source for the hilula in Tiberias appeared in the year 5633 in an article in the newspaper "Havatzelet."[113] It reads: "It is now six years that the people of Tiberias are jealous of them [people of Meron] ... also in our midst there are graves of righteous Tannaim of which we are proud and why should we be inferior to them? And they finished by saying that one makes a bonfire by the grave of the Tanna Rabbi Meir Baal Haness on Pesach Sheni, the fourteenth of Iyar." One can say that the people of Tiberias wanted to make a hilula in their city and they relied on the opinion that Rabbi Meir Baal Haness died on Pesach Sheni[114] and also that the fact that the hilula in Meron on Pesach Sheni had ceased, and therefore they decided to make a hilula by the grave of Rabbi Meir Baal Haness in Tiberias on the day of his death.

In the customs of the Jews of Tunisia today[115] there is a mention of the connection that existed between Meron and Pesach

Sheni. According to their custom: "Anyone who survives an illness or danger, then he or his relatives make an oath to organise by him a meal in honor of Rabbi Meir or in honor of Rabbi Shimon Bar-Yochai (sometimes two meals are held)." These meals are held during the daytime of Pesach Sheni. It is possible that the reason why the meal for Rashbi is held on Pesach Sheni stems from the original custom of going up to Meron on Pesach Sheni and not on Lag b'Omer.

Summary

Pesach Sheni: From the thirteenth century there was the pilgrimage to the graves of Hillel and Shammai in Meron to pray for the rains. After the increase of people going up to the hilula at the tomb of Rashbi on Lag b'Omer, the going to the graves of Hillel and Shammai ceased and instead there is a going up to the grave of Rabbi Meir Baal Haness in Tiberias.

Lag b'Omer: Pilgrimage to the grave of Rashbi in Meron. The earliest source for this hilula is to be found in the writings of Rabbi Chaim Vital, a student of the Ari.

The day of Shmuel Haramati's death: The earliest source for the going up to his grave on the day of his death is in the middle of the fifteenth century.

References

Abbreviations
SA =Shulchan Aruch
OC = Orach Chaim
(1)   R' Eliakim Carmoly, *Eile haMassaot*, (Brussles, 5601), p.20
(2)   R' Yehudah David Eisenstein, *Ozar Massaoth*, (New York, 5687/1926), p.65
(3)   *Luach Davar b'Ito* (Every Thing in its Proper Time), year 5748 (Achiezer: Bnei Brak, 5748), p.501
(4)   MS Paris, heb.312/2 folios 235b-236a
(5)   *Sefer Hamaasaot*, op. cit.

(6)   Rabbi Simcha Assaf, "Totsaot Eretz Yisrael", *Yerushalayim* edited by
        Avrahan Moshe Lunz, (Jerusalem, 5688), pp.51-52
(7)   Ibid., p.52 footnote 1
(8)   *Cosmographie de Chems-ed-din Abou Abdallah Mohammed
        ed-Dimichqui*, (Saint Petersbourg, 1866), p.118
(9)   Bernard Lewis, "An Arabic Account of the Province of Safed -I", *Bulletin
        of the School of Oriental and African Studies*, University of London, vol.
        xv, 1953, pp.477-78
(10)   Ibid., pp.480-81
(11)   *Qalqashandi*, Subh Al Asha, (Cairo, 1914), vol.4, p.78
(12)   Yaakov ben Moshe Chaim Baruch, *Shivchi Yerushalayim*, (Levorno,
        5629), p.56
(13)   Biblical Bamidbar, chap.9 verses 9-14
(14)   *Megillat Ta'anit*, (Warsaw, 5634), chap.2 month of Iyar, p.6 (11)
(15)   Talmud Bavli, Rosh Hashanah 19b
(16)   Rabbi Chaim Hezekiahu Medini, *Sdei Chemed*, (Bet Hasofer: Bnei Brak,
        5723), vol.5, p.378
(17)   *Luach Davar b'Ito*, op. cit., pp.500-501
(18)   Rabbi Chaim Elazar Spira, *Shaar Yissachar*, (Munkacz, 5698), vol.1
        p.52b; Rabbi Chaim Elazar Spira, *Darkei Chaim v'Shalom,* edited by
        Rabbi Yechiel Michel Gold, (Munkacz, 5700), minhagei yemai haomer,
        pp.204-05
(19)   Rabbi Menachem ben Meir A'sh (Eisenstadt), *Zichron Yehuda*, (Munkacz,
        5660), p.35
(20)   Rabbi Yissachar ibn Mordechai Sussan, *Tikkun Yissachar – Ibur Shanim*,
        (Venice, 5339), p.23b
(21)   Ibid., p.2b
(22)   Rabbi Tzvi ben Dov Cohen, *Bein Pesach l'Shavuot*, (Bnei Brak 5745),
        pp.203-04
(23)   Rabbi Yisrael Yaakov ben Yom Tov Elgazi, *Shalmei Tzibur*, (Salonici
        (Thessaloniki), 5555), dinei nefilat apayim, par.10
(24)   Rabbi Chaim Elazar Spira, *Nimukei Orach Chaim*, (Slovakia, 5690),
        p.37
(25)   Rabbi Yosef David, *Bet David*, (Salonici (Thessaloniki), 5500), p.13b
(26)   Rambam, Hilchot Kiddush Hachodesh, chap.3 halacha 12, chap.5
        halacha 6
(27)   *Bet David*, op. cit., p.14
(28)   Ibid.
(29)   *Luach Davar b'Ito*, op. cit., p.502
(30)   *Likut Dinim v'Hanhagot m'Maran haChazon Ish,* (published by Rabbi
        Meir Grainaman: Bnei Brak, 5748), vol.1 OC p.47
(31)   Rabbi Yosef Ginsberg, *Itim leBina*, (Warsaw, 5647), article 15, p.126
        (251)
(32)   Rabbi Yechiel Michel Tucazinsky, *Sefer Eretz Yisrael*, (Jerusalem, 5726),
        chap.18 par.1 (p.64); *Sdei Chemed*, op. cit.
(33)   Rabbi Avraham David of Buczacz, *Eshel Avraham*, OC chap.131

(34) Rabbi OvadiaYosef, *Chazon Ovadia*, (Jerusalem, 5727), vol.2 hilchot Pesach im Hagadah shel Pesach, pp.175-76

(35) *Shivchei R' Chaim Vital*, (Lemberg, 1863), p.2b; *Shivchei haAri*, (Lemberg, 5610), p.2b; Rabbi Yitzchak Luria (Ari), *Sefer haKavanot uMaase Nissim*, (Kushta (Constantinople), 5480) p.2; Rabbi David Kahana, *Even Negef*, (Vienna, 5633), pp.23-24

(36) Rabbi Chaim Vital, *Shaar haKavanot*, (Salonici (Thessaloniki), 5612), gateway 6, p.127

(37) Rabbi Chaim Elazar Spira, *Divrei Torah*, third edition, (Munkacz, 5690), chap.46

(38) *Pesikta* of Rab Kahana, (Lyck, 1868), piska 10 vayehi beshalach, p.87b

(39) Rabbi Yehudah haLevi Levi, *Zemanei Hayom Bahalacha*, (Leo Levi, Jewish Chrononomy), (Gur Aryeh Institute for Advanced Jewish Scholarship: Brooklyn New York, 5727), p.11

(40) Rabbi Aharon Refoel Chaim Pereira, *Me'il Kodesh v'Bigdei Yesha*, (Jerusalem, 5648), Bigdei Yesha, Torat Emet v'Shalom, p.59b

(41) *Shivchi Yerushalayim*, op. cit.

(42) e.g. Talmud Bavli, Pesachim 4a

(43) *Shaar haKavanot*, op. cit.

(44) Rabbi Chaim Vital, *Pri Etz Chaim*, (Korzec, 5545), gateway 22, gate sefirat haomer, chap.6, p.108

(45) Rabbi Yaakov Chaim Tzemach, *Nogid u'Mitzvah*, (Amsterdam, 5472), p.64

(46) *Shivchei R' Chaim Vital*, op. cit., pp.14b, 16a, 16b(

(47) Ibid., p.2b

(48) Ibid., p.16b

(49) Jewish Chrononomy, op. cit.

(50) *Shivchei R' Chaim Vital*, op. cit., p.18b

(51) Yosef ben Yitzchak Sambari, *Divrei Yosef*, MS Paris "Kol Yisrael Chaverim" folio 115b (published by Mercaz Zalman Shazar and Merzaz Dinor, Jerusalem 5741); MS Bodleian, catalogue Neubauer 2410, folio 108a

(52) Letter of Shimshon Bak dated 19 Tammuz 5342 from Jerusalem, published by R' Avraham Moshe Lunz, in his book *Yerushalayim*, second year, (Jerusalem, 5647), pp.146-47

(53) Jewish Chrononomy, op. cit.

(54) Rabbi Shabtai miRashkov, *Sefer Tefillah miKol haShana im Kavanat haAri* , Korzec, 5554), p.26a

(55) Rabbi Avraham ben Natan haYarchei, *Sefer haManhig*, (brought out by R' Yitzchak Refoel, Mossad Harav Kook: Jerusalem, 5728), vol.2, p.538 (with the variant versions)

(56) Talmud Bavli, Yevamot 62b

(57) Rabbi Menachem ben Shelomo – "Meiri", *Beit haBechirah*, Masechet Yevamot, (Institute for the Complete Israeli Talmud: Jerusalem, 5728), p.229

(58) letter from Rabbi Yoesef Kapach to Chaim Simons, Adar 5751

(59) *Sefer haManhig*, op. cit.

(60) Teshuvot haGeonim, *Shaarei Teshuva*, (Leipzig, 5618), chap.278, p.26

(61) Rabbenu Yerucham, *Sefer Toldot Adam v'Chava*, (Venice), Chava, section 22 part 2, (p.186b)

(62) Rabbi Chaim Hakohen, *Tur Bareket*, (Amsterdam, 5414), OC chap.493, (p.89)

(63) *Nogid u'Mitzvah*, op. cit.

(64) Rabbi Moshe ben Machir, *Seder haYom*, (Berdyczow(?), 5566), Seder Kesirat uSefirat haOmer

(65) SA OC chap.493 par.1-2

(66) Ibid., Rema

(67) The Rema writes in connection with Lag b'Omer: "One does not recite Tachanun" (Rema "SA OC chap.493 par.2). The Ba'er Heteiv adds: (OC chap.493 par.5)" And the Ashkenazi custom is that in shacharit one does not recite tachanun, but this is not the case in the evening when one does say tachanun. Rabbi Yaakov Emden (Mor uKetzia chap.493) explains the words of the "Ba'er Heteiv "the evening" mean the day before as in the case of erev Shabbat or erev Yom Tov, but in the minchah service of Lag b'Omer of course one does not make a difference from the shacharit service. However, the "Noheg kaTzon Yosef" p.62 (Hanau, 5478) writes that one says tachanun at minchah of Lag b'Omer (and likewise on 15 Shevat), and it is only in shacharit that one does not say tachanun.

(68) Rabbi Ovadia Yosef, *Yechave Daat*, (Jerusalem, 5740) vol.3 chap.31

(69) Mishnah Berurah SA OC chap.493 par.8

(70) *Ateret Zekenim*, SA OC chap.493

(71) Rabbi Chaim Yosef David Azulai ("Chida"), *Tov Ayin*, (Husyatin, 5664), chap.493, (p.50)

(72) Rabbi Avraham ben Shmuel Zacuto, *Sefer Yuchasin Hashalem*, (London, 1857), fifth paper, p.228

(73) *Masa Meshulam miVolterra b'Eretz Yisrael biShenat 5241*, (Mosad Bialik: Jerusalem, 5709) p.74

(74) Menachem Hartom, "Harav Ovadia Bartinura v'igrotov", *Yehudim b'Italia – Mechkarim*, (Hebrew University Jerusalem, 5748), p.101 and variant readings

(75) Rabbi Gedaliah of Simiatitz, *Shaalu Shalom Yerushalayim*, (Jerusalem, 5723), p.25

(76) Rabbi Chaim Palagi, *Ruach Chaim*, (Izmir, 5641), OC chap.580 par.1

(77) MS London, British Museum 1074.3

(78) Hartom, op. cit., p.49

(79) MS Moscow catalogue Günzburg 333

(80) Senior Sachs, "Zwei Briefe Obadjah's aus Bartenuro", *Jahrbuch fur die Geschichte der Juden und der Judenthums,* (Leipzig, 1863), vol.3, p.193

(81) MS Warsaw 281/4

(82) MS Bet Hamidrash l'Rabonim New York, Mic 3617

(83) Hartom, op. cit., p.101

(84) MS Moscow catalogue Günzburg 333 folio161a; Senior Sachs, op. cit., p.222

(85) *haMasa l'Eretz Yisrael Bishnat 5247-5248*, (Klal-Verlag: Berlin, 5682) p.50

(86) *Darchei Tzion*, (Warsaw, 1895), p.21

(87) *Darchei Tzion*, op. cit., later editions: Pietrikov, 5688/1928, p.26; Jerusalem, 5691/1931, p.24; Slovenia, 1934, p.28

(88) Rabbi Ascher Zelig Margoliot, *Hilula d'Rashbi*, (Jerusalem 5701) p.13b

(89) SA OC chap.580 par.2

(90) Megillat Ta'anit, op. cit., appendix, p.19b

(91) Talmud Bavli, Shabbat 13b

(92) Magen Avraham, SA OC chap.580 introduction

(93) Chaim Simons, Ha'im haKeta "Yamim Shemitanim bahem hu Perek Batra shel Megillat Ta'anit?" *Sinai*, (Mosad Harav Kook; Jerusalem), vol.107 p.58

(94) Rabbi Yaakov Emden, *Siddur Shaarei Shamayim*, (Warsaw, 5641), Shaar haYesod, p.267a

(95) Rabbi Shlomo Netter, *Siddur Derech haChaim im Nehora haShalem*, (Vienna, 5628), vol.2, p.38b

(96) *Itim leBina*, op. cit., p.126

(97) e.g. *Luach l'Eretz Yisrael* for the year 5666, arranged by R' Avraham Moshe Lunz, (Jerusalem, 5665), p.20

(98) *Sefer Yuchasin*, op. cit.

(99) Rabbi Yosef Hacohen, *Haadrot haKedoshot*, (Amsterdam, 5468), introduction

(100) MS Bodleian F105, folio 6a

(101) Rabbi Menachem ben Aharon ibn Zerach, *Tzeda laDerech*, (Warsaw, 5640), fifth paper, first principle, chap.8, (p.142a)

(102) MS London British Museum, catalogue Margoliouth 1168 folio 290b; MS Bodleian, catalogue Neubauer 893/2 folio 216b

(103) Rabbi Yechiel Avraham Zilber, "keviyot hataaniyot", *Otzrot Yerushalayim*, (Jerusalem), vol.56, pp.887-888

(104) *Shivchi Yerushalayim*, op. cit.

(105) Rabbi Shraga Weiss, Hilula d'Rabi Meir Baal Hanes – 14 Iyar, *Orayta*, (Kolel Avreichim Tiferet Netanya "Yad Moshe), no.13, Iyar 5743, p.112

(106) *Yalkut Minhagim miMinhagei Shivte Yisrael*, arranged by Asher Wassarteil, (Misrad haChinuch vehaTarbut: Jerusalem, 5740), p.265

(107) Yalkut Minhagim, op. cit., p.307

(108) Rabbi Avraham Yitzchak Sperling, *Ta'amei haMinhagim u'Mekorei haDinim*, (Eshkol: Jerusalem), p.462

(109) Rabbi Rahamim Yosef Chaim Ofelatka, *Masei h'Yerach,* (Jerusalem, 5636), p.5

(110) *Luach l'Eretz Yisrael*, Lunz, op. cit., year 5656, (Jerusalem, 5655)

(111) e.g. *Luach Yerushalayim lishnat 5701*, edited by R' Dov Natan Brinker, (Jerusalem, 5700), p.28

(112) e.g. Luach for year 5751, Luach Dinim u-Minhagim, (brought out by the Ministry of Religions, Department for Synagogues); Luach for year 5751, (brought out by the chain of Porat Yosef Yeshivot in Israel)
(113) Habazeleth, Jerusalem, 18 Sivan 5633, p.238
(114) Rabbi Chaim Kneller, *Dvar Yom b'Yomo*, (Przemysl, 5667), p.56
(115) *Yalkut Minhagim*, op. cit., p.307

✿ ✿ ✿ ✿ ✿ ✿ ✿ ✿ ✿ ✿ ✿

APPENDICES

A number of years after the journal "Sinai" had published the above paper, certain Manuscripts became accessible and I accordingly, at different periods of time, wrote two appendices which were subsequently published in "Sinai"

Appendix 1

After a brief summary of what I wrote in my paper concerning the date of the death of the prophet Shmuel, I continued ...

When I wrote my paper the Moscow Manuscript [of the letter by Ovadia Bartinura] was not accessible. The only copy was a printed copy made by Senior Sachs in 1863. According to Sacks the date [of the death of the prophet Shmuel] that appeared on this manuscript was 28 Iyar and this is the date that appears in the Rabbinical literature. In my paper I suggested that "One cannot exclude the possibility that when Senior Sachs printed out this manuscript, he corrected the date from 21st to 28th of Iyar in accordance to what is written in the Shulchan Aruch and the other Rabbinical Literature where it states that 28th of Iyar is the date of his death."

Today following the dissolution of the Soviet Union, a microfilm of this manuscript (MS Moscow Günsburg 333) has arrived in Jerusalem and can be found in the Israel National Library (microfilm no. 43029). I have checked the relevant part (folio 161a) and found that the date 28th Iyar appears on it, which agrees with the date that Sacks published.

If this is the case, what is the date that Rabbi Ovadia Bertinura wrote in his letter - 21st or 28th of Iyar? From what I wrote in my paper it seems that the proof tends towards the 21st, but it is impossible to be completely certain.

Appendix 2

In the year 5714 "Mossad Harav Kook" published the book "Sefer ha-Chezyonot" written by Rabbi Chaim Vital. The book was prepared by Dr. Aharon Eshkoli. He utilised a manuscript that was in the possession of Rabbi Dr. Elio. Toaff, (the Chief Rabbi of Rome), and he wrote that the manuscript is in the handwriting of R' Chaim Vital himself.

On the first page appears the date that the Ari arrived in Eretz Yisrael. According to Dr. Eshkoli it is written in this manuscript, that in "the year 5330 a clever woman speaks about the future... namely that a great wise man will come during this year to Safed from the south, for example from Egypt... and this occurred, for in that year my teacher, my he be remembered for blessing, came from Egypt" – namely that during the year 5330 the Ari arrived to Eretz Yisrael, but it is impossible to determine the exact date during the year 5330 that he arrived.

However, from all the editions of the book "Shivchei R' Chaim Vital" (e.g. Ostroh 5586, Lemberg 5623, Lemberg 5625, Jerusalem 5626, Lublin 5660, Lublin 5687, Jerusalem 5713, Jerusalem 5726, Ashdod 5748) that I was able to check, including the edition which was printed from the manuscript (Jerusalem 5626) of the grandson of R' Chaim Vital, it states "basof" (at the end of) and not "betoch" (during the course of) that in the year 5330 the Ari arrived in Eretz Yisrael. (This book is an abridgement of "Sefer ha-Chezyonot.")

In addition, there are a number of manuscripts of "Sefer ha-Chezyonot" (MS Jerusalem Machon Schocken 11585/2, MS Cambridge Trinity College F12.43, MS Jerusalem Machon Ben Zvi 2234, MS Jerusalem Hebrew University 8^0 5300) and in all of them appears the word "basof" It should be mentioned that in one of them (MS Cambridge Trinity College F12.43) it states that it was copied from a manuscript written by R' Moshe Vital who had copied it from a manuscript of his grandfather R' Chaim Vital.

Therefore I decided to check the manuscript which Dr. Eshkoli had used. At present this manuscript is not accessible, but to our good fortunate a photocopy of the page which I required was reproduced in the book of Dr. Eshkoli. The difference in writing the words "sof" and "toch" is just the letter "samech" instead of the letter "tof," and a final "pai" instead of a final "kof." I made an enlargement of this photocopied page and then studied this word very carefully, and also the formation of the letters "samech," tof," final "pai," and final "kof" as they appeared elsewhere on this page. My conclusion (and also that of others) is that in this manuscript it seems the scribe had written "basof." I say "it seems" because the writing is not very clear, and it is therefore difficult to arrive at an absolute conclusion!

After that, I weighed up what appears in the manuscript that Dr. Eshkoli used, and also the versions which appear in all the printed editions and manuscripts, and from all this one can conclude with almost certainty that the correct version is "basof."

However the question remains how to explain that the Ari arrived in Eretz Yisrael at the end of the year 5330, in view of the fact that he was a pupil of Rabbi Moshe Cordovero who died on 23 Tammuz 5330. Sambari has written that the period that the Ari was a pupil of R' Moshe Cordovero was for a "very short period" (MS Paris Kol Yisrael Chaverim H 130 A folio 99b) – in fact a very very short period! Perhaps it was only for a few weeks and that the Ari arrived in Eretz Yisrael in Tammuz 5330. It appears that the intention of the words "sof hashanah" (end of the year) was not precisely the month of Elul but relative to the end of the year.

Appendix 3

Follow on to publication of my paper.

The material below is taken from my unpublished autobiography.

After my paper had appeared in the journal "Sinai," a professor wrote a letter to that journal criticising a number of the points I had made in my article. (I am not the only person who has been criticised by this professor!) Although I would regularly see this professor at the Minchah Minyan at the Jewish National Library, he never said anything to me on this matter. It is likely that he did not connect my face with my article.

It was about eight years later that he mentioned the matter to me! He had obviously by then found out who I was. This was the first I knew of his letter of criticism and I immediately went to look it up. Even though a long time had elapsed since he had written his criticism and because there is no "statute of limitations" on this subject, I at first decided to prepare an answer. I carefully studied his letter line by line and prepared the draft of an answer. However, I then decided that since so many years had passed, I would not send it up to the journal "Sinai."

I have in my files my draft answer and will give here the main points. The professor was critical that I would repeatedly describe a traveler to Eretz Israel mentioned in my paper as "the anonymous traveler from the year 5282" and not as R' Moshe Basula. (He had likewise in the past criticised another writer on this very same point.) Actually I had added in brackets that "some say he is R' Moshe Basula."

I utilised this term "the anonymous traveler" since this was the term that had been used to describe him for over four hundred years. In fact, his identity is not so clear cut. At first he had been identified as R' Moshe Basan from Noveira. At a later date it was Yitzchak Ben-Zvi (who was later a President of the State of Israel) who decided that it was R' Moshe Basula. It is quite possible that in the future, some other historian will decide that Ben-Zvi was wrong, especially as his reasoning is specious.

The professor also claimed that I made a number of errors in the chronology of the Ari. He was critical that I had stated that "there are differences in opinion regarding the year that the Ari died." He held that the opinion of R' Chaim Vital that he died in the year 5332 was the correct one. But the fact is that in addition to R' Chaim Vital, there are opinions that he died in 5333 or 5334. The historian Shlomo Schechter saw in these different opinions sufficient importance to quote them in his book.

Another chronological fact that I quoted from a source, but which the professor objected to, was that I stated that the Ari arrived in Israel at the *end* of 5330. He claimed that this source was not accurate and that the correct version appeared in an edition of "Sefer ha-Chezyonot" of R' Chaim Vital, prepared by Dr. Aharon Eshkoli and published by Mossad Harav Kook. This edition stated *during* 5330 (in place of *end* of 5330) and it was taken from a manuscript written by R' Chaim Vital himself, which was in the possession of Rabbi Dr. A. Toaff of Rome.

I felt that this point merited further research. I went through all the other available manuscripts of "Sefer ha-Chezyonot" and the printed editions of "Shivchei R' Chaim Vital" including one taken from a manuscript written by R' Chaim Vital's grandson. In *every* case it stated *end* of 5330.

I then decided to study the manuscript used by Dr. Eshkoli. Unfortunately, a microfilm of it was not in the Jewish National Library, and Mossad Harav Kook which did have a copy was at the time closed for reorganisation.

However fortunately, the book by Dr. Eshkoli reproduced one sample page of this manuscript and it was the page I required. (Usually according to Murphy's Law, it is the opposite which is the case!) I photocopied it and made an enlargement of the photocopy. The difference between "sof" (end) and "toch" (during) especially in handwriting which isn't of the best is small. I, and also others studied the enlargement carefully and also compared it with how the writer formed his letters. We came to the conclusion that this manuscript said "sof" and not "toch" and was thus in complete agreement with the other manuscripts and published editions.

I felt that this piece of research merited publication and I sent it up to the journal "Sinai" and they published it. In my appendix which I sent to the journal "Sinai," I did not mention the professor's letter. However, the next time I saw the professor, I told him about this research and asked him to read it and let me have his comments. However, I never met him again, and so I don't know his observations on it.

I wrote in my paper that R' Chaim Vital began to learn with the Ari at the end of 5331. On this the professor wrote that there is no basis for this. However, we know that R' Chaim Vital only went *once* on Lag b'Omer to see the Ari in Meron. As his outstanding pupil, he would surely have gone *every* Lag b'Omer possible. In other words, he was *not yet* a pupil as at Lag b'Omer 5331. This is further supported by the evidence of R' Yonatan Sagish. In fact, in one of his *own* papers, this professor wrote that he was a pupil for "a year and a few months" - about 14 months - which is in good agreement with my assessment!

In my paper I commented that there were errors that had crept into "Shivchei R' Chaim Vital," in particular in connection with dates. I gave as an example "Leil Shabbat 5th day of Chanukah." I pointed out that the 5th day of Chanukah cannot occur on Shabbat! On this the professor wrote that in "Sefer ha-Chezyonot" it states instead "Leil Shabbat 8 Tevet" and so all my "pilpulim" (argumentation) are "superfluous." What the professor "overlooked" was that 8 Tevet also cannot occur on Shabbat!

In his letter, the professor stated that I wrote that Shmuel Hanavi died on 21 Iyar. Anyone reading my paper would see that this is not so. I in fact brought down the various opinions regarding the date of his death, *one* of them being 21 Iyar. The professor also claimed that this date 21 Iyar only appeared in one manuscript. In fact, it appears in a number of places. He also claimed that 28 Iyar is the "accepted" date. Also this is not correct! Pilgrims go on *both* 28 and 29 Iyar to pray at his grave, since it is held that either of these dates could be correct.

Another point that I mentioned in my draft did not arise from his letter, but from some articles he wrote. The "anonymous traveler" wrote "15th Iyar which is called Pesach Sheni." The professor "corrected" this by adding "apparently this should be the 14th." It would seem that the professor was not aware of the Rabbinic literature on the subject which I brought down in detail in my paper that the date of Pesach Sheni continues into the 15th Iyar and thus the "anonymous traveler" did not err.

I ended my draft by thanking the professor, who is one of the authorities on the Ari, for reading my paper and writing his comments.

THE REASONS FOR THE FAST ON
THE NINTH OF TEVET

In the Shulchan Aruch there is a list of more than twenty days "on which troubles occurred to our ancestors and it is proper to fast on these days."[1]

This list appears in a number of manuscripts, as an appendix to Megillat Ta'anit (even though this appendix only appears there from the sixteenth century onwards), and in the Rabbinical literature from the period of the Geonim (great Rabbis who lived between about the 6th to the 11th centuries) to the present day. It is also to be found in a liturgical form in some selichot.

The reasons for all the fasts appearing in the above-mentioned list is given, but with one exception, namely, the ninth day of the month of Tevet. Regarding this fast, in almost all the sources (except those who do not mention it!) it is written that the "Rabbis did not write the reason for it." Only in a few cases was the reason mentioned, namely that it is the day that Ezra and Nehemiah died.

As we shall see below, a number of Rabbis tried to find reasons for this fast. In this paper, we will try to research and review these reasons, and also to determine why the Rabbis avoided giving a reason for it.

1) The Death of Ezra the Scribe

It is written in the Shulchan Aruch: "It is not known what misfortune occurred on that day."[2] In his commentary to the Shulchan Aruch, the Taz[3] (Rabbi David Halevi, 16th-17th centuries, Poland), was very surprised at this statement because in the selichot for the tenth of Tevet it states that on the ninth of Tevet, Ezra the Scribe died. In addition, other commentaries on the Shulchan Aruch, for example the "Magen Avraham"[4] (Rabbi Avraham Abele Gombiner Halevi, 17th century, Poland), and the "Ba'er Heteiv"[5] (Rabbi Yehuda Ashkenazi, 18th century, Germany), refer to these selichot.

Rabbi Baruch Frankel[6] (18th-19th centuries, Poland), explains that the above commentators are referring to the selichah "Ezcara Matzok" which is the first selichah in the Ashkenazi rite for the selichot for the tenth of Tevet. There it states: "On the ninth [of Tevet] I was denounced with reproach and shame. My robe of majesty and diadem were stripped off, and the giver of goodly words was forcibly torn away. He is Ezra the Scribe."

From the last line of the selichah we can see that the author's name is "Yosef" and apparently,[7] that he is Rabbi Yosef ben Shmuel Bonfils Tov Elem, who was one of the greatest Tosafists in France in the mid-eleventh century.

Rabbi Avraham Meir Israel[8] (Head of the Bet Din in Hunyad in Brooklyn 20th century), explains why the author of the Shulchan Aruch was not aware of this selichah. He considered the reason was that the author of this selichah was an Ashkenazi, but the author of the Shulchan Aruch was a Sefaradi and he had therefore not seen this selichah, and hence was not aware of the tragedy which occurred on the ninth of Tevet.

In addition to the selichot of Rabbi Yosef Bonfils, there are also selichot in the Sefaradic rite who mention the death of Ezra. In the selichah "Shaah Elyon" it is written, in contrast to the selichah "Ezkara Matzok," that the date of Ezra's death was the tenth of Tevet. The author of this selichah, apparently,[9] was Rabbi Yechiel Mondolfo, who lived in the seventeenth century (after the death of Rabbi Yosef Karo).

Ezra's death is also mentioned in the selichah "Yoshev Bashamayim" whose authorship and date of composition are unknown, and which follows immediately after "Shaah Elyon," and states that Ezra died on the tenth of Tevet and not on the ninth.

Indeed, from the Rabbinical literature we see that Ezra died on the tenth of Tevet. It is stated by Rabbi Avraham Zechut, (15th-16th centuries, Spain), in his book "Sefer Hayuchasin":[10]

"Then Ezra the scribe died on the tenth of Tevet, and Haggai and Zechariah and Malachi and then prophecy departed from Israel." About a century later, Rabbi Gedalia ben Yosef ibn Yechia, one of the sages of Italy in the sixteenth century, and the author of "Shalshelt haKabbalah"[11] wrote: "The prophets Haggai, Zechariah and Malachi died on the tenth of Tevet." In addition, he wrote[12] that "it states in a Midrash that Ezra is the prophet Malachi." This identification that Ezra is Malachi also appears in the Targum (translation of) Yonatan ben Uziel.[13] [In the Tractate Megillah[14] there are differences of opinion on the identification of Malachi; some say that he is Mordechai whilst others say he is Ezra.] In addition, in two manuscripts,[15] one can see that Ezra died on the tenth of Tevet, and not on the ninth.

The question to be asked is, if Ezra died on the tenth of Tevet, why fast on the ninth? Rabbi Asher ben Yosef David (17th–18th centuries, Salonika) in his book "Bet David"[16] answers this: "Because Ezra died on the tenth of Tevet one should have to fast on it ... but people already had to fast on it because the king of Babylon had begun a siege on Jerusalem and therefore it would not be recognised that the fast was (also) to commemorate the death of Ezra ... Therefore, they advanced the fast to the ninth in the same way as a fast occurring on Shabbat would be postponed to the next day, although in this case it would be advanced." However, one needs to investigate his comments because there is a principle in the Talmud[17] that "we do not hasten the approach of trouble," and therefore the fast should be postponed to the eleventh Tevet and not be advanced to the ninth.

According to the Bet David,[18] the reason for the observance of the fast on Ezra's death on a date that was not the day of his death was because the Rabbis did not know why one fasts on the ninth of Tevet. In contrast to this, Rabbi Yonatan Eybeschutz[19] said in a sermon he delivered in Hamburg on the ninth of Tevet 5518 (1751) that Ezra in his generation was like Moshe, and he thus had the qualifications of Moshe to receive the Torah. Since Ezra was compared to Moshe, and in the same way as the Torah intentionally did not reveal the burial place of Moses, it was necessary to conceal something from Ezra's death, and the Rabbis chose not to reveal the date of his death. It is possible that there is a support for the words of Rabbi Eybeschutz in the book "Tzeda laDerech" by Rabbi Menachem ibn Zerach,[20] (14th century, Spain) who wrote that the Rabbis did not want to write on what occurred on that day [ninth of Tevet]. From this one can see that even though the Rabbis knew the reason for the fast, they intentionally "did not want" to reveal it.

There are several sources who state that the date of Nehemiah's death was identical to that of Ezra's. However, the earliest reference which gives identical dates is to be found in the books "Kol Bo"[21] (authorship unknown), and "Orchot Chaim"[22] (Rabbi Aharon Hakohen m'Lunel, 14th century). [The date of the composition of a lamentation[23] (which was discovered in the Cairo Genizah) linking the death of Ezra to the death of Nehemiah, is unknown; However, the day of their deaths were included in the events of the month of Kislev!]

Rabbi Shlomo Haas, (19th century) in his book "Kerem Shlomo"[24] did indeed write: "The ninth [of Tevet] ... Ezra died and the 'Halachot Gedolot [BaHaG]' adds and also Nehemiah ben Hacaliah." However, it is clear that that the "Kerem Shlomo" copied this from a printed edition of the BaHaG (who lived in the period of the Geonim) in which the words regarding "Nehemiah" were a later addition. This can be seen in one of the manuscripts[25] of the BaHaG in which there is no reference at all to the deaths of Ezra and Nehemiah. In another manuscript,[26] the date of their death only appears as a marginal addition. A further proof can be found in Rabbinical literature which was written hundreds of years after the BaHaG, and in addition, a list of all the fast days throughout the year (including the ninth of Tevet) appears in the following books: Seder Rav Amram Gaon,[27] (9th century), Sefer Raviah (Rabbenu Eliezer ben Yoel haLevi,[28] about the twelfth century) and "Tzeda laDerech (Rabbi Menachem ibn Zerach, 14th century)."[29] These authors do not add the comment on the deaths of Ezra and Nehemiah. Also, the Tur (Rabbenu Yaakov ben Asher,[30] 13th-14th centuries), who begins this list with the words: "Written by the BaHaG" does not include these words regarding Ezra and Nehemiah.

Even though Rabbi Yaakov Emden (Yaavetz)[31] (18th century, Altona) and Rabbi Elia Shapira (Eliya Rabba)[32] (17th-18th centuries, Prague), quote the death of Nehemiah together with the death of Ezra, others reject it. Rabbi Shlomo Yehudah Rapoport (Shir)[33] who was the Rabbi of the city of Prague in the middle of the nineteenth century, writes: "It would really be a great coincidence if these two died on the same date." Rabbi Avraham Naphtali Zvi Roth[34] from Frankfurt am Main (20th century) comments: "It is possible that the subject of the death of Nehemiah ben Hacaliah in connection with the fast of the ninth of Tevet is incidentally mentioned together with the death of Ezra the scribe." However, it is not clear whether Rabbi Roth wants to determine that Nehemiah's death was not on the same date of Ezra's death. It is interesting to note that Rabbi Moshe Rivkash[35] known as the "Be'er Hagolah" (17th century, Amsterdam), and Rabbi Yosef Molcho,[36] the author of "Shulchan Gavoha" (18th century, Salonika), both write: "In the 'Kol Bo' it states that Ezra died on [9 Tevet]." However, they both omit that the "Kol Bo" also brings that Nehemiah died that day. Furthermore, the "Aruch Hashulchan,"[37] who had obviously seen the books of the Rishonim (great Rabbis who lived approximately between the 11th and 15th centuries) and the Acharonim (great Rabbis who lived from about the 16th century) that mentioned that Nehemiah's death was on the same day of Ezra's death, wrote: "It was later discovered that that on it [ninth of Tevet] Ezra died" and he does not mention Nehemiah.

The two books "Orchot Chaim" and "Kol Bo" are similar. However, there is a difference between them on the subject which is before us. According to the "Orchot Chaim":[38] "Our ancestors did not write about the ninth of Tevet, the reason being that it is a secret which is hidden from us, and on it [the same day] Ezra the Kohen and Nehemiah son of Hacaliah died." From the language "and on it" ["ubo" in Hebrew] means that there is an additional reason for the fast of ninth of Tevet, other than the death of Ezra and Nehemiah. In addition, there are three extant manuscripts[39] of the "Orchot Chaim" on which appear the words "and on it." On the other hand, according to the language of the Kol Bo[40] in place of the words "the secret which is hidden from us, and on it" ["bo" in Hebrew] appears "the secret is found on it." From the language "on it," namely the death of Ezra and Nehemiah is the only reason for this fast. In a manuscript[41] of the "Kol Bo" this point is emphasised even more, and instead of the word "is on it" there is the word "that is on it" ["shebo" in Hebrew]. In the "Siddur Minhag Romania"[42] from the year 5280 (1520) also appears the word "shebo" in Hebrew, and it states "and it is found 'shebo' that Ezra the Kohen and Nehemiah the son of Hacaliah died on [ninth of Tevet].

In apparently the oldest manuscripts[43] of the "Orchot Chaim" (from the year 1524), instead of the words and "the secret which is hidden" it states "and the secret is given," and this language is closer to the language of the "Kol Bo" namely, "the secret is found on it." However, in his commentary to the Shulchan Aruch, Rabbi Chaim Yosef David Azulai[44] known as the "Chida," (18th century), in his book "Birkei Yosef" he corrected the language of the "Kol Bo" to that of the "Orchot Chaim." (However, according to Rabbi Roth[45] the exact opposite is the case.)

In contrast to other commentators, Rabbi Yehudah Leib Gordon, (19th-20th centuries), in his commentary "Iyun Tefillah"[46] on the Siddur "Ozar haTefilot," completely rejects the reason for fixing a fast on the day Ezra and Nehemiah died: "for we do not find that any fast was decreed for the death of a great person with the exception of Gedaliah son of Ahikam, since then the kingdom of Israel ceased altogether and it was thus regarded like the destruction of the Temple." Rabbi Roth[47] expresses astonishment at this supposition: "Indeed one fasts on the seventh of Adar on the death of Moshe Rabbenu." In addition, it is expressly written by Rabbi Yehudah ben Shmuel of Regensburg (12th-13th centuries, Germany), in his book "Sefer Hasidim":[48] "One must fast on the death of a great man." One can see that on some of the dates appearing on this list in the Shulchan Aruch are fasts due to the deaths of great people.

2) The Killing of HaNagid R' Yehosef Halevi

In the book "Sefer Hakabalah"[49] written by Rabbi Avraham Ibn Daud (known as the Ra'avad and who died in 1180) is a passage about the about the vizier of Granada, haNagid Yohosef Halevi who was killed in 1066. On him the Ra'avad writes: "Of all the fine qualities which his father [R. Shmuel haNagid] possessed, he only lacked one, namely since he had been raised in wealth and thus never having to bear a burden of responsibility in his youth, he grew haughty, and this led to his destruction. The Berber princes became so jealous of him that he was killed on Shabbat the ninth of Tevet in the year 4824 [should be 4827] (1066) along with the community of Granada and all those who had come from far off lands to see his learning and power... and from the days of the early Rabbis who had written Megillat Ta'anit had decreed a fast on the ninth of Tevet but the reason for it was not known. From this incident (the murder of R' Yehosef and the community of Granada] we see that they had arrived by 'ruach hakodesh' (prophetically) to this very day."

Therefore, because this event had not yet occurred when the Rabbis instituted the fast on the ninth of Tevet, they did not write the reason for it.

The killing of R' Yehosef and the community of Granada on the ninth of Tevet, is also quoted by Rabbi Moshe Ibn Ezra (11th-12th centuries, Spain) in his book "Kitab al-Muhadara wa al-mudhakarah"[50] who writes: "He was killed together with the community of Granada on Shabbat the ninth of Tevet 4827 (1066)."

[There is some confusion[51] about the date of this killing, because in the translation by Ben-Zion Halper,[52] of Rabbi Moshe Ibn Ezra's book from the Arabic original, the date of the murder is given on Shabbat, the twentieth of Tevet. Because we also have the day of the week and also the date, it can be calculated that the twentieth of Tevet was a Wednesday and not a Shabbat. In addition, in a later translation by Avraham Helkin[53] who brings the original Arabic text, one can see that Rabbi Ibn Ezra wrote "the ninth of Tevet" which accords with version of the author of "Sefer haKabbalah."]

This killing is also mentioned in Arab sources. One of them is in "Abdullah's Memoirs,"[54] which states: "G-d declared their downfall on Shabbat, the tenth day [of the Moslem month of] Safar). Also in the summary of the event brought in a book by Reinhart Dozy[55] the date given is Shabbat, 10 Safar 459 (year of Moslem calendar). According to the fixed calendar of the Moslems, 10 Safar 459 was on Sunday and not on Shabbat; The corresponding date in the Hebrew calendar was the tenth Tevet, and not the ninth of Tevet. How can these contradictions be resolved?

The Moslems did not always follow a pre-fixed calendar. In some Arab countries they determined the new months by the first sighting of the new moon.[56] The average molad of the month of Tevet in the year 4827 was on Tuesday (December 19, 1066) in the afternoon at 5 hours 19 minutes and 11 hourly divisions; therefore it is indeed possible that in the year 459 (in the Moslem calendar), the first day of the month of Safar was determined by a lunar observation to be on Thursday (30 Kislev); therefore the tenth of Safar was indeed on Shabbat (ninth of Tevet), and this is in accordance with the Jewish sources.

The "Birkei Yosef"[57] brings the conclusion of the Ra'avad that the Rabbis arrived prophetically to this date in order to explain the question of why the Rabbis did not write a reason for the fast, since they already knew that Ezra had died that day. The "Birkei Yosef" comments: "There is a secret in the matter" and he brings evidence from the words of the "Orchot Chaim":[58] "For the ninth of Tevet, the Rabbis did not write the reason, and it is a hidden secret," and he hints that the solution of this secret is to be found in the words of the Ra'avad that they had "arrived by 'ruach hakodesh' (prophetically) to that date [9 Tevet] for the murder of Rabbenu Yehosef Halevi," namely, when they established the date of the fast, the event has not yet occurred.

However, by arriving at such a conclusion, there are difficulties, namely how can a fast be determined for an event that has not yet occurred? For example, the book "Igrot Shir"[59] describes it as "a remote prophecy," and Rabbi Aharon veYirmash[60] (the Rabbi of the city of Metz in the nineteenth century, and author of the book "Meorei Ohr") describes it as "forced." Rabbi Roth[61] from Frankfurt hints that the reason that the Taz and the Magen Avraham did not bring the words of the Ra'avad was: "We never found a case where a decree was made for an event which would occur in the future." Also in two manuscripts[62] of "Tzeda laDerech" it states: "The Rabbis did not write the reason about an event which had occurred on the ninth of Tevet," namely one which had occurred in the past. Although this is not conclusive evidence, it is possible that this expression was used in order to emphasise that this is an event that occurred in the past, and not an event that would occur in the future.

One finds that the Rabbis who cite this reason of the Ra'avad do not state this as the only reason, but in addition to the death of Ezra. Rabbi Yaakov Emden in his Siddur "Shaarei Shamayim"[63] writes among various other things that occurred in the month of Tevet: "Ninth [of Tevet] fast for the death of the righteous. Ezra Hakohen and Nehemiah the son of Hacaliah (and the murder of the Granada community together with R' Yehosef haNagid)." The parentheses are included in the text of Rabbi Yaakov Emden. Rabbi Shlomo Netter (19th century) in his Siddur "Derech haChaim with Nehora haShalem"[64] quotes the same words, but without the parentheses. Among the others who cite this reason as an additional reason, are Rabbi Nachman Kahana,[65] "Kerem Shlomo"[66] and the "Tosafot Chadoshim" to Megillat Ta'anit.[67] The only source that cites only the reason of the murder of R' Yehosef haNagid is Rabbi Chaim Palagi (19th century, Izmir), in his book "Ruach Chaim,"[68] and even he directs the reader to other sources.

3) Taking of Esther to the King's Palace

"And Esther was taken to the King Ahachvairosh' palace on the tenth month, which is the month of Tevet, in the seventh year of his reign."[69]

In his commentary on the Book of Esther, entitled "Mechir Yayin," Rabbi Moshe Isserlis[70] (the "Rema," 16th century, Krakow), comments on this verse that the month in which Esther was taken is seemingly unimportant to the reader, but he replies that the Book of Esther mentions the month to inform us that we should fast on the ninth of the month in order to remember this event.

This interpretation raises a number of questions:

a) Why fast because Esther was taken to the king's palace?
b) How does one know that it was on the ninth of the month?
c) Why did the Rabbis refrain from writing this as a reason for fasting?

From the Book of Esther itself it is difficult to find an answer to these questions, but from various Midrashim one can find an answer.

a: In "Midrash Rabbah"[71] one finds that Esther was forcibly taken to the king's palace, and it must be remembered that at that period Esther was a married woman, namely she was married to Mordechai.[72] In the Midrash "Shocher Tov"[73] one can see that Esther wept to G-d saying that even though the Matriarch Sarah was only with Pharaoh for one night, Pharaoh and all his household were punished. In contrast, Esther complained that she had been forced to live with the wicked king for many years. She thus asked G-d why he had not performed a miracle for her and begged "why have you forsaken me?" From this, it seems there is a sufficient reason to declare a fast on account of Esther being taken to the king's palace.

b: The date of Esther being taken appears in the Midrash "Lekach Tov."[74] On this verse in the Book of Esther, the Midrash writes that the day that the rumour of the exile of Yehoyakim arrived was the very day that Esther was taken before the king. From the book of Ezekiel[75] one can see that the day that the rumor came was on the fifth of Tevet: "And it came to pass in the twelfth year [of our captivity], in the tenth month [Tevet] on the fifth day of the month that one who had escaped from Jerusalem came to me saying the city is smitten."

214

The question arises, why did they declare the day of fasting on the ninth of Tevet instead of the fifth? The following explanation can be offered: The girls taken to the king were usually rejected after one night with the king.[76] Esther, on the other hand, was not rejected, and after three nights with the king (sixth, seventh and eighth of Tevet), the Rabbis in accordance with the principle of "presumption" came to the conclusion that the king intended to take her for a wife. Therefore, on the eighth day of Tevet, the Rabbis decided to establish a fast for the next day, and in the following years, the Jews continued to fast on the ninth of Tevet, to commemorate this event.

However, the question arises: Why in the subsequent years, did the Jews not fast on the fifth of the month? An answer to this question can be found in "Perush haEshel" of Rabbi Avraham Eliyahu Bornstein[77] (20th century), in connection with the Fast of Gedaliah: "Even though he [Gedaliah] was killed on Rosh Hashanah, and even on the second day still no-one knew about it. Therefore, on the first year they fasted when they heard the report, namely on third of Tishrei. Similarly, it is stated in Masechet Rosh Hashanah that the day of hearing the report is like the day of the burning [of the Temple] and therefore they fixed it [the Fast of Gedaliah] perpetually on the third of Tishrei." This principle, which Rabbi Bornstein brings, can also be extended to the establishing of the fast for future generations on the ninth of Tevet in commemoration of taking Esther to the king's palace.

c: One can find a reason from the Gemara[78] why the Rabbis refrained from writing this reason for this fast. There it states: "Esther sent to the Wise Men saying 'Commemorate me for future generations.' They replied: You will incite the ill will of the nations against us. She sent back a reply: I am already recorded in the chronicles of the kings of Media and Persia." It appears from this Gemara that the Rabbis publicised only things which already had appeared in the chronicles. On the other hand, the ninth of Tevet fast was not among these things, so the Rabbis refrained from publishing it in order that the non-Jews would not talk about the dissatisfaction of the Jews, at the period when a Jewish girl was chosen as the queen.

4) The Death of Shimon haKalfus - The "Apostle Peter"

In his commentary on the Shulchan Aruch, Rabbi Baruch Frankel[79] writes: "I found in a manuscript that on the ninth of Tevet, Shimon haKalfus who delivered the Jewish people from a great misfortune during the period of the 'lawless ones' died. It was fixed that the day of his death should be commemorated as a fast in perpetuity in Jerusalem."

Rabbi Frankel did not identify the manuscript, nor did he give further details about the identity of Shimon haKalfus. However, from the fact that the Rabbis refrained from writing the reason for the fast, it can be said that the actions of Shimon haKalfus were sensitive, and therefore it was advisable not to publish them to non-Jews, since it was associated with the establishment of Christianity.

Over the course of last centuries, a number of manuscripts have been found in Hebrew that deal with the birth of Jesus and the apostles. Usually this material is called "Toldot Yeshu" or "Teliyat Yeshu." One of these manuscripts was published in 1705 by a Christian named Johann Jacob Huldricus.

The last chapter of the manuscript of Huldricus[80] speaks of a man named "Shimon haKalfi" and the trick he used to get the "lawless ones" out of Jerusalem. [The expression "lawless ones" ("pritzim" in Hebrew) which appears in this literature refers to Jewish-Christians who lived at the early days of Christianity.] Regarding the death of Shimon, it is written in the manuscript: "Shimon died and the children of Israel mourned for Shimon, and declared the day of his death to fast every year and this is the ninth day of the month of Tevet."

It can be determined with a high degree of probability that the manuscript that was before Rabbi Baruch Frankel was indeed this manuscript, and it is also possible to explain why the Rabbis refrained from publishing the reason.

The only difference between the manuscript of Huldricus and the comments of Rabbi Frankel is the small difference in name; in the manuscript of Huldricus[81] the name appears as "haKalfasi" while Rabbi Frankel writes "haKalfus." However, such a change is common in the Hebrew language; the real name is "haKalfus," and Shimon, who was an important member of his family, was called the "haKalfasi."[82] Furthermore when Rabbi Shlomo Zvi Schick,[83] (19th-20th centuries, Hungary), quotes this part of the Huldricus manuscript, he writes "haKalfus and not "haKalfasi."

In his book "Meorei Ohr," Rabbi Aharon veYirmash[84] writes that he saw in the Book of Remembrances a mention of the death of Shimon haKalfoni who succeeded in making a complete separation of the Jews from Christianity. Even though he writes the name "haKalfoni," it does not mean that he saw another manuscript. For the difference between "haKalfus" and "haKalfoni" is "s" instead of "ni"; a mistake that can easily be made by reading in a manuscript "ni" instead of "s." Unfortunately, the original manuscript of Huldricus is not extant, and therefore, one cannot see how the writer wrote the letter "s."

A question which arises, is that are books such as "Toldot Yeshu" or "Teliyat Yeshu," the sort of books that the Rabbis used in their commentaries on the Shulchan Aruch? The following examples show us that the answer is definitely positive.

a) Rashi in his commentary on Masechet Avodah Zara,[85] in a passage that was deleted by the censor, writes a comment which corresponds to the passage in the manuscript of Holdricus,[86] and he concludes: "Everything is as explained in the story "Teliyat Yeshu."

b) In the book "Takanot uTefillot" by Rabbi Shlomo Schick,[87] a letter is brought that the author received from the Rabbi of Beregsas, which states: "The pamphlet [Manuscript Huldricus] copied by my grandfather [Chatam Sofer] is from the book by a Christian named Johann Jacob Huldricus.

c) The manuscript itself[88] concludes with these words: "These are the words of our Rabbi, Rabban Yochanan ben Zakkai in Jerusalem." It is obvious that this is not conclusive evidence that Rabban Yochanan ben Zakkai was the author, but it is a possibility. Alternatively, it can be said that it was material based on the words of Rabban Yochanan ben Zakkai.

Another question that arises is related to the character of Shimon haKalfus, also known as Shimon Keifa, and who was the apostle Peter, and afterwards the first Bishop in Rome. [Keifa is his name in Aramaic, and in Greek it is Petros ("Peter")]. Were the Rabbis of the opinion that his death merited a fast every year for the Jews? In order to answer this question, one needs to know more details about his life and work. One can learn details of this from the various manuscripts[89] on Shimon haKalfus (Keifa).

In the early days of Christianity, even before there was a difference between Judaism and Christianity, the "Christians" persecuted the Jews and would torture and slaughter them, and gradually the "Christians" strengthened and thus established their power. At a meeting called by the elders of the Jews, to try to find a solution to these problems, one of those elders called Shimon Keifa said: "Listen to me my brothers and my people. If it is acceptable to you, I will separate these people from the Jewish race and they will have no place amongst the Jews, provided that you will take upon yourselves the transgression." The elders answered: "We will take upon ourselves the transgression, but you must do what you have spoken." After Shimon had put the Ineffable Name in his flesh, he went among the Christians and shouted: "Anyone who believes in Jesus should come to me as I am his messenger," and he performed the same signs that Jesus had done in his time to gain their trust. He told the Christians that "Jesus was a hater of Israel and their teachings," and commanded them "that you do no more harm to any Jew." Instead of celebrating the Jewish holidays, you will celebrate holidays related to Jesus' life and death. [The Rabbi from Munkatch[90] writes that Shimon said: "From now on their Sabbath will be on Sunday and not on Saturday." However, this does not appear in the version which one has today. The source of this is from a different version on the life of Shimon Keifa which identifies him as a "wise man called Eliyahu."[91]] The Christians answered together: "Everything you have spoken will be done provided that you stay with us." Shimon agreed and they built him a tower in which he lived for six years until the day he died on the ninth of Tevet. In the tower, he ate only bread and water. Every day he was alone and he studied Torah by day and by night, and served G-d with all his soul.

This account presents Shimon Keifa in a positive light. Furthermore, in his comments to "Machzor Vitry," Rabbi Shimon Halevi Ish Horowitz[92] (19th century), writes that Rabbenu Tam wrote that "he never believed in the new faith [Christianity], and that everything he did was for the sake of Heaven." Also "Sefer Hasidim"[93] describes him as a "righteous person."

The various accounts about the life of Shimon Keifa conclude that he composed liturgical poems, and in one of the accounts[94] it is written that these liturgical poems were said every Shabbat in the prayers in the synagogue, their intention being the prayer "Nishmat" which is mentioned in Masechet Pesachim.[95]

In his commentary on the Siddur "Avodat Yisrael," Rabbi Yitzchak Baer[96] (19th century), writes: "I found an old commentary in a siddur manuscript from the year 5187 (1427) according to the customs of Troys, and there it stated that I heard from R' Yehudah ben Yaacov that Shimon ben Keifa was the author of Nishmat. Also in the Kabbalistic commentary on the Siddur "Meleah Aretz De'ah," Rabbi Naftali Hirtz Treves[97] (16th century), writes: "Nishmat. I found that Shimon ben Keifa composed it, but some say that it was Shimon ben Shetach."

Rashi, however, vehemently rejects any possibility of attributing the "Nishmat" prayer to Shimon Keifa. Rabbenu Simcha who was a pupil of Rashi writes in his book "Machzor Vitry":[98] "There are those who say concerning the scoundrel Shimon Peter Chamor [donkey], the abomination of Rome, that he was the author of this prayer [Nishmat] along with other prayers. But G-d forbid no such thing should occur in Israel, and anyone who says this thing, when the Temple is rebuilt will bring a sin offering."

In contrast to this, the author of "Machzor Vitry"[99] quotes afterwards Rabbenu Tam (12th century), who writes: "From the days of Shimon Keifa who wrote a liturgical item entitled 'Eten Tehilah' which is recited on Yom Kippur."

One should note that all that is written above is not conclusive evidence. However, it can certainly be assumed that Shimon haKalfus (Keifa), who was the Apostle Peter, was a person who on account of his actions merited to have the day of his death declared as a fast for generations, and the reason for not publicising it is obviously clear.

5) The Day on which the Jesus was born

The Dayan Rabbi Yehudah Leib[100] in his commentary entitled "Tosafot Chadoshim" on Megillat Ta'anit writes in connection with the fast on the ninth of Tevet: "And I heard in the name of a great man that it was on that day that 'that man' was born." The phrase "that man" appears in the books of the Rabbis[101] as an alias for Jesus. [However, Rabbi Bornstein[102] in his commentary "Haeshel," uses the word "Jesus" instead of the alias "that man"]

Rabbi Yehudah Leib in his "Tosafot Chadoshim" does not reveal the identity of "that man," but Rabbis who lived in the nineteenth and twentieth centuries, such as "Shir"[103] and Rabbi Yehuda Leib Gordon[104] suggested the "Sefer haIbur"[105] as a source for what is written in the "Tosafot Chadoshim." The author of "Sefer haIbur" was Rabbi Avraham bar Chiyah who lived in the eleventh century in Spain, and who writes: "And he [Jesus] was born according to them [the Christians] in the year 3761 after the Creation of the World on the twenty-fifth day of the month of Deigber [December], and that was Shabbat the ninth day of Tevet. About two hundred years later in the year 1310, Rabbi Yitzchak ben Yosef haYisraeli, who also lived in Spain, quotes in his book "Yesod Olam"[106] the same date in both the Jewish and secular calendars for the birth of Jesus. From the tables[107] which give the corresponding dates for the Jewish and secular calendars, one can confirm that the ninth of Tevet, in the year 3761, occurred on Saturday, 25 December, 1 BCE. [In fact, in those days the Jewish calendar was fixed by the evidence of witnesses who testified on seeing the new moon. Therefore, the parallel between the dates is only theoretical.]

But these two authors[108] continue that Christians counted the year 1 CE from the first of January before the birth of Jesus.

The question arises, is there therefore a parallel of dates with regards to 25 December 3761, for the year 1 CE?

[There are several methods in counting the years in the Hebrew calendar, with two of them being relevant to our case. According to tradition, the first day of the creation of the world was 25 Adar or 25 Elul. According to one method (molad "baharad") and this is the method which is used today, the first year began on the first of Tishrei before the creation of the world. In the second method (molad "veyad") the first year began on the first of Tishrei after the creation of the world.][(109)]

If these two authors used the "veyad" method, then 25 December 3761, corresponds to the year 1 CE. However, this will not occur on Shabbat, the ninth of Tevet, but on Sunday, the twenty-first of Tevet; it follows that the two authors (quoted above) did not use the "veyad" method. In addition, it appears from the "Sefer haIbur"[(110)] that the author, who lived in Spain, used the Western method, namely the method of "baharad."

The problem can be resolved by the fact that the Christians changed their method to count the years. The previous method was according to the calendar of Philochlos from the year 354, and according to this calendar, December 25 of the year of the birth of Jesus is called the year 1 CE. At a later date they changed to the Gregorian system (which is the system used today), and according to this system the first of January after the birth of Jesus is called 1 CE. In practice this method was not used in most of Spain until the fourteenth century. If this is indeed the case, it is understandable why Rabbi Avraham bar Chiyah and Rabbi Yisraeli used the calendar of Philochlos.

There is a tradition[(111)] that holds that Jesus was born on the equinox of Tevet. The equinox that one is referring to is that of Shmuel and this equinox occurred on the Julian calendar on 24-25 December. [Today, one uses the Gregorian calendar, and the equinox falls (during the twentieth and twenty-first centuries) thirteen days later, namely, on 6-7 January.]

In a marginal note on the commentary "Tosafot Chadoshim" there are additions, brought by Rabbi Raphael Gordon[(112)] which states that in the year 3761, the equinox of Tevet fell on a Friday, the eighth of Tevet - December 24 - at ten-thirty in the morning, and Jesus was born on Saturday, the ninth of Tevet which was December 25.

[Rabbi Schick[113] and Rabbi Moshe David Hoffman[114] write that in the "Sefer haKabbalah" of the Ra'avad it is written that Jesus was born on the ninth of Tevet. However, this does not appear in the version of "Sefer haKabbalah" which we have today. It is difficult to believe that two Rabbis could make the same mistake within ten years of each other. How then can one explain this? In the "Sefer haKabbalah" there are two passages[115] about Jesus. One can see from a number of editions and manuscripts (for example: Basel edition of 1580, Amsterdam edition of 1711, Prague edition of 1795, and Moscow manuscript[116] from the seventeenth century), that these passages were censored. It is also known that at the end of the nineteenth century, there were attempts[117] to reconstruct the censored passages. It is possible that during the restoration a mistake was made and these two Rabbis thus saw an inaccurate wording.]

Is there a support in the Jewish tradition to declare the day of Jesus' birth as a day of fasting, namely a day of mourning?

Although the fast of ninth of Tevet is not observed today, (nor was it observed during the period of the "Beit Yosef,"[118]), there is a Jewish tradition not to study Torah on the eve before the secular date of the birth of Jesus, the night of December 24, namely Christmas Eve. This is called in Yiddish "Nital Nacht." This custom is especially observed among Hasidim in the Diaspora, and even if it occurs on the night of Shabbat, Torah lessons are cancelled.[119] [After the publication of this paper (in Hebrew) in the journal "Sinai," I received a letter from Yonah Heinrich Hagar who lives in the city of Emanuel in Samaria. In it he writes that in his city of Frankfurt am Main "this 'prohibition' was very strictly observed."[120]]

There are several reasons for this custom. One of them is from Rabbi Nathan Adler, (18th century, Frankfurt), and it is to be found in a responsum of his student the "Chatam Sofer."[121] In this responsum he writes: "On the question of the prohibition of learning Torah on the night of their festival [Christmas] ... he

[Rabbi Adler] says that the reason is because of mourning." Further evidence for the reason of Rabbi Adler comes from the fact that on that night the mikvahs were closed and sexual relations were forbidden. This reasoning of Rabbi Adler is compatible with proclaiming a fast on the day Jesus was born.

Several Rabbis in recent times, including Rabbi Bornstein,[122] Rabbi Gordon,[123] Rabbi Hoffman,[124] and Rabbi Schick[125] explain that the Rabbis' reason for not publishing the fast on the day of the birth of Jesus was to prevent hostility and trouble towards the Jews. Rabbi Hoffman added that today times have changed and we are able to write things without repercussions.

There is support in Masechet Ta'anit[126] not to declare a fast on non-Jewish festivals. While the Temple existed, the Jews were divided into twenty-four "groups," in connection with the daily sacrifice. People on the "group" on duty fasted each day during the week with the exception of Friday, Shabbat, and Sunday. Among the various explanations for the reason for not fasting on Sunday was "because of the Christians." Rabbenu Gershom[127] (10th-11th centuries, Mainz), comments on this: "Because Sunday was their festive day and if the Jews were to fast, it would anger them [the Christians]"

However, a chronological problem arises here. For during the period that the Temple was standing, Christianity had only just begun and Sunday, apparently, had not yet been their festive day. So how can it be that the groups did not fast on Sunday "because of the Christians"?

In his answer, we see that Rabbi Menachem Meiri[128] (13th-14th centuries, Perpignan Catalonia), instead of the word "Christians" writes "Netzarim" which means Babylonians, taken from the word "Nebuchadnezzar" king of Babylon. The festive day of the Babylonians was on Sunday. In addition, in various manuscripts of "Masechet Soferim,"[129] on this subject, there are several versions of this word namely, Egyptians, nochrim, netzarim, minim, Christians. (On the other hand, the Rabbi of Munkatch[130] agreed with the version which states "Christians," and claimed that this reason was written right at the beginning of Christianity.)

However, the conclusion from all this, is the principle not to anger non-Jews by establishing fasts on their festivals.

From all the above, it is possible to understand the Rabbis' refusal to write down this reason for fasting on the ninth of Tevet.

A Copyist's Error

In "Machzor Vitry"[131] is written: "On the ninth [Tevet] the [king] of Babylon began the siege on Jerusalem." This is obviously an error, since it is clearly stated in the Bible[132] that this event took place on the tenth of Tevet.[133] However, this was not just a printer's error since these words appear in two extant manuscripts[134] of Machzor Vitry. In addition, this appears in two manuscripts[135] of "Siddur Rashi" (although in another manuscript, [136] the ninth of Tevet is completely omitted).

"Machzor Vitry" was written by Rabbenu Simcha who was a student of Rashi and therefore it is probable that this part of "Machzor Vitry" was copied from a manuscript of the "Rashi Siddur." But it is very surprising that no-one attempted to correct this obvious error.

Summary

I found five reasons for the fast of the ninth of Tevet.:

* 1) The death of Ezra the Scribe
* 2) The Killing of HaNagid R' Yehosef Halevi
* 3) Taking of Esther to the King's Palace
* 4) The Death of Shimon haKalfus - the "Apostle Peter"
* 5) The Day on which Jesus was born

For each of these reasons, I wrote the background, the Rabbis' comments and the reason they refrained from writing the reason for this fast. However, because the Rabbis consistently wrote "On the ninth of it [Tevet] our Rabbis did not give the reason," it would be improper for me to give an opinion as to which of these reasons is the correct one.

References

Abbreviations
SA = Shulchan Aruch
OC = Orach Chaim

(1)   SA OC chap.580 par.1
(2)   Ibid., par.2
(3)   Taz SA OC chap.580 par.1
(4)   Magen Avraham, SA OC chap.580 par.6
(5)   Ba'er Heteiv SA OC chap 580 par.6
(6)   Comments from Rabbi Baruch Frankel, SA OC chap.580
(7)   Yisrael Davidson, *Otzar haShira v'haPiyut,* vol.1, (Ktav: New York, 1970), p 108
(8)   Rabbi Avraham Meir Israel, *Hamaor,* year 30, pamphlet 2 (p.241) Kislev-Tevet 5738, New York, p.7
(9)   Davidson, op. cit., vol.3, p.502
(10)   Rabbi Avraham Zacuto (Zechus), *Sefer Yuchasin,* (Zalkava, 5559), p.9b
(11)   Rabbi Gedaliah ibn Yechia, *Shalshelet haKabbalah,* (Venice, 5346), p.23a
(12)   Ibid., p.21b
(13)   *Targum Yonatan ben Uziel,* Malachi, chap.1 verse 1
(14)   Tractate Megillah 15a. See also: Rabbi Don Yitzchak Abarbanel, commentary on Neviim Malachi 1
(15)   MS Bodleian, catalogue Neubauer 882; MS Bodleian, catalogue Neubauer 902
(16)   Rabbi Yosef David, *Bet David,* (Saloniki (Thessaloniki), 5500), chap.316
(17)   Tractate Megillah 5a
(18)   Bet David, op. cit.
(19)   Rabbi Yonatan Eybeschutz, *Ya'arot Devash,* (Lemberg (Lvov), 5623), vol.2, pp.74b-75a
(20)   Rabbenu Menachem ben Aharon ibn Zerach, *Tzeda laDerech,* (Warsaw, 5640), article 5, principle 1, chap.8 (p.142b)
(21)   *Kol Bo,* chap.63
(22)   Rabbi Aharon haKohen m'Lunel, *Orchot Chaim,* (Firenze (Florence), 5510), p.214
(23)   MS Cambridge, T.S. H 11.32
(24)   Rabbi Shlomo Haas, *Kerem Shlomo.* (Pressburg, 5603), SA OC chap.580
(25)   MS Vatican, 304
(26)   MS Cambridge, T.S. NS 329.432
(27)   *Seder Rav Amram Gaon,* (Kiryat Ne'emana: Jerusalem, 5725), p.34b
(28)   Rabbenu Eliezer ben Yoel haLevi, *Sefer Raviya,* R' Avigdor Aptowitzer edition, (Machon Harry Fischel: Jerusalem, 5724), chap.889, (vol.3 p.672)
(29)   *Tzeda laDerech,* op. cit.
(30)   Tur, OC chap.580
(31)   Rabbi Yaakov Emden, *Siddur Shaarei Shamayim,* (Warsaw, 5641), Shaar Yechonya, p.90b
(32)   Rabbi Elia Shapira, *Elia Raba,* (Sulzbach, 5517), OC chap.580 par.5

(33) Rabbi Shlomo Yehudah Rapoport ("Shir"), *Igrot Shir,* (Przemysl, 5645), letter 33, (p.202)

(34) Rabbi Avraham Naphtali Zvi Roth, *Hamaor,* year 30, issue 4, (p.243), Nisan-Iyar 5738, (New York), p.36

(35) Be'er Hagolah, SA OC chap.580

(36) Rabbi Yosef ben Avraham Molcho, *Shulchan Gavoha,* (Salonika (Thessaloniki), 5516), SA OC chap.580 par.3

(37) Aruch Hashulchan, OC chap.580 par.3

(38) *Orchot Chaim,* op. cit.

(39) MS New York Bet Hamidrash l'Rabbonim, Rab 667 Adler collection 1770 folio 122b; MS New York Bet Hamidrash l'Rabbonim, Rab 666 folio 195b; MS Bodleian, catalogue Neubauer 2366 folio 114b

(40) *Kol Bo,* op. cit.

(41) MS Moscow, catalogue Günzburg, 72/1 folio 141b-142a

(42) *Siddur Tefilot Hashanah l'Minhag Kehilot Romania,* (Venice, 5280), p.230

(43) MS New York, Bet Hamidrash l'Rabbonim, Rab 666, folio 195b

(44) Rabbi Chaim Yosef David Azulai (Chida), *Birkei Yosef,* OC chap.580 par.1

(45) Rabbi Roth, op. cit.

(46) *Siddur Ozar haTefilot,* Nusach Sefarad part 2, commentary *Iyun Tefillah* by Rabbi Aryeh Leib ben Shlomoh Gordon, (Nehora d'Oraita: Jerusalem, 5720), p.108

(47) Rabbi Roth, op. cit.

(48) Rabbi Yehudah ben Shmuel, *Sefer Hasidim,* (Mosad Harav Kook: Jerusalem, 5717), chap.231 (p.206)

(49) Rabbi Avraham ibn Daud Halevi (Ra'avad), *Sefer haKabbalah,* (Manitoba, 5284), pp.16-17; Ra'avad, *Sefer haKabbalah,* English translation by Gerson D Cohen as *The Book of Tradition,* (Jewish Publication Society of America: Philadelphia,1967-5728), pp.75-76

(50) Rabbi Moshe ben Yakov ibn Ezra, *Kitab al-muhadarah wa al-mudhakarah - Liber Discussionis et Commemorationis (Conversations and Recollections* - translated into Hebrew as *"Shirat Yisrael"),* translated by R' Avraham Shlomo Halkin, (Jerusalem, 5735) p.66

(51) *Sefer haMoadim,* collected and arranged by Yom-Tov Levinsky, (Agudat Oneg Shabbat, 7th edition: Tel-Aviv, 5731), vol.7 p.94

(52) *Kitab al-muhadarah,* op. cit., translated by R' Ben-Zion Halper, (Leipzig, 5684), p.68

(53) *Halkin,* op. cit. p.67

(54) Abdallah ben Belkin, *Les Memoires de Abdallah,* (Lévi Provençal: Cairo, 1955), p.54

(55) Reinhart Dozy, *Recherches sur l'histoire et la litterature de l'Espagne pendant le moyen age,* (Paris, 1881), summary, appendix xxvi, p.lxii

(56) *Luach l'Sheshet Alafim Shanah,* arranged by Avraham Aryeh Akavia, (Mosad Harav Kook: Jerusalem, 5736; Sherrard Beaumont Burnaby, Elements of the Jewish and Muhammadan Calendars, (London, 1901), p. 380,

(57) *Birkei Yosef*, op. cit.
(58) *Orchot Chaim*, op. cit.
(59) *Igrot Shir*, op. cit.
(60) Rabbi Aharon bar Avriel veYirmash, *Meorei Ohr,* vol.8, *Od laMoed*, (Metz, 5582), p.110
(61) Rabbi Roth, op. cit., p.37
(62) MS London British Museum, Catalogue Margoliouth 1168, folio 290b; MS Bodleian, catalogue Neubauer 893/2 folio 216b
(63) *Siddur Shaarei Shamayim*, op. cit.
(64) *Siddur Derech haChaim* with *Nehora haShalem*, (Shlomo Zalman Netter, Vienna, 5628), vol.2 p.104a
(65) Rabbi Nachman Kahana, *Orchat Chaim*, (Sighet Romania, 5658), SA OC chap.580, par.5
(66) *Kerem Shlomo*, op. cit.
(67) Rabbi Yehudah Leib ben Menachem, (Dayan in Krotoshin), *Tosafot Chadoshim*, (Warsaw, 5634), Megillat Ta'anit, final chapter, p.20a (39)
(68) Rabbi Chaim Palagi, *Ruach Chaim,* (Izmir, 5641), OC chap.580 par.5
(69) Biblical book of Esther, chap.2 verse 16
(70) Rabbi Moshe Isserles, *Mechir Yayin*, (Kraków, 5641), p.19
(71) *Midrash Rabba*, Esther, parashah 6
(72) Tractate Megillah 13b
(73) *Midrash Shocher Tov on the book of Psalms*, (Jerusalem, 5728), Psalm 22, p32 (63-64)
(74) *Midrash Lekach Tov,* Esther chap.2 verse 16
(75) Biblical book of Ezekiel, chap.33 verse 21
(76) Biblical book of Esther, chap.2 verse 14
(77) Rabbi Avraham Eliyahu Bornstein, *Perush Haeshel* on *Megillat Ta'anit*, (Jerusalem, 5668), final chapter, p.132
(78) Tractate Megillah 7a
(79) Comments of Rabbi Baruch Frankel, op. cit.
(80) Johann Jacob Huldricus, *Sefer Toldot Yeshu Hanotzri*, (Historia Jeschuae Nazareni), (Leiden, 1705), pp.125-26
(81) Ibid., pp.101, 107, 125
(82) see Biblical Book of Numbers, chap.26
(83) Rabbi Shlomo Zvi Schick, *Takkanot uTefillot*, (Munkatsch, 5650), p.97b
(84) *Meorei Ohr*, op. cit.
(85) Rabbi Raphaeol Natan Nata Rabbinowitcz, *Dikdukei Soferim,* (New York, 5737), vol.2, Avodah Zarah, p.5 footnotes
(86) Huldricus, op. cit., pp.108, 115-16
(87) *Takkanot uTefillot*, op. cit.
(88) Huldricus, op. cit., p.127
(89) Adolf Jellinek, *Bet Ha-Midrasch*, (Bamberger and Wahrmann: Jerusalem, 5698), part 5 pp.60-62, part 6 pp.155-56
(90) Rabbi Chaim Elazar Shapira (Spira), *Sefer Divrei Torah,* second edition (Munkatsch, 5689), chap.47
(91) *Bet Ha-Midrasch*, op.cit., part 6 p.11

(92) Rabbenu Simcha, *Machzor Vitry,* notes by Rabbi Shimon Halevi Ish Horowitz, (Machon Publishers: Jerusalem, 5727), chap.325, footnote 5 (p.362)

(93) *Sefer Hasidim,* op. cit., chap.191 (p.188)

(94) *Bet Ha-Midrasch,* op. cit. part 6 p.11; [see also R' Yitzchak Rafael "maamar al Agadot Shimon Kaifa", "Davar" 28 Tishrei 5700]

(95) Tractate Pesachim 118a

(96) *Seder Avodat Yisrael,* ed. Rabbi Yitzchak ben Aryeh Yosef Dov (Seligman Baer), (Schocken: Berlin, 5697), p.206

(97) *Siddur Hirts,* commentary by Rabbi Naftali Hirts Treves, (Tihinigin, 5320), end of the Nishmat prayer

(98) *Machzor Vitry,* op. cit., p.282 (and footnote 5)

(99) *Ibid.,* chap 225 (p.362)

(100) *Tosafot Chadoshim* on *Megillat Ta'anit,* op. cit.

(101) *Teshuvot haRadak l'Notzrim,* appendix to *Chesronot haShas,* (Sinai: Tel-Aviv), p.44

(102) *Perush Haeshel,* op. cit., p.134

(103) *Igrot Shir,* op. cit.

(104) commentary *Iyun Tefilah,* op. cit.

(105) Rabbi Avraham bar Chiyah haNasi, *Sefer haIbbur,* (Filipowski: London, 5611), third paper, third chapter, p.109

(106) Rabbenu Yitzchak ben Yosef haYisraeli, *Sefer Yesod Olam,* (Berlin, 5608), fourth paper, chap.17 (p.31)

(107) *Luach l'Sheshet Alafim Shanah,* op. cit., p.316

(108) *Sefer haIbbur,* op. cit.; *Sefer Yesod Olam,* op. cit.

(109) *Commentary of Rabbenu Ovadiah,* on *Rambam, Kiddush haChodesh,* chap.6 halacha 8; *Chumash Torah Sheleima,* edited by Rabbi Menachem Mendel Kasher, (New York, 5714), vol.13 p.110

(110) *Sefer haIbbur,* op. cit., pp.96, 99

(111) Rabbi Avraham Yitzchak Sperling, *Taamei haMinhagim uMekorei haDinim,* (Eshkol: Jerusalem), p.500

(112) Rabbi Refoel Gordon, footnote to *Tosafot Chadoshim* on *Megillat Ta'anit,* (Vilna), Yemei haTzomot, p.22b

(113) *Takanot uTefillot,* op. cit., p72a

(114) Rabbi Moshe David Hoffmann, *Toldot Elisha ben Abuyah,* (Vienna, 5640), p.13

(115) Ra'avad, *Sefer haKabbalah,* op. cit., pp.4, 9

(116) MS Moscow, catalogue Günzburg, 179/13 folio 100a, 101b

(117) *Chesronot haShas* from the year 5625; *Dikdukei Soferim* from the years 5627-5646

(118) Beit Yosef on Tur, OC chap.580 first words "kol hayomim"

(119) *Each Thing in its Proper Time – Calendar 5758 (Luach Davar b'Ito – 5748),* (Achiezer: Bnei Brak, 5748), pp.289-291

(120) Letter from R' Yonah Heinrich to Chaim Simons, 4 Adar 5751

(121) MS Montefiore London, 450.2 folio 16

(122) *Perush Haeshel,* op. cit.

(123) Rabbi Refoel Gordon, op. cit.

(124) *Toldot Elisha ben Abuyah*, op. cit.

(125) *Takanot uTefillot*, op. cit.

(126) Tractate Ta'anit 27b

(127) Rabbenu Gershom ben Yehudah on Ta'anit 27b

(128) Rabbi Menachem Meiri on Ta'anit 27b

(129) Masechet Sofrim, edited by Dr. Michael Higger, (New York, 5697), p.301, variations in texts

(130) *Sefer Divrei Torah*, op. cit.

(131) *Machzor Vitry*, op. cit., p.230

(132) Biblical books: Kings II chap.25 verses 1-2; Ezekiel chap.24 verses 1-2

(133) According to the Jewish calendar, the tenth of Tevet cannot occur on Shabbat. According to the opinion of the Abudraham (Jerusalem, 5723, p.254) even if it would occur on Shabbat, one would not be able to postpone the fast to another day because it states regarding this fast "on that same day" just as is stated with Yom Kippur. On the other hand, the Rambam (Hilchot Ta'anit chap.5 halacha 5), and the Shulchan Aruch (OC chap.550 par.3) write that if one of the four fasts falls on Shabbat, one would postpone it to Sunday. From this language one can see that if the tenth of Tevet would occur on Shabbat, the fast would be postponed to the next day During the period of after the destruction of the first Temple, and after the destruction of the second Temple until the establishment of a fixed calendar, there was the possibility that the tenth of Tevet could occur on Shabbat. Therefore, one needs to investigate whether they would have fasted on Shabbat or postponed the fast to Sunday.

(134) MS London British Museum, catalogue Margoliouth, 655 folio 154a; MS Warsaw, 240/1

(135) MS Munich, 28/5 folio 321b; MS Cambridge, 786/1 Or. folio 42a

(136) MS Parma, Catalogue De Rossi, 858 folio 255a

Appendix

Follow on to publication of my paper.

The material below is taken from my unpublished autobiography

In June 2005, I received an e-mail from y.y.f. (that is all he disclosed of his name) stating "I'm very interested in the story that st. peter was really as good jew that did it to save the jews..." [no capital letters in original]. He enclosed a paper which had written in 2001 by Rabbi Shlomo Shmuel Fleishman from Nachalat Har Chabad in Kiryat Malachi., which inter alia quoted from a newspaper article in the Agudah newspaper "Hamodia" that in addition to Peter, John and Paul were good Jews.

I telephoned Rabbi Fleishman to ask him for more details of this article in "Hamodia" but all he could tell me was that it was written by Gerlitz and had appeared several years earlier.

IS THE CHAPTER "DAYS ON WHICH ONE FASTS" THE LAST CHAPTER IN MEGILLAT TA'ANIT?

(This chapter is an appendix to the paper "Reasons for the Fast on the Ninth of Tevet")

The book "Megillat Ta'anit" is composed of thirteen chapters. The content of the first twelve chapters are days throughout the year on which one is not allowed to fast. In contrast, the last chapter is composed of the days throughout the year when one is obligated to fast. The dates in this last chapter are quoted in the Shulchan Aruch under the heading "Days on which one Fasts."

The question before us is whether this last chapter appeared in the *original* Megillat Ta'anit, or was it added at a later period?

To find an answer, we will study the following five questions. [From now on, for the purpose of the discussion, the list of fasts, which *now* appears in the last chapter of Megillat Ta'anit, will be referred to as the "list".]

1) Is it possible to bring evidence from the Talmud that this chapter appeared in the original Megillat Ta'anit?
2) Are there internal contradictions between the first twelve chapters of Megillat Ta'anit and the "list"?
3) Does the "list" appear in the various books and manuscript of Megillat Ta'anit?
4) Is it possible to bring proofs from the various manuscripts of the "list," which state that their origin is in another composition?
5) Is it possible to bring evidence from the Rabbinical literature written during the period of the Geonim (great Rabbis who lived between about the 6th to the 11th centuries), the Rishonim (great Rabbis who lived approximately between

the 11th and 15th centuries) and the Acharonim (great Rabbis who lived from about the 16th century) of the origin of this "list"?

The Talmud

Megillat Ta'anit is mentioned several times in the Gemara (both in the Talmud Bavli[1] and also in the Talmud Yerushalmi[2]), and it is even mentioned in the Mishnah.[3] This proves that a book by the name of Megillat Ta'anit was already known during the period of the Tannaim (Rabbis whose views are recorded in the Mishnah). The question which then arises is, is this the same book that we have today?

From the various places in the Gemara one can get information about the contents of the book "Megillat Ta'anit" which is mentioned in the Talmud. For example: the twelfth of Adar is Torianus day;[4] the thirteenth of Adar is Nicanor day;[5] the third of Tishrei is the day that the Greeks abolished the prohibition of writing the Divine name on documents;[6] the twenty-eighth of Adar is the day that good news arrived to the Jews, that they will not be prevented from studying the Torah.[7] All these dates together with the reasons for not being permitted to fast on them appear in the text of Megillat Ta'anit which we have today. All this proves that the Megillat Ta'anit that we have today is the same book which is referred to in the Talmud.

Even though there are several sources in the Talmud of the days mentioned in Megillat Ta'anit on which one is forbidden to fast, there is not even a single hint in the entire Talmud regarding the "list." Therefore it seems that the sages of the Talmud did not know about it.

Internal contradictions

As mentioned above, there are in the various editions of Megillat Ta'anit which are extant today, days when it is forbidden to fast, and also days on which one is mandatory to fast. A study of these dates shows us that there are some days when fasting is both forbidden and also mandatory!

First of Nissan
Forbidden to fast: From the first to the eighth day of Nissan the daily sacrifice in the Temple was inaugurated.[8]

Mandatory to fast: On the first of Nissan the sons of Aharon died.[9]

Tenth of Nissan
Forbidden to fast: From the eighth day of Nissan until the end of the Festival of Pesach, a holiday was declared.[10]

Mandatory to fast: On the tenth day Miriam the prophetess died and the Well of Miriam ceased to give water.[11]

Seventh of Elul
Forbidden to fast: The seventh of Elul was the day of the dedication of the wall of Jerusalem.[12]

Mandatory to fast: On the seventh of Elul, the people who gave an evil report of Eretz Yisrael died in a plague.[13]

Seventh of Kislev:
Forbidden to fast: On the seventh of Kislev a Festival commemorating the death of Herodes.[14]

Mandatory to fast: On the seventh of Kislev, Yehoyakim burned the megillah.[15] (there are a number of different versions regarding the exact date of this event and we brought the version that appears in Megillat Ta'anit)

Ninth of Adar
Forbidden to fast: On the eighth and ninth days of Adar they sounded blasts of a horn for rain.[16]

Mandatory to fast: On the ninth day a fast was decreed as a result of disputes between Bet Hillel and Bet Shammai.[17]

Needless to say, it is impossible throughout history, to prevent both a victory and a disaster from happening on the same day. In such a case, however, if the twelve chapters of Megillat Ta'anit and the "list" were in the same composition, it is clear that the author of Megillat Ta'anit would have commented on this matter. The absence of any such comments most likely proves that the "list" was not part of the *original* Megillat Ta'anit.

Books and Manuscripts of Megillat Ta'anit

The first printed edition of Megillat Ta'anit was in Mantoba, in the year 5274 (1514). In this edition this "list" is to be found after Chapter 12, and this was so in all printed editions of Megillat Ta'anit. In every edition the chapters are labelled chapter 1, chapter 2 ... chapter12 (or similar wording). However, the "list" has the heading: "The Last Paper" or there is no title at all. Were the "list" to be the last chapter of Megillat Ta'anit, it would have been headed chapter13; however there is no edition with such a heading. Moreover, in the Hamburg edition of the year 5517 it is written: "From here onwards this is an addition from 'Ba'al Halachot Hagedolot'" namely it is an explicit recognition that the "list" is taken from "Halachot Gedolot" and is not part of the Megillat Ta'anit.

The printing era began in the middle of the fifteenth century, and before that, books were written by hand. We have several manuscripts of Megillat Ta'anit, written before the era of printing. The earliest manuscript[18] that is extant is from the year 5105 (1344), and the "list" does *not* appear in it. It is impossible to say that this manuscript in our possession is missing its end, because immediately after chapter twelve it is written: "The completion of Megillat Ta'anit, with the help of Heaven."

A manuscript written in the year 5269 (1509), (which was before the printed first edition of Megillat Ta'anit) was destroyed during the Second World War. However, there are two sources from which we know a little about the contents of the destroyed manuscript: The first is that in 1875, Joel Muller[19] compared the destroyed manuscript and the Mantoba printing. The second one is that there is a manuscript from the nineteenth century[20] which opens with the words: "Baraita of Megillat Ta'anit was copied from a manuscript written on parchment from the year 5269," namely it had been copied from the manuscript which was destroyed. From these two sources we can see that the "list" already appeared in the year 5269 at the end of Megillat Ta'anit.

From the above we can see that between the years 5105 and 5269 the "list" was "annexed" to Megillat Ta'anit. It is not possible to be more exact since we do not have manuscripts of Megillat Ta'anit which were written between these two dates.

Fourteen Manuscripts

There are at least fourteen manuscripts in which appears *only* the "list" and they do not give any connection to another source. In two of them there is a heading which could hint of the source of this "list.":

In the first of them the heading is "Mankol m'Ba'al Halachot Hagedolot"[21] namely the source is the book "Halachot Gedolot" which was written at the period of the Geonim.

The second is in a manuscript[22] which was written in the fourteenth century, and there is a heading "Megillat Ta'anit." However, this is not conclusive evidence that the author copied it from Megillat Ta'anit, the reason being that because this title appears *only* in this manuscript, it is possible that the scribe himself *added* the title to describe the content of the material.

Books of the Sages

Until the era of the Acharonim, this "list" appeared in a number of books of the Sages *without stating* that the source is Megillat Ta'anit.

The first is "Ba'al Halachot Gedolot" (BaHaG),[23] written either by Rabbi Yehudai Gaon (in the 8th century) or by R. Shimon Kayyara (in the 9th century).

Rabbi Yaakov ben Asher known as the "Tur"[24] (in the 14th century) begins the "list" with the words: "Written by the

BaHaG," namely the Tur did not find the "list" at the end of Megillat Ta'anit.

There are also Rishonim who mention these lists of fasts without elaborating, and among them is Rabbi Yechiel ben Yetutiel Anau author of the book "Tanya Rabati,"[25] (13th-14th centuries) who wrote: "I found that the Geonim were asked regarding the twenty-two fasts, or twenty-four, and thirty-six, if they occurred on Rosh Chodesh or on days when one does not fast on them. In such a case, may one or may one not fast on them? He replied to them. I do not know who instituted the twenty-four fasts which you wrote about, whether it was by the Rishonim or by someone else. Therefore one should not fast on them on Rosh Chodesh since Rosh Chodesh is from the Torah. But Megillat Ta'anit was written by Hananiah ben Garon..."

The question which now arises: Is the "Tanya Rabati" speaking about the "list"? In answer, it is possible to bring positive proof from a manuscript[26] (date unknown), in which twenty-two days from the "list" appear under the heading: "Twenty-two fasts in the twelve months of the year." (The number "twenty-two" is the first number which appears in "Tanya Rabati.") At a later date someone added to this manuscript additional dates at the side of the list and also at its end, and from this one can understand the other numbers quoted by the "Tanya Rabati." The "Tanya Rabati" also mentions that there are fasts whose dates are on Rosh Chodesh. (In the "list" Rosh Chodesh Nissan and Rosh Chodesh Av appear as fast days.) One can conclude from the above that "Tanya Rabati" is speaking about the "list." In addition, Rabbi Avraham Gombiner[27] known as the "Magen Avraham" (17th century) identifies the words of the "Tanya Rabati" with the "list."

Another question that one can ask is: Did the "Tanya Rabati" see the "list" at the end of Megillat Ta'anit or elsewhere? From his words: "But Megillat Ta'anit," the word 'but' proves that he did not find the 'list' in Megillat Ta'anit. Another proof from his language: "I do not know who instituted them." If the fasts had appeared in Megillat Ta'anit, he would not have written these words, because Megillat Ta'anit was written by the sages of the Talmud.

The book "Shibbolei Haleket"[28] (Rabbi Zedekiah ben Avraham Anaw, 13th century) is similar to "Tanya Rabati" and has almost the same wording when he writes on this subject.

In the book "Tzeda laDerech"[29] (written by Rabbi Menachem ben Aharon ibn Zerach, one of the sages of Spain in the fourteenth century) it is written: "And these are the days when the early Rabbis wrote that the pious people would fast on them." After that appears the "list" and he continues: "And these are the days written in Megillat Ta'anit," and he brings the days when fasting is forbidden. The words of the "Tzeda laDerech" are divided into two: Firstly, the words of the "early Rabbis" who established the fasts that appear in the "list," and secondly the words of "Megillat Ta'anit" which state the days when fasting is forbidden. Therefore, it is not clear whether or not the "Tzeda laDerech" saw the "list" in Megillat Ta'anit.

In the "Mishneh Torah" of the Rambam (Maimonides) the "list" is not found. Were it to have been in Megillat Ta'anit, namely it would then be a decree of the sages of the Talmud, and thus obvious that the Rambam would have quoted it. One of the "days in which a fast was decreed" was the twenty-eighth day of Iyar, the day of the death of Shmuel Haramati. However, on that *very day* the Rambam made a day of feasting and rejoicing due to a personal event. Rabbi Chaim Yosef David Azulai[30] known as the "Chida," (18th century), writes the reason for the event: "The Rambam wrote a copy of the Torah from a copy of the Tanach which was now in Egypt, and which had been in Jerusalem from the time of the Tannaim but had recently been brought to Egypt. He then heard that in the kingdom of Burgundy (province of east central France) there was a Sefer Torah which had been written by Ezra Hasofer the High Priest, and the Rambam went there and he found that all the open and closed paragraphs of this Torah corresponded to what he had copied from the Tanach that was then in Egypt. The Rambam therefore wrote that I was very happy and accepted upon myself to make a day of feasting and rejoicing on that day every year, namely the twenty-eighth day of the month of Ziv [Iyar]." From this account Rabbi Yitzchak Palagi[31] (19th-20th centuries) wrote in his book "Yafe laLev": "Perhaps the Rambam was not accustomed to fast on these fasts which included the twenty-eighth of Iyar the day when the prophet Shmuel died." However, the "Yafe laLev" continues that maybe the Rambam was indeed accustomed to fast on them, and he brings a proof of the version which appears in the "list" that the death of the prophet Shmuel was on the *twenty-ninth* of Iyar "and for that reason he did not fast on the twenty-eighth," and apparently from his words, the Rambam did fast on the twenty-ninth of Iyar.

However, there are Rishonim who mention this "list" *together* with the words "Megillat Ta'anit." The "Sefer haKabbalah" of the Ra'avad"[32] (Rabbi Avraham Halevi ben David, 12th century) is one of the sources and he writes: "From the days of the early Rabbis who wrote Megillat Ta'anit they decreed a fast on the ninth of Tevet..." [the ninth of Tevet is one of the fasts in this list. [33]] However, if we study his wording carefully, we can see that he does not write that this fast *appears in Megillat Ta'anit*, but he holds that the date of the establishing of these fasts was at the same period as the composition of Megillat Ta'anit. It is possible that the word "Torah" in the language of the BaHaG,[34] namely: "These are the days that one fasts as decreed from the Torah" his intention is that they were from the time of the sages of the Talmud.

Chapter 63 of the book called "Kol-Bo"[35] whose contents is that "list" and is headed "Din Seder Megillat Ta'anit" (The laws concerning Megillat Ta'anit) and he finishes this chapter with the words "the end of Megillat Ta'anit." However, the author gives a title to each chapter in his book which describes its contents. Also, in some chapters he concludes with the words: "end ....", or a similar wording. Therefore it is possible to assume that the use of the words Megillat Ta'anit is just coincidental, and it does not hint that the author of the book "Kol-Bo" actually saw the "list" in Megillat Ta'anit.

The "list" was copied several times by various writers, and in the process, mistakes were made on the dates of the fasts. Rabbi Yeshayahu ben Avraham Halevi Horowitz,[36] known as the "Shelah" (16th-17th centuries) made a comparison between the version found in the Tur in the "Kol-Bo", and in the BaHaG, but *not* in Megillat Ta'anit. It seems that had by the period of the Shelah the "list" would had already appeared at the end of Megillat Ta'anit, he would have included it in his comparison.

We saw above that already by the beginning of the sixteenth century, the "list" appeared at the end of Megillat Ta'anit. Many of the Acharonim who saw the printed editions of Megillat Ta'anit thought that Megillat Ta'anit was the origin of this "list."

One of them was the "Magen Avraham,"[37] who wrote his commentary in the middle of the seventeenth century. He was amazed at the "Tanya Rabati," "Shibbolei Haleket" and the Beit Yosef who had written that they had not seen or heard of anyone who actually fasted on these days. Afterwards he wrote: "But I saw that it is written in Megillat Ta'anit, and thus they were decreed in the days of the sages of the Talmud (within the words of this book). (The parentheses are those of the "Magen Avraham"). The "Magen Avraham" mentions that the Shelah had compared the versions found in the Tur, the "Kol-Bo" and the BaHaG, and he continues: "And I saw in Megillat Ta'anit a few other versions and it seems to me that one could use them since they are the source of the material, and even in the Amsterdam newly printed version is written in the margin variant versions as an addition by a commentator, who had taken it from the Shulchan Aruch, but in the older books these other versions are not to be found." The Magen Avraham a lived from about 1637 until about 1683, and therefore he speaks of the Amsterdam edition from the year 5419 (1659), in which four of the dates have different versions. But from earlier printings: Mantoba (5274 /1514), Venice (5305/ 1545), and in Basel (5340 /1580), the "other versions" are not included. According to his words: "In the older printings the variant versions are not to be found," we can learn that he saw at least one of these printings.

In a footnote to an edition of Megillat Ta'anit, at the beginning of the chapter on these fasts, Rabbi Yaakov Emden[38] known as the "Yaavetz" comments: "From here onwards was added from the BaHaG, and the Magen Avraham had erred in chapter 580 thinking that the Sages of the Talmud had written them." (The same words appear also in Megillat Ta'anit that was printed in Hamburg in the year 5517, namely the period when Yaavetz lived.) The Vilna Gaon[39] (Gr'a) holds that the source of these fasts is Megillat Ta'anit. According to his language: "as is found in Megillat Ta'anit."

About two hundred years after the Shelah, the "Eliyahu Zuta"[40] Rabbi Eliyahu Spira, 17th-18th centuries), made a comparison of the different versions, but he also included the version of Megillat Ta'anit in his comparison because the "list" had by then already appeared in the printings of Megillat Ta'anit. The commentaries on the Shulchan Aruch, namely the Ba'er Heteiv[41] (Rabbi Yehudah ben Shimon Ashkenazi, 18th century), and the Mishnah Berurah[42] apparently also held that the source of the "list" is Megillat Ta'anit. One needs to remember that the Sages in the former centuries did not have access to the manuscripts which are to be found in libraries throughout the world, and so they needed to rely only on the printed books.

However, not all the Acharonim hold that Megillat Ta'anit is the source of this "list." In the sixteenth century, Rabbi Azaria ben Moshe Del Rossi Min haAdumim[43] known as the "Meor Einayim" writes: "At the end of Megillat Ta'anit, according to the version found in the printed editions, there are the days when one is obligated to fast." From his wording "according to the version found in the printed editions" there is a hint that the "list" did not appear in the *manuscripts* of Megillat Ta'anit. The "Be'er Hagolah"[44] (Rabbi Moshe Rivkash), who lived at the same period as the Magen Avraham, and Rabbi Yosef ben Avraham Molcho[45] known as the "Shulchan Gavoha" who lived in the eighteenth century wrote that the source of these fasts was the "Tur in the name of the BaHaG" and they made no mention of Megillat Ta'anit.

The Aruch Hashulchan[46] begins the "list" with the words: "The Geonim wrote," namely he holds that the "list" is from a later period than the Talmud.

Summary

From all of the above, we can see, without a doubt, that the "list" did not appear in the *original* Megillat Ta'anit. It appears only in extant manuscripts and printed books from the sixteenth century onwards. It seems that the first source of the "list" was the BaHaG who lived in the eighth or ninth centuries.

References

Abbreviations
SA = Shulchan Aruch
OC = Orach Chaim
(1) e.g. Talmud Bavli: Shabbat 13b, Rosh Hashanah 18b-19b, Ta'anit 17b-18b
(2) Talmud Yerushalmi: Ta'anit chap.2 halacha 12, Megillah chap.1 halacha 4, Nedarim chap.8 halacha 1
(3) Mishnah Ta'anit chap.2 mishnah 8
(4) Talmud Bavli Ta'anit 18b
(5) Ibid.
(6) Talmud Bavli Rosh Hashanah 18b
(7) Talmud Bavli: Rosh Hashanah 19a, Ta'anit 18a
(8) *Megillat Ta'anit*, (Warsaw: 5634), chap.1, (p.3a)
(9) *Megillat Ta'anit*, op. cit., ma'amar ha'acharon, (p.19b)
(10) *Megillat Ta'anit*, op. cit., chap.1, (p.4a)
(11) *Megillat Ta'anit*, op. cit., ma'amar ha'acharon, (p.19b)
(12) *Megillat Ta'anit*, op. cit., chap.6, (pp.11a-11b)
(13) *Megillat Ta'anit*, op. cit., ma'amar ha'acharon, (p.19b)
(14) *Megillat Ta'anit*, op. cit., chap.9, (p.13b)
(15) *Megillat Ta'anit*, op. cit., ma'amar ha'acharon, (p.20a)
(16) *Megillat Ta'anit*, op. cit., chap.12, (p.16b)
(17) *Megillat Ta'anit*, op. cit., ma'amar ha'acharon, (p.20b)
(18) MS Parma calalogue de Rossi 117/4
(19) Joel Muller, "Der Text der Fastenrolle," *Monatsschrift fuer Geschichte und Wissenschaft des Judenthums*, (Breslau, 1875), vol.xxiv, pp.143-44
(20) MS New York Bet Hamedrash l'Rabbonim 3200 folio 1a
(21) MS New York Bet Hamedrash l'Rabbonim L896/3 folio 114b
(22) MS Vatican 299.5 folio 23
(23) *Sefer Halachot Gedolot, (BaHaG)*, (Ezriel Hildesheimer, Hevrat Mekize Nirdamim: Jerusalem, 5732), part 1, hilchot Tisha b'Av, pp.396-98
(24) Tur, OC chap.580
(25) *Tanya Rabati*, (Warsaw, 5639), chap.62, p.66b
(26) MS Vienna catalogue Zechariah Schwartz 97/1
(27) Rabbi Avraham Gombiner, *Magen Avraham* on SA OC chap.580 introduction

(28) Rabbi Zedekiah ben Avraham Anaw, *Shibbolei haLeket Hashalem*, (Vilna, 5647), chap.278, p.132

(29) Rabbenu Menachem ben Aharon ibn Zerach, *Tzeda laDerech*, (Warsaw, 5640), article 5, principle 1, chap.8, (p.142b)

(30) Rabbi Chaim Yosef David Azulai, (Chida), *Sefer Shem haGedolim*, (Vilna, 5613), part 1, ma'arechet gedolim, p.76b

(31) Rabbi Yitzchak Palagi, *Yafe laLev*, (Izmir, 5636), part 2 OC chap.580 par.1

(32) Rabbi Avraham Halevi ben David (Ra'avad), *Sefer haKabbalah l'haRa'avad,* (Mantoba, 5284), p.16

(33) see: Chaim Simons, "Sibot l'Ta'anit Tisha b'Tevet," journal *Sinai*, (Mossad Harav Kook: Jerusalem), vol.106, p.138

(34) *BaHaG*, op. cit., p.396

(35) *Kol Bo*, chap.63

(36) Rabbi Yeshayahu ben Avraham Halevi Horowitz, *Shney Luchot Habrit*, (Shanghai, 5704?), Masechet Ta'anit, p.290

(37) *Magen Avraham*, op. cit.

(38) *Megillat Ta'anit*, (Vilna, 5685), footnote by Rabbi Yaakov Emden (Yavetz), Perek Yemai Hatzomot, (p.21b)

(39) Vilna Gaon, *Beur haGra*, SA OC chap.580 par.1 (

(40) *Eliyahu Zuta,* Levush, SA OC chap.580 par.3

(41) Rabbi Yehudah ben Shimon Ashkenazi, *Ba'er Heteiv*, SA OC chap.580 paras.2, 3, 4, 5, 7

(42) Mishnah Berurah, SA OC chap.580 paras.5, 11, 14

(43) Rabbi Azaria ben Moshe Del Rossi Min haAdumim, *Meor Einayim,* (Vilna, 5626), Imrei Bina, chap.7, (p.130)

(44) Rabbi Moshe Rivkash, *Be'er Hagolah*, SA OC chap.580 par.1

(45) Rabbi Yosef ben Avraham Molcho, *Shulchan Gavoha*, (Salonika, 5516), SA OC chap.580 par.1

(46) *Aruch Hashulchan*, OC chap.580 par.1

BLOWING THE SHOFAR DURING
THE MONTH OF ELUL

Part One

Introduction

The earliest source for blowing the shofar in the month of Elul is in a book of Midrashim and Aggadot called "Pirkei Rabi Eliezer" which is attributed to the Tanna Rabbi Eliezer Hagadol the son of Hyrcanus. In this book he writes: "Rabbi Yehoshua ben Karcha says for forty days [Moshe] was on the mountain ... and after forty days he took the Tablets of Stone and went down to the camp and on the seventeenth of Tammuz he broke the Tablets ... and on Rosh Chodesh Elul G-d said to him come up unto Me in the mountain and sound a shofar in the camp, and thus Moshe ascends the mountain in order that the people will no longer err again with idol worship..., as it says, G-d went up with the sound of a shofar, and therefore the sages instituted the blowing of the shofar on Rosh Chodesh Elul every year."[1]

However, this is not the only reason for blowing the shofar during the month of Elul. There are several additional reasons which can be found in the Talmudic Encyclopedia, and they are:

- To arouse people to repent
- "Blow the Shofar on the New Moon," namely during an entire month
- To confuse the Satan (the Devil)[2]
- Furthermore, the Maharam (Meir) of Rothenburg (13th century) gives another reason, namely to teach and educate people how to blow the shofar.[3]

A number of questions are asked, such as on what day or days is the shofar blown during the month of Elul, and at what hours of the day is it blown? These questions will be discussed in accordance with the reasons for blowing the shofar during the month of Elul.

Moshe on Mount Sinai

Moshe ascended Mount Sinai for the third time on a certain date, and remained there for forty days and forty nights, and on the day he ascended the shofar was blown in the camp.[4]

It is not only a historical matter to know exactly on what day Moshe ascended and heard the sound of the shofar, but it is the date, when we today blow (or begin to blow) the shofar.

There are many discussions and differing opinions regarding these dates and we will bring some of them:

It is written that Moshe ascended the mountain on Rosh Chodesh Elul and came down on Yom Kippur, and he was on Mount Sinai for forty days and forty nights. Had there been thirty days in the month of Elul, there would be no problem to arrive at the number forty, namely thirty days in Elul and ten days in Tishrei. However, even prior to the date of fixing the Hebrew calendar, it is written in the Gemara: "From the days of Ezra onwards we did not find the month of Elul of length thirty days."[5] To resolve this contradiction, some write that in the year that Moshe Rabbenu went up Mount Sinai, there were in fact thirty days in the month of Elul.[6]

Another suggestion is that Moshe went up on the first day of Rosh Chodesh Elul, namely, on the thirtieth day of the month of Av.[7] Yet even if he went up on the morning of the thirtieth of Av, and descended during the daytime of Yom Kippur there would not be forty days and forty nights. In order to complete forty days and forty nights some have suggested that Moses descended from Mount Sinai towards the conclusion of Yom Kippur,[8] and others say that he in fact came down on the day after Yom Kippur.[9]

There is a further discussion by the Bach (Rabbi Yoel Sirkis, 16th-17th centuries), and he brings two different explanations on Rashi's comments,[10] but one of them he immediately rejects.

On the comments which he accepts, the Bach writes: "What Rashi wrote in the parashah of Ki Tisa, was that it was on Rosh Chodesh that Moshe was told to ascend in the morning. The explanation is that on the night of Rosh Chodesh he was told to prepare himself, so that in the morning he would be on the summit of the mountain, but at the night he would still be on the slopes of the mountain, and it was only on the morning that he reached the summit. The decree to blow the shofar was not in the evening but in the morning of thirtieth of Av, since it was then that he reached the summit of the mountain, and it was at that time that they blew the shofar in the camp.[11]

However, in the second explanation of Rashi's comments, the one that the Bach rejects, he writes that it seems that this was on the second day of Rosh Chodesh. However, this is difficult since according to this commentary, from the second day of Rosh Chodesh until the morning of Yom Kippur would be only thirty-eight days and nights.[12]

The Bach also quotes the commentary of Rabbi Eliyahu Mizrahi known as the Re'em, who lived in the fifteenth to sixteenth centuries: Moshe's ascent was on the twenty-ninth day of Av, but why then should one count the forty days from the thirtieth of Av and not from the twenty-ninth day when he actually went up the mountain? The reason could be because the twenty-ninth day of Av did not include the night, but the forty days included the nights, and thus one does not count forty days but from the thirtieth of Av which was Rosh Chodesh Elul, because the night of the thirtieth of Av completed the first complete day. This was because the first day comprised the daytime of twenty-ninth of Av and its subsequent night which was the night of the thirtieth of Av. According to this, one finds that the forty days and nights finish on the daytime of Yom Kippur. It was then that G-d accepted the Jews repentance and gave Moshe the second set of the Tablets of Stone.[13]

On the words of Rabbi Eliyahu Mizrahi, the Bach argues that the wording does not prove this, and furthermore, if so, why does one not blow the shofar on the twenty-ninth of Av?

There are commentators who state that Elul was thirty days and Moshe went up on the thirtieth of Av and came down during the day of Yom Kippur, which would make it forty days and forty nights

Another suggested possibility to resolve the problem, is that he went up at the beginning of the night of the thirtieth of Av, and came down towards the end of Yom Kippur, thus Yom Kippur would be included in the forty days.[14] Also Rabbi Yosef ben Meir Teomim known as the "Pri Megadim," (18th century), writes of the descent of Moses: "It is possible that it was on Yom Kippur towards the evening."[15]

In most sources it is written that Moshe went up Mount Sinai on the thirtieth of the month of Av. However, there is a source which holds that the ascent was on the first day of the month of Elul, that is, the second day of Rosh Chodesh Elul. The Magen Avraham (Rabbi Avraham Abele Halevi Gombiner who lived in Poland in the seventeenth century) quoted the Tosafot in the name of the Midrash Tanchuma that Moshe went up on the second day of Rosh Chodesh Elul, but in that year, Elul had thirty days.[16] In the words of the Tosafot which the Magen Avraham brings, the day Moshe went up the mountain for the forty days of his last of three sojourns on Mount Sinai, was on the first of Elul.[17] On this there is the comment of the Bach: "His last sojourn of forty days began on the first of Elul."[18] The Midrash Tanchuma writes: "The entire month of Av was included in the forty-day second sojourn. He went up on Rosh Chodesh Elul [for his third sojourn]... And Moshe ascended early in the morning and was there for the entire month of Elul and the ten days of Tishrei and he descended on the tenth.[19]

In his commentary on the Torah, the Ramban (Rabbi Moshe ben Nachman (Nachmanides), 13th century) writes: "On the Day of Atonement the replacement Tablets of Stone were given and on the next day Moshe descended and spoke with the children of Israel,[20] namely, it was only on the eleventh day of Tishrei that Moshe descended from Mount Sinai. In addition, the "Daat Zkenim" write that Moshe went down from Mount Sinai on the day after Yom Kippur.[21]

In conclusion, according to the opinions that Moshe went up Mount Sinai on the thirtieth of Av, emphasising that it was on the beginning of the night of the thirtieth, and came down on Yom Kippur, towards the end of the day, he was there forty days and forty nights, and this is even more so, according to the opinion that he came down on the day after Yom Kippur. However, according to the opinion that he went up only on the first day of Elul, even if he came down the day after Yom Kippur, one needs to investigate how to arrive at forty days and forty nights.

From all the above debates and differing opinions, we need to arrive at the halacha (practical ruling) on what day we blow the shofar.

Rabbi Binyamin Aharon Slonik (who lived in the 16th century and was a Rabbi in the "Council of Four Lands" (Central Body of Jewish Authority in Poland) and the head of a Bet Din in Poland), wrote that some people blow the shofar on the first day of Rosh Chodesh, and others on the second day, and after a discussion on the subject he came to the conclusion: "Those who begin to blow the Shofar on the second day of Rosh Chodesh are making a mistake."[22]

The Acharonim (great Rabbis who lived after about the 16th century), however, do not agree with his view that "they are making a mistake" to start blowing the shofar on the second day of Rosh Chodesh. In his commentary on the Siddur "Derech HaChaim," Rabbi Yaakov Lorberbaum of Lissa (who lived in Poland in the 18th-19th centuries) wrote: "On the second day of Rosh Chodesh Elul one begins to blow the shofar."[23] Also Rabbi Moshe ben Avraham of Przemysl (who lived in the 16th century in Poland) in his book "Mateh Moshe,"[24] and the "Kitzur Shulchan Aruch[25] and the "Aruch Hashulchan"[26] all write that one blows on the second day of Rosh Chodesh, and they do not even mention the first day of Rosh Chodesh. After a discussion on the subject, the Magen Avraham concludes "One should not change the custom of beginning to blow the shofar on the second day [of Rosh Chodesh]."[27]

In the various calendars (luchot) that are published every year, it states that one begins to blow the shofar only on the second day of Rosh Chodesh Elul. These calendars include: Luach l'Eretz Yisrael (Rabbi Tucazinski),[28] Luach Dvar b'Ito,[29] Luach Belz,[30] Luach Hechal Shlomo,[31] and Luach Hachayal.[32] Furthermore, in Luach Belz it is specifically written: "One does not blow the shofar [on the 30 Av].[33]

On the other hand, Rabbi Avraham Danzig in his book "Chayei Adam" (published in the early 19th century) writes: "And some begin from the first day of Rosh Chodesh whilst others and from the second day."[34] Similar wording is to be found in the book "Mateh Ephraim" written by Rabbi Ephraim Zalman Margoliot (who lived in the 18th-19th centuries in Brodie),[35] and also in the "Be'er Heteiv" (Rabbi Yehuda Ashkenazi who lived in the first half of the 18th century.)[36]

In the "Shulchan Aruch Harav" (written by Rabbi Shneur Zalman of Liadi, who lived in the 18th-19th centuries) an original opinion was found on the subject. There it is written: "On the first day of Rosh Chodesh Elul one blows the shofar in order to learn, and one begins to blow after the service of the second day of Rosh Chodesh."[37]

Until now, we have not written that one blows the shofar every day during the month of Elul because according to the "Pirkei Rabi Eliezer," one only blows it for one day, that is, the day that Moshe ascended Mount Sinai.

According to the version before us of the "Pirkei Rabi Eliezer," the blowing of the shofar in the month of Elul is limited to just one day, either the thirtieth of Av or the first of Elul. There are some Rishonim (great Rabbis who lived approximately between the 11th and 15th centuries) who bring this version, and they include Rabbenu Yerucham (14th century),[38] Machzor Vitry (early 13th century),[39] and the Rosh (Rabbenu Asher, 13th-14th centuries).[40]

However, there are some Rishonim who had a different version: They include the Raviah (Rabbi Eliezer ben Yoel Halevi, 12th century),[41] Tzeda laDerech (Rabbi Menachem ben Aharon ibn Zerah, 14th century),[42] and the Tur,[43] who quote from the "Pirkei Rabi Eliezer" but they write: "And therefore the Rabbis decreed that one should blow during all the month of Elul." Namely they added the words "all the month."[44]

Even though we saw that the Rosh had a version without the words "all the month" he added "but it is customary for Ashkenazim to blow during the whole month of Elul.[45] In addition, all the calendars (mentioned above), as well as the commentary "Derech HaChaim" on the Siddur, specifically state that one blows throughout the month of Elul.

Moshe was on Mount Sinai for forty days, that is, until Yom Kippur, so why do we blow only in the month of Elul, and not until Yom Kippur? Rabbi Eleazar of Garmiza (who lived in the 12th-13th centuries) in his book "Harokeach" wrote about this: "It was decreed that one should blow for forty days until Yom Kippur corresponding to the forty days that Moshe was on the mountain, and we blow every day in order that people should not again err with idol worship." However, he concludes that one only blows until Rosh Hashanah.[46]

The commentary "Ketze Hamateh" written by Rabbi Chaim Zvi Ehrenreich (19th–20th centuries, Hungary) on the book "Mateh Efraim" writes that in the Book of Psalms there is a hint for blowing the shofar in the month of Elul. He bases this on the writings of the "Beit Yosef" who writes that the word "halelu" appears twelve times in Psalm 150, and this corresponds to the twelve months of the year, and the repetition of the last verse corresponds to the thirteenth month in a leap year.[47] On this, the "Ketze Hamateh" adds that the sixth month in the year is Elul and the sixth "halelu" in this Psalm is "haleluhu bteika shofar."[48]

To Arouse People to Repent

It is written in Rambam regarding the blowing of the shofar: "Wake up you sleepy ones from your sleep and you who slumber, arise. Inspect your deeds, repent, remember your Creator."[49]

In a similar statement about "Israel's custom is to begin to blow the shofar from Rosh Chodesh Elul," Rabbi Tzvi Hirsch Kaidanover (who was the head of a court in Frankfurt and lived in the 17th-18th centuries), writes in his book "Kaf Hayashar": "Everyone arise to the sound of the Shofar which proclaims, arouses and warns one to repent."[50]

The question to be asked is, if the blowing of the shofar is to arouse the public to repent, why does one not blow the shofar during the Ten Days of Repentance?

It can even be said that it is more relevant to blow the shofar during these days than to blow it in the month of Elul. This can also be seen from the words of the Rambam, who writes: "Even though repentance and calling out [to G-d] are desirable at all times, during the ten days between Rosh Hashanah and Yom Kippur, they are even more desirable and will be accepted immediately."[51]

There is also a chapter in "Shulchan Aruch" entitled "Sidrei Yemai Teshuvah." There it is written: "On all the days between Rosh Hashanah and Yom Kippur, one increases one's prayers and supplications ... and every day a prayer confessing one's sins is recited three times before dawn,"[52] namely the selichot (penitential prayers) recited on these days. There is also a chapter in the Shulchan Aruch whose content is that one needs to be more particular on those days on the observance of mitzvot: "Even one who is not particular to eat non-Jewish baked bread, during these Ten Days of Repentance should eat only Jewish baked bread, and every person should investigate and scrutinise his deeds and do repentance during these Ten Days of Repentance."[53]

The recital of selichot in particular during these ten days is already mentioned by the Geonim (great Rabbis from about the year 600 to about the year 1000): "Rav Kohen Tzedek (9th century) stated that the custom of the 'two yeshivot' [the two most famous academies in Babylon] was to recite penitential and supplication prayers during these Ten Days of Repentance. Rav Amram Gaon (9th century) said that during these ten days one gets up early to go to the Synagogue each day... and Rav Hai Gaon (10th-11th centuries) said that the custom is to say supplication prayers on these ten days."[54]

Today, even though Ashkenazim begin to say selichot in the latter days of the month of Elul, the selichot recited during the Ten Days of Repentance are much longer than the selichot said in the days of Elul (with the exception of those recited on the eve of Rosh Hashanah).[55]

From all this, why does not one blow the shofar during the Ten Days of Repentance, namely until Yom Kippur? And furthermore, why not even until Hoshana Raba?

The author of the book "Shibbolei Haleket" (Rabbi Zedekiah ben Avraham Anav, 13th century), wrote of Hoshana Raba: "And the day of Arava [Hoshana Raba] is the day of the sealing of the judgment of Yom Kippur."[56] There is a source for this from the Zohar: "On the seventh day of Sukkot the judgment of the world is finally sealed and the edicts are sent forth from the King."[57] One could therefore ask why not blow the shofar every day until Hoshana Raba in order to arouse the people to do repentance?

We see from the Rabbinic literature that Hoshana Raba is considered a kind of mini-Yom Kippur. The person conducting the service in the Synagogue on Hoshana Raba wears a kittel,[58] some of the tunes in the service are those sung on Yom Kippur,[59] and there are even some who recite the additions in the amidah recited during the Ten Days of Repentance; an example of such an addition is "Remember us unto life,"[60] and furthermore some even say "the Holy King" and "the King of Judgment"[61] which are changes which are made in the prayers during these ten days. There is even a version of the liturgical hymn "unetanne tokef" [recited at Mussaf on Rosh Hashanah and Yom Kippur] which includes the words: "On Hoshana Raba the decree is sealed."[62] In the Rome congregation they would say "unetanne tokef" on Hoshana Raba.[63]

Is there a source for blowing the shofar on Hoshana Raba? The answer is yes, and there is even a source in the Jerusalem Talmud where it is written: "Yet they seek me daily with the shofar and willow."[64] Rabbi David Frenkel, who was the Rabbi of the Berlin congregation in the eighteenth century, and wrote the commentary "Korban haEdah" on the Jerusalem Talmud, comments on this: "They are the Festival of Rosh Hashanah and the seventh day of the willow (Hoshana Raba), days when everyone hears the sound of the shofar..."[65] It is written in the book "Kaf haChaim" on the chapter on Hoshana Raba: "It is customary to start with a request for mercy from Rosh Chodesh Elul, whether as a result of the sounds on the shofar or in the

251

selichot, every one according to his custom, and from the first day of Elul until Hoshana Raba are fifty-one days."[66] According to this book's language, it is possible that one blows the shofar every day until Hoshana Raba.

Blowing the shofar on Hoshana Raba is not just theoretical. There are communities that blow the shofar on Hoshana Raba. Among them are several communities of Hasidim and they blow tekia – shevarim teruah – tekia between the various hoshanot.

Hasidei Sanz blow even more than this, namely tekia – shevarim teruah – tekia, tekia – shevarim – tekia, tekiah – teruah – tekia.[67] In the book "Seder Hayom" by Rabbi Moshe ben Machir (16th century) it is written that the reason for these blowings is: "To arouse mercy and every place should follow its own custom."[68]

A man known as "Charif Echad" sent a question to Rabbi Meir Horowitz, (who lived in the late 19th century), and authored the book "Imrei Noam." This person wrote to him regarding the blowing of the shofar on Hoshana Raba: "Because this custom has no source and is not found in any book it should be forbidden because of the principle of 'not separating into different groups'." Rabbi Horowitz rejected the words of the "Charif Echad" and wrote that a number of Rabbis such as the Tzaddik of Apta and the Tzaddik of Rapshitan followed this custom, and therefore according to those who observe it "one has on whom to rely on." Rabbi Horowitz concludes: "Without any doubt, it is permissible to blow the shofar on Hoshana Raba ... and that one may not abolish this custom which has a holy source."[69]

In addition to the customs of the Hasidim, there are synagogues of the Sefaradim from Spain and Portugal who blow the shofar after every circuit with the arba'at haminim (the four species taken on Sukkot). For example, in the Shearit Yisrael Synagogue in Manhattan, New York, which was established in the seventeenth century, they blow the shofar on Hoshana Raba.[70]

On the night of Hoshana Raba, some read the book "tikun leil Hoshana Raba, and it is written that in the Sefaradic communities during the reading of this book, they blow the shofar at specific places.[71]

Another question which can be asked is, why are the three types of notes blown every day in the month of Elul, namely tekia, shevarim and terua? The sound of the shofar is to arouse people to repent, and therefore why is not one of these notes sufficient? The "Ben Ish Chai" (Rabbi Yosef Chaim who lived in Baghdad in the 19th century) explains: The note tekia hints at joy and good things; shevarim and terua hint at pain and suffering since terua is a weeping sound and shevarim a groaning sound.[72] Therefore, all of them are necessary for repentance.

Blow the Shofar on the New Moon

"Blow the Shofar on the New Moon" is part of a verse in the book of Psalms,[73] and the Rabbis learned from it that one blows the shofar for a month, and the length of a month is thirty days. However, the month of Elul is in fact only twenty-nine days, and furthermore one does not blow the shofar on the day preceding Rosh Hashanah, therefore there are only twenty-eight days of blowing during Elul. However, in addition, one blows on the two days of Rosh Hashanah thus making the thirty days.[74] However, in practice one does not blow for thirty days because one does not blow on the four Shabbatot in the month of Elul, and likewise if the first day of Rosh Hashanah occurs on Shabbat. Therefore, in practice there are only twenty-six days, or in some years only twenty-five days when one blows.

Regarding the blowing on Shabbat during the month of Elul, the Aruch Hashulchan writes: "And one blows [during the month of Elul] tekia – shevarim terua – tekia with the exception of Shabbat ..."[75] Why does one not blow on Shabbatot during the month of Elul?

Moreover, in years when Rosh Hashanah occurs on Shabbat one does not blow the shofar.[76] The reason is: "A decree lest one takes it [the shofar] in his hand and goes to an expert in order to learn how to blow it and he carries it four cubits [about two metres] in a public domain."[77]

The question to be asked is whether it is possible to give the same opinion not to blow on Shabbatot in the month of Elul?

It seems not. The reason lies in the fact that blowing the shofar on Rosh Hashanah is a mitzvah from the Torah, but blowing in the month of Elul is only a custom. It could be compared to the performing of tashlich (prayer recited alongside a body of running water) on Rosh Hashanah which is only a custom. What happens when the first day of Rosh Hashanah falls on a Shabbat? Here there are different opinions. Some perform tashlich that year on Shabbat, whilst others postpone it until the second day of Rosh Hashanah.[78]

According to the opinion that one performs tashlich on Shabbat, because it is just a custom, one may ask why not blow the shofar on Shabbat in the month of Elul because likewise it is also just a custom? The question to be asked: Is it a forbidden labour to blow the shofar on Shabbat? We can see from the Gemara that this is not the case: "Blowing of the shofar ... is a kind of skill and not a labour."[79] Therefore, why should one not blow the shofar on Shabbat? It is written in Rambam: "It is forbidden to sound musical tones on Shabbat, whether using a musical instrument such as a harp or lyre or other object."[80] The reason for forbidding this is a precaution that he might come to transgress the work category of "makeh b'patish" (putting the finishing touches to an object); in the words of the Rambam: "a person repairing an object in any manner."[81] This does not only apply on Shabbat, but also on Yom-Tov it is forbidden to play musical instruments.[82] For example, on the Yom Tov of Pesach, which occurs on weekdays, it is forbidden to play musical instruments. In addition, even on Rosh Hashanah which occurs on a weekday, one is limited to blowing the shofar only to perform the mitzvah, and "one may not blow additional notes for no reason."[83]

In conclusion, one may not blow on Shabbatot during the month of Elul, even though this will result in blowing the shofar for less than thirty days!

Confuse the Satan

The role of Satan (the Devil) is to prosecute the Jewish people on Rosh Hashanah. That is the reason that one does certain actions that aim to hide from the Satan when Rosh Hashanah falls, and among them: One does not announce on the Shabbat prior to the month of Tishrei (the month when Rosh Hashanah occurs) the day on which Rosh Chodesh Tishrei will occur, and this is unlike the other months during the year.[84] In the book of Psalms it is written about Rosh Hashanah "in the time appointed ("bakese" – covered) for our day of Festival,[85] and the meaning is that the moon is covered, at the time when Rosh Hashanah occurs;[86] (it is on the first day of the month of Tishrei, when the moon is covered, namely there is a new moon). In connection with the blowing of the shofar, one blows every weekday of Elul but does not blow on the eve of Rosh Hashanah, the reason being "to confuse the Satan in order that he should not know on which day is Rosh Hashanah,"[87] namely to confuse him into thinking that Rosh Hashanah has already passed, and thus he has lost the opportunity to prosecute the Jewish people.

It is also written that the shofar is not blown on the eve of Rosh Hashanah in order to "make a break between the customary blowing [in the month of Elul] and the obligatory blowing [on Rosh Hashanah]."[88] Therefore, the question arises what happens in the years when the first day of Rosh Hashanah falls on Shabbat and there is no shofar blowing on that day? Is this Shabbat also considered a break between the blowing on the last day of Elul and the second day of Rosh Hashanah? The Poskim (Rabbinical arbiters) write: "And even when the first day of Rosh Hashanah occurs on Shabbat, one also does not blow on the day before Rosh Hashanah."[89] The reason is that since we say in the prayers a "memorial of the Shofar," it is as if one blows the shofar.[90]

On the other hand, there is a Siddur that came out in Britain at the end of the nineteenth century with the authorisation of the then Chief Rabbi of Britain, Rabbi Natan Adler, and this Siddur subsequently had many editions. In some of them it is written: "If that day [eve of Rosh Hashanah] be a Friday, the shofar is sounded."[91] Also, in a Siddur authorised by the then Chief Rabbi of Britain, Rabbi Yosef Zvi Hertz, the identical language appears.[92] One needs to investigate where is the source for this!

The Siddur that came out under Rabbi Natan Adler is based on the Siddur "Avodat Yisrael." However, in this Siddur it is written: "There is no blowing the shofar on the eve of Rosh Hashanah"[93] and it does not add that one blows if it occurs on Friday.

There is another custom of not to stop blowing just one day before Rosh Hashanah, but instead three days. On this the Maharam of Rothenburg explains: "And the reason not to blow three days before Rosh Hashanah is so that the Satan will not be accustomed to blowing for thirty consecutive days."[94]

Some of the places where it is customary not to blow on the three days prior to Rosh Hashanah are Frankfurt,[95] Piorda (the city of Fürth located in the state of Bavaria in Germany),[96] and Wurzburg (in the state of Bavaria in Germany).[97]

However, there is a situation in which it is permitted to blow the shofar on the eve of Rosh Hashanah. It is written in the book "Elia Rabba": "It is permissible to blow the shofar in order to learn on the eve of Rosh Hashanah ... I found in the book Amcharel which had been written on parchment, that a person who wants to blow the shofar on the eve of Rosh Hashanah in order to learn how to blow should blow in a mikvah (ritual bath) or in a closed room."[98] On the other hand, it is stated in a commentary to the Siddur "Beit Yaakov": "[One blows] every weekday [in the month of Elul] with the exception of the eve of Rosh Hashanah, even if one blows in one's house in order to learn."[99]

To Learn how to Blow the Shofar

On Rosh Hashanah there is a commandment from the Torah to blow the shofar, and in the month of Elul it is customary to blow almost every day. Blowing a shofar is a skill,[100] and by blowing in the month of Elul there is an opportunity to learn how to blow. The Maharam of Rothenburg writes about this: "It is customary to blow the shofar ... from the month of Elul to three days before Rosh Hashanah in order to learn and to instruct in the mitzvah which states that one asks on the laws of a Foot Festival [Pesach,

Shavuot, Sukkot] thirty days before the Festival.[101] (It seems that the Maharam of Rotenburg believed that the rule thirty days before the Festival[102] also applies to Rosh Hashanah.)

It is also written in the sermons of Rabbi Chaim Ohr Zarua (who lived in the second half of the thirteenth century): "It is customary to blow the shofar throughout the month of Elul, so that one would be accustomed to blowing and would not err."[103]

Rabbi Avraham Yitzchak Hakohen Kook (19th-20th centuries) also thinks so and wrote: "One begins to study the laws of Pesach thirty days before the Festival and the same law applies to other Festivals, therefore one should study the laws thirty days before Rosh Hashanah, and concerning the blowing of the shofar one needs to learn the arrangements and details of the notes, therefore one needs to start to blow from Rosh Chodesh Elul which is thirty days before Rosh Hashanah."[104]

From this, the reason for blowing during the month of Elul is to learn how to blow on Rosh Hashanah, and for this reason there is no need to blow after Rosh Hashanah.

There are three types of notes that one blows on the shofar, namely: tekia, shevarim and terua. Each of the notes has a different method on how to produce the sound. Every day during the month of Elul, one blows at least the notes tekia – shevarim terua - tekia, (henceforth: TASHRAT) and by doing so, the blower of the shofar has the opportunity to have practice in producing all these notes.

Most congregations during the month of Elul blow only TASHRAT,[105] and the majority of them belong to Ashkenazi Jewry,[106] and these include the Perushim (disciples of the Vilna Gaon) in Eretz Israel[107] and Lithuania,[108] a number of groups of Hasidim,[109] who include Belz,[110] Breslov,[111] Bobov,[112] Amshinov,[113] Vizhnitz[114] and Lelov.[115]

However, there are communities that during the days of Elul, in addition to the order of TASHRAT, also blow tekia – shevarim – tekia (henceforth: TASHAT), and tekia – terua – tekia (henceforth: TARAT).[116] These include Frankfurt,[117] Chabad,[118] Sanz,[119] Stolin-Karlin,[120] Chernobyl,[121] and Satmar.[122] Hasidei Kretshnif blow the shofar before the morning service, and apparently, blow every day TASHRAT, TASHAT and TARAT.[123]

The source for including TASHAT and TARAT in the month of Elul is given by the Bach in his commentary on the Tur where he writes: "Any person who blows the shofar in a congregation, should only blow TASHRAT, TASHAT, TARAT, because if he were to blow only TASHRAT or TASHAT or TARAT, the listeners would come to the conclusion that this was gist for the blowing and they would therefore come to err [on Rosh Hashanah] ... and thus in the blowing of the shofar even throughout the entire month of Elul, one should not blow less than the orders TASHRAT, TASHAT, TARAT." However, the Bach then concludes that "this is not the custom" and he gives a reason not to blow all these orders, namely: "and perhaps according to Rabbenu Tam (Rabbi Yaakov ben Meir, 12th century) who decreed that if one were to blow just TASHRAT, one should have no apprehension, and we follow the custom of Rabbenu Tam."[124]

However, we see from the customs mentioned above, as well as from the books "Darkei Chaim v'Shalom" (Rabbi Chaim Elazar Spira of Munkatch, 19th-20th centuries),[125] and "Shevet Hakehati" (Rabbi Shammai K'hat Hakohen Gross, 21st century),[126] that in practice there are some who do not follow the conclusion of the Bach, but blow TASHRAT, TASHAT, TARAT throughout the month of Elul.

There are discussions on whether it is necessary on Rosh Hashanah to blow shevarim-terua in one breath or in two breaths? On this one can ask the question: "In the month of Elul, in the order TASHRAT does one blow shevarim-terua in one breath or in two breaths?" The answer given by the Rabbis of "Machon Halacha Chabad was: "In one breath."[127] Obviously this is "lechatchila" (from the outset) because even on Rosh Hashanah there are differences of opinion whether to blow shevarim-terua in one breath or in two breathes.[128]

Despite all the preparatory exercises performed during the month of Elul, it is written in the Shulchan Aruch that there are cases in which the person blowing the shofar does not succeed on being able to blow on Rosh Hashanah, and it rules: "If a person begins to blow but is not able to complete the blowing, another person should complete them, and even three or four people, and the berachah (blessing) made by the first person is sufficient ... and even if the first person recites the berachah but does not succeed in blowing at all, the second person continues without reciting a berachah, and the berachah [recited by the first person] is not a berachah recited in vain.[129] The "Shulchan Aruch haRav" adds that this is not limited to three or four people but to an indefinite number.[130]

There have been cases where an expert shofar blower did not succeed in blowing even a single note on Rosh Hashanah. One such incident occurred at a minyan where Rabbi Aharon Yehuda Leib Shteinman was praying in the city of Bex Les Baines in Switzerland during the Second World War.[131] Another incident that happened around the same period occurred in the village of Kfar Pines in Eretz Israel.[132]

Part Two

Blowing the Shofar in Elul by Sefaradim

Sefaradim have a different custom from Ashkenazim regarding the blowing of the shofar in the month of Elul. Sefaradim begin to say selichot on the second day of the month of Elul, and blow the shofar whilst reciting the selichot. Rabbi Mordechai Eliyahu (a former Chief Rabbi of Israel) mentions three different customs in this context: During the recital of the selichot, there are some Sefaradim who blow the order TASHRAT whilst saying the "shelosh esrai middot" [the thirteen Attributes of G-d's Mercy], others who blow during the kaddish before the words 'taanu v'taharu,'[133] and there are yet others who do not blow the shofar at all.[134] It would seem from the language of Rabbi Eliyahu, that the forementioned are three distinct alternatives.

However according to the language of the Kaf Hachaim, one blows on both occasions (namely, shelosh esrai middot and kaddish), and this is addition to blowing at the end of the morning service according to the custom of the Ashkenazim: "And those who follow the (Beit Yosef in the) Shulchan Aruch, (namely the Sefaradim), there are those who also blow the shofar in the selichot when reciting the 'shelosh esrai middot' and also before 'kaddish titkabel' in order to keep every opinion."[135]

Among the Sefaradic communities in the various countries, there are different customs regarding the blowing of the shofar in the month of Elul. Rabbi Shem Tob Gagin (who died in England in 5713, and was the author of the book "Keter Shem Tov" which was published in seven volumes, and whose contents include in detail the customs of the Sefaradim from both the Eastern and Western countries). It is written in this book that the Sefaradim in Eretz Yisrael, Syria, Turgama (Turkey), and Egypt blow the shofar when the man conducting the service is saying the "shelosh esrai middot" and before kaddish titkabel at the end of the selichot.[136] It is not clear from his wording on which days they would blow the shofar during the selichot, and also if it included the Ten Days of Repentance.

On the other hand, in the book "Yalkot Minhagim" it is written about the Sefaradic Jerusalemites: "They were not accustomed to blow the shofar during the month of Elul in order not to becloud the mitzvah of blowing the shofar on Rosh Hashanah, but they would not prevent a person who wanted to blow the shofar on these days. If they blew the shofar, it would be when saying the words in the selichot 'Kel Melech Yoshev al Kisei Harachamim'?"[137] It seems possible to explain the contradiction between "Keter Shem Tov" and "Yalkut Minhagim," that the custom in Jerusalem was not like other places in Eretz Yisrael, and also to suggest, as it is written in "Yalkut Minhagim" in Jerusalem as elsewhere in Israel, that they would blow the shofar.

In addition, the "Keter Shem Tov" wrote that the custom in London and Amsterdam is to blow only at the end of selichot, only during the Ten Days of Repentance.[138] It seems from his words that they did not blow in the month of Elul.

He also writes: "In Aram Tzova they did not blow at all on the days of the selichot, and this is surprising to me, because in

Aram Tzova all the customs are according to the customs of Eretz Yisrael, and in Eretz Yisrael they blow."[139]

Rabbi Raphael Moshe Daluya (21st Century) wrote about the Moroccan custom of blowing in the month of Elul: "During the selichot they are accustomed to blow TASHRAT, TASHAT, TARAT every time they say the word 'vayavor' (namely the shelosh esrai middot). He added that he heard this from Rabbi Avraham Hafuta and Rabbi Shalom Gabay.[140] To this, Rabbi Maor Kayam added, "also in kaddish titkabel," and he also writes that there is this custom among the people who came from Egypt and Bukhara, and Libya.[141]

On this subject, Rabbi Kalfon Moshe Hakohen in his book "Brit Kehuna" (20th century) writes about the custom of Djerba (an island, which is part of Tunisia and is situated close to it): "Our customs in all the synagogues that after the selichot they blow TASHRAT, TASHAT, TARAT, provided they have a person who can blow a shofar and have a shofar available."[142]

The custom of Kurdistan is: that in a number of synagogues, it is customary to blow TASHRAT, TASHAT, TARAT, after the selichot. In other synagogues they do not blow, in order not to detract from the blowing of the shofar on Rosh Hashanah and Yom Kippur."[143] It would seem that the meaning of Yom Kippur refers to the Jubilee Year or the blowing of the shofar after Yom Kippur.

The book "Shtilei Zeitim" written by Rabbi David Mashraki, a Yemenite Rabbinic authority who lived in the eighteenth century, brings the customs of Yemen and he wrote: "And our customs are to blow the shofar on these days after the selichot until the twenty-eighth of Elul."[144] Rabbi Yitzchak Ratzabi (Yemenite Rabbinic authority in the 20th-21st centuries) adds that some of the Shami group (one of the three groups of Yemenite Jewry) are accustomed to blow also in the "shelosh esrai middot."[145] However, in the book "Halichot Teman" by Rabbi Yosef Kapach (a Yemenite Rabbinical authority in the 20th century), he does not mention the custom of blowing the shofar during the selichot during the month of Elul but he does mention the custom of the children: "During the entire month of Elul, they do not stop blowing the shofar by day and by night, and they are accustomed to compete with each other on who is able to blow the longest and most pronounced tekia. Because of this there were many experts in blowing the shofar."[146]

Part Three

So far we have discussed the various reasons for blowing the shofar in the month of Elul and the implications arising from these reasons. We will now discuss general matters related to blowing the shofar in the month of Elul.

The Time of Day during the Month of Elul to Blow the Shofar

The Rema writes: "From Rosh Chodesh [Elul] onwards one begins to blow the shofar after shacharit (morning service) and there are places where they also blow at ma'ariv (evening service),"[147] the intention being right at the end of the service. However, the Mishnah Berurah writes: "It is customary in our countries that from Rosh Chodesh Elul until Yom Kippur to recite every day after the end of the service the Psalm beginning with the words 'Mizmor leDavid HaShem ori' in the morning and in the evening."[148] Therefore, when should one blow the shofar – before or after reciting this Psalm? In fact, there is no contradiction between the words of the Rema and those of the Mishnah Berurah. The custom to recite this Psalm appears in writing for the first time in the book "Shem Tov Katan" which was published at the beginning of the eighteenth century,[149] that is, after the period of the Rema. The custom today is to blow the shofar before reciting this Psalm.

The time for blowing the shofar during the services in Elul is already mentioned by the Rishonim. The Rosh wrote: "The Ashkenazi custom is to blow throughout the month of Elul in the morning and in the evening."[150] One could derive from the wording of the Rosh that in *every location* they blow twice every day, namely, once in the morning and once in the evening.

One can understand this because he does not write "*there are some locations* where they also blow in the evening." Furthermore, there is a similar wording by the Tur.[151]

However, a different wording is to be found in the book "Chayei Adam": "And there are places where they blow the shofar also at minchah (afternoon service).[152] The question arises is why does he use the word "also"? Does he mean that the blowing at minchah is in addition to the "evening" or is it instead of the "evening"?

According to Rabbi Moshe Leib Tziltz in his book "Mili D'Avot," the words of the "Chayei Adam" are difficult since all the Poskim write that one blows after ma'ariv. He writes that perhaps in his opinion, the Chayei Adam does not mean that one should blow the shofar at minchah throughout the month of Elul, but only on the first day of Rosh Chodesh Elul.[153]

Rabbi Moshe Feinstein (20th-21st centuries) also discusses the intention of the use of the word "evening" by saying: "It seems that evening could also mean when it was actually day, or that it was after minchah, or it was after having recited ma'ariv whilst it was still day, because it does not seem that one should blow from the beginning of the night until after midnight, a period that is not an appropriate time [to request Heavenly intervention]."[154]

Rabbi Chaim Kanievsky (20th-21st centuries, son of the "Steipler") was asked if in a case when the congregation forgot to blow the shofar at shacharit in the month of Elul, is it possible to blow at minchah? He replied: "one can do so."[155] In a similar situation, Rabbi Feinstein wrote that it is possible to make up not blowing at shacharit by blowing in the evening.[156]

However, we see from the responsa of Rabbi Menashe Klein, (Rabbi in the city of Ungvar Ukraine, author of the Mishnah Halachot responsa), that although it is possible, or alternatively, one is able to blow the shofar later in the day: "It is not necessary to make up and blow the shofar after minchah."[157]

Rabbi Feinstein also believed that the custom is to blow only once a day, and the reason that blowing is done in shacharit is that everyone is present in the synagogue for shacharit, but this is not so for minchah, and also, shacharit is recited before people go out to work.[158] However, in practice many of the worshippers at shacharit "go out to work" before the end of the service and thus miss the shofar blowing! Therefore, according to this reasoning, it is possible to suggest (and this is only a suggestion!) that it is better to blow the shofar during the repetition of the amidah, maybe after the berachah "t'ka beshofar gadol." This would be similar to the blowing during the mussaf (additional service) amidah of Rosh Hashanah.

Among the communities that blow the shofar both in the morning and evening are Prague,[159] Frankfurt[160] and Piorda (the city of Fürth located in the state of Bavaria in Germany).[161] The city that once had a custom of not blowing at all in the month of Elul, was the city of Kotzk, a town located in eastern Poland, and it was an important centre for Hasidim. Apparently, at a later period, Rabbi Yitzchak Meir Alter of Gur, instituted that one should blow the shofar in that city during Elul.[162]

Another question regarding the blowing of the shofar during the month of Elul is whether it is possible to postpone the blasts until a later hour in the morning when there is a brit milah (circumcision) in the synagogue immediately after shacharit. Rabbi Gavriel Zinner, (21st century), in his book "Nitei Gavriel" discusses this issue and wrote: "When there is a brit milah in the month of Elul, one performs the brit milah and afterwards blows the shofar. But then he adds that it is not customary to do so since one should one blow immediately after shacharit.[163]

Regarding the times in the year for the blowing of the shofar, it seems that the "Shibbolei Haleket" brings a novella: "The latter generations decreed to blow the shofar on the nights of Rosh Chodesh Elul and on the termination of Yom Kippur in remembrance of the joy on receiving the latter set of Tablets of Stone.[164] Are these the only occasions when the shofar was blown, or is it his intention that they were in addition to the blowing of the shofar during the month of Elul?

Do communities that are accustomed to blow in the evenings of the month of Elul also blow on "motzaei Shabbat" (termination of the Sabbath)? This is an actuality because according to the book "Mili d'Avot" in the city of Prague they blew on motzaei Shabbat.[165]

Regarding blowing on motzaei Shabbat there are some questions which need to be asked. Should one wait until the stars come out, namely, the time of motzaei Shabbat, or is it already possible to blow at twilight? Is it permitted to blow before reciting havdalah (separation ceremony at conclusion of the Sabbath), namely in the amidah (i.e., the addition 'ata chonantanu') or at least does one have to wait until at least the person blowing the shofar has made havdalah?

There is one occasion every year that all places blow the shofar in the evening, and that is on motzaei Yom Kippur. There are two opinions on when to blow – immediately after ne'ila (concluding service) or after ma'ariv. The Tur brings two opinions: "some are accustomed to blow immediately after the ne'ila service before ma'ariv; even though one has not yet made havdalah there is no apprehension since blowing the shofar is not regarded as work, but it is more correct to wait until after ma'ariv even if one has not yet made havdalah (on the cup of wine) since one has made it in the amidah of ma'ariv."[166] However, the Bach writes on the Tur: "Some are accustomed to blow immediately after ne'ila ... such is the custom today."[167] The Mishnah Berurah writes that even if the time for the termination of Yom Kippur has not arrived, "even if it is twilight it is permitted since (the blowing of a shofar) is permitted since it is just a Rabbinical restriction which is being performed for a mitzvah, but if it is still daytime it is forbidden to blow."[168] The "Luach l'Eretz Yisrael" writes that one blows twenty-one minutes after sunset and this is twelve minutes before the termination of Yom Kippur.[169] Also the "Mateh Ephraim"[170] and the "Aruch Hashulchan"[171] write that one blows after ne'ila. However, "Machzor Vitry" brings the two opinions quoted above.[172] Further investigation needs to be made that according to the opinion that one blows after ma'ariv, is it necessary that it is already after the termination of Yom Kippur.

Rabbi Shmuel ben Natan Neta Halevi Loew Kelin, who lived in the eighteenth century in Boskovice, Czech Republic, discusses in his book "Machazit haShekel" whether it is permissible to blow the shofar at the termination of Yom Kippur before making havdalah in the amidah. On this he wrote: "Even though on motzaei Shabbat before making havdalah in the amidah, it is even forbidden to move the candle, nevertheless one can be lenient here [motzaei Yom Kippur].[(173)] Further investigation needs to be made regarding what is the halacha when Yom Kippur occurs on Shabbat?

A question to be asked is whether it is possible to use the same conclusions arrived at for blowing the shofar on motzaei Yom Kippur, for the motzaei Shabbat in the month of Elul? However, it seems that the subject of blowing the shofar on motzoei Shabbat in the month of Elul is hardly mentioned by the Poskim.

The few sources in which it is mentioned are the book "Milei d'Avot" and the commentary "Elef Lamateh" to the book "Mate Ephraim," and there it is written that the blowing the shofar on motzaei Shabbat in the month of Elul should be done by a person who has already made havdalah in the amidah.[(174)] However, it is not written if it is permitted by someone who has prayed ma'ariv during the twilight period.

A Minor Blowing the Shofar during Elul

The question arises, may a minor (boy under the age of Barmitzvah) blow the shofar during the month of Elul?

In the laws of Rosh Hashanah there is a chapter entitled "Who may blow the shofar" and there it is written: "Anyone not obligated for something is unable to fulfill others' obligations."[(175)] Amongst those who are exempt are a minor.[(176)]

A question that can be asked is whether this halacha also applies to the month of Elul? In the book "Shevet Hakahati" it is written: "One needs to be punctilious ... that a minor does not blow [in the month of Elul]."[(177)] In contrast to this, one can see from other authorities who state that according to the letter of the halacha a child may blow the shofar during the month of Elul, but they add that if a grown up knows how to blow, he should do so out of respect to the congregation.[(178)]

Regarding the first day of Rosh Hashanah, what is the definition of "minor" who cannot blow? On this the Mishnah Berurah wrote: "Even he is thirteen years old, as long as it is not known that he grown two body hairs he cannot blow in order to make others fulfill the mitzvah."[179] Because blowing the shofar on Rosh Hashanah is a mitzvah from the Torah, one cannot rely on the presumption that he has reached this state of physical maturity.[180] Even though the second day of Rosh Hashanah is Rabbinical, Rabbi Tzvi Pesach Frank (Ashkenazi Chief Rabbi of Jerusalem, 20th century) is doubtful if one can rely on presumption, since the two days of Rosh Hashanah are regarded as "one long day."[181]

The blowing in the month of Elul is only a custom, so there is room to investigate if a child who has reached the age of thirteen can blow, namely if one can rely on the presumption of physical maturity.[182]

Furthermore, not everyone can enable others to fulfill their obligation of blowing the shofar on Rosh Hashanah. Among them are a woman,[183] an androgynous, and a person whose sex is undetermined.[184] Probably the same laws apply to them as to a minor.

According to the opinions that allow a minor, a woman, etc. to blow the shofar in Elul, do they limit it to humans or does it also include certain animals. For example, if a monkey is taught to blow the shofar, can he blow it in the month of Elul? This is not ridiculous because there are monkeys who have learned to do various actions for the disabled.[185] There are cases which appear in the halachic texts that a monkey can help in the observance of some mitzvot. For example, some permit monkeys to pour water on a man's hands for washing before eating bread.[186] Also a monkey can be a courier for mishloach manot (sending food items to a person on Purim),[187] and regarding extending the area permitted to enter on Shabbat.[188] Therefore, there is room to investigate if (theoretically!) a monkey could blow the shofar in the month of Elul!

Another question is whether in the month of Elul it is possible to hear the sound of the shofar via a recording. If the reason is to stir up people to do repentance, a recording will also help. However, if the reason is to learn how to blow, probably a recording will not help!

Use of an Invalid Shofar in the Month of Elul

On Rosh Hashanah there are a number of defects that can invalidate a shofar for the observance of the mitzvah. These things can be a defect in the actual horn of the shofar, for example a crack, and it can also be related to the history of the shofar, for example a shofar used for idol worship.[189] The question is whether these defects which disqualify the shofar for Rosh Hashanah also do so for the month of Elul? Rabbi Yom Tov ben Avraham Asevilli who was known as the "Ritva," (13th-14th centuries) wrote explicitly about this: "One is not particular for any defect neither with the sounding of the notes nor in the state of the actual shofar."[190] This point was also mentioned in the book "Nitei Gavriel."[192] However, on the other hand, the author of the book "Shevet Hakahati" writes: "Care must be taken ... that the shofar is not invalid and the laws lechatchila are the same as for Rosh Hashanah."

There are a number of details which will be studied regarding blowing the shofar during the month of Elul. We will look at the laws of the shofar on when it is possible to use a certain shofar in the month of Elul in accordance with the opinion that during Elul one is allowed to use shofars with defects.

It is written: "The shofar .... of a cow is invalid,"[193] and the reason is "because it is called a horn and not a shofar."[194] Therefore it is obvious that even in the month of Elul one cannot use a shofar from a cow.

Even though the Rema rejects a shofar from an unclean animal,[195] the Mishnah Berurah writes on this statement: "If one has no other shofar, one should use such a shofar but should not say the berachah over it since it may be forbidden by the halacha.[196] According to these conditions it is possible to use such a shofar during Elul.

It is written: "If one steals a shofar and blows on it, one has fulfilled the mitzvah."[197] The reason is "that the mitzvah of the

268

shofar is only hearing it being blown and there is no law of theft by hearing a sound ... but one should not recite a berachah."[198] Therefore, also in the month of Elul it is possible to use a stolen shofar, but the issue of a berachah is irrelevant, since one does not recite a berachah on blowing a shofar in the month of Elul.

It is written in the "Aruch Hashulchan" that the size of the shofar is that one should be able to hold the shofar in one's hand and see it jutting out on both sides of one's hand, and the reason for this is that someone should not suggest that one is making the sound by blowing into one's hand.[199] One needs to study further whether because blowing the shofar in the month of Elul is just a custom, there will be no apprehension regarding its size.

If on Rosh Hashanah one blows on a shofar owned by a Jew and it is an object of idol worship, or is made from an animal that was sacrificed for idol worship, one does not fulfill the mitzvah of shofar blowing.[200] The reason is that such a shofar must be burned and if they are burned, even the minimum size of the shofar will not remain.[201] Therefore, if one does not have a shofar of the minimum size in Elul, then it is open to investigation whether such an undersized shofar can be used.

The physical defects that usually disqualify the shofar on Rosh Hashanah are certain holes or cracks.[202] Since there are different opinions on whether a particular shofar is kosher (fit) it seems that even those Poskim who hold that one should not use a possul (non-fit) shofar during Elul, would allow it.

It is written in the Shulchan Aruch: "If one stuck broken shofars to each other to make a shofar, it will be possul."[203] The reason is: "G-d says one shofar and not two or three shofars."[204] For the same reason "if one attached to a shofar either shofar material or non-shofar material, the shofar will be possul."[205] One needs to investigate if it is permissible to use this shofar in the month of Elul.

It is written: "If one inverted the shofar and then blow on it, one does not fulfill the mitzvah of shofar."[207] The reason is that it states in the Torah "v'havarta shofar terua" (and you shall pass a shofar), namely one must blow on the shofar at the mouthpiece and not from the other end. Apparently, this also applies to blowing in the month of Elul because otherwise it is not called blowing at all.

Praying Privately

Is the blowing of the shofar in the month of Elul limited to congregational prayers, or does one who "prays privately in the month of Elul blow the shofar, or was the decree instituted only to a congregation?" This question was submitted to Rabbi Chaim Kanievsky and he answered "it would seem that it is not," but he suggests that in such a case: "The proper thing is to go to another place when he has not heard in it in his synagogue."[208] Rabbi Menashe Klein also believes that it is a congregational duty, but he does not demand that the person go to hear the shofar after he has finished praying. He writes: "The blowing of the shofar in the month of Elul was instituted only for a congregation, and therefore one who prays privately does not need to blow after his prayers."[209]

In addition, a number of other Poskim have discussed this issue. Some of them arrived at the conclusion that it is only a congregational obligation, whilst others believed that it is also an individual obligation. The origin of the dispute is in the way to interpret the words of the Magen Avraham namely: "In his house he is permitted to learn how to blow [on erev Rosh Hashanah]."[210]

Rabbi Tzvi Pesach Frank thinks that blowing the shofar in the month of Elul is an obligation on an individual since the Magen Avraham permits blowing in one's house on erev Rosh Hashanah in order to learn how to blow;[211] from this wording it is not necessary *to stop* blowing in one's house on erev Rosh Hashanah, namely during the entire month of Elul he has been blowing in his house. From this one can see that also an individual is included in the regulation to blow during the month of Elul.

However, Rabbi Eliezer Yehuda Waldenberg (20th century) in his book "Tzitz Eliezer" rejects the words of Rabbi Frank, and believes that this was not the intention of the "Magen Avraham" because he uses the word "to learn" that *is only to learn* to blow the shofar in the month of Elul. Therefore, blowing the shofar in the month of Elul is not an obligation for an individual, but only for a congregation, adding: "We have never heard that even the most meticulous person will be strict in the month of Elul when as result of necessity he has to pray privately, that a shofar be brought to blow for him in order to keep the decree to blow in the month of Elul."[212]

The "Nitei Gavriel" at first held that an individual does not need to blow the shofar during Elul, but later changed his opinion and ruled: "One who prays privately during the days of Elul should also blow the shofar."[213] On the other hand, Rabbi Yosef Shalom Elyashiv (20th-21st centuries) believed that it was not necessary for him to blow."[214]

In one of his responsa, Rabbi Ephraim Greenblatt (20th-21st centuries) in his book "Rivevot Ephraim" discusses the subject, and brought a number of Poskim whose opinion on this subject that he personally accepted. One of them was his father, Rabbi Avraham Baruch, who wrote to him: "It seems that this is only for a congregation but if an individual wants to blow, it seems that one does not prevent him from doing so." Also, Rabbi Yitzchak Bernstein, the Principal of the Knesset Hezekiah Yeshiva in Kfar Hasidim said: "According to all the reasons for this custom one should not make a distinction between an individual and a congregation." On the other hand, Rabbi Greenblatt brings the opinions of Rabbi Moshe Sachshevsky and Rabbi Yaakov Yitzchak Ruderman, the Principal of the "Ner Yisrael" Yeshiva in Baltimore, USA, who held that blowing in the month of Elul was only for a congregation and not for an individual.[215]

A Distortion in the Sound of the Shofar

In order to observe the mitzvah of Shofar on Rosh Hashanah, one must hear the sound of the shofar without any distortion in the sound. On this the Shulchan Aruch writes: "If one sounds a shofar in a pit or in a cave those standing in the pit or the cave fulfill their obligation. Concerning those standing outside, if they hear the sound of the shofar they fulfil their obligation ... but those who hear the echo do not."[216]

The mitzvah is to hear the *actual sound* of the shofar, and there is a responsum by Rabbi Avraham Yitzchak Hakohen Kook regarding the observance of the mitzvah of blowing the shofar on Rosh Hashanah in a situation where a person covers his ears with their tallit during the blowing. Does such a person observe the mitzvah of blowing the shofar? After an in-depth discussion, he writes: "There is no apprehension about it ... other than being diligent in its observance; I am also accustomed to removing anything that covers my ears."[217]

One can also investigate the situation regarding a person who covers his ears with a tallit when hearing the blowing of the shofar during the month of Elul. Because even on Rosh Hashanah not covering one's ears is only for one who is diligent in the mitzvah, it seems that in the month of Elul there is no problem.

Nowadays, there are additional questions on the subject, because today there are different types of electronic devices for hearing voices, and the Poskim have written many responsa to discuss whether it is permissible to hear the sound of the shofar through them on Rosh Hashanah. If they are allowed on Rosh Hashanah, surely they are allowed in the month of Elul. On the other hand, if they are forbidden on Rosh Hashana, it needs to be ascertained whether they are allowed in the month of Elul.

One of the machines for hearing is the telephone, and the question arises as to whether the mitzvah is observed if the sound of the shofar is heard on Rosh Hashanah through the telephone. (Because dialing on the telephone is forbidden on Yom Tov (Rosh Hashanah), a non-Jew should dial on the telephone.[218])

Rabbi Neta Shlomo Schlissel, a Rabbi in Hungary in the twentieth century, in his book "Neta Shaashuim" wrote a long responsum on the subject and ruled that in an exceptional case like a person who was in prison or in a place where under no circumstances can a shofar be obtained, then "there is room to allow a non-Jew to dial on the telephone on Rosh Hashanah, and to permit the blowing of the shofar."[219] He explains that since one can hear the person speaking, there is no break [in the transfer of the sound] and one is actually hearing the voice of the speaker even from a long distance away such as hundreds of kilometres at exactly the same instant that he speaks and there is a direct connection between the person hearing and the person speaking even if the words from the voice of the speaker is not always clear, but he adds that in contrast, the sound of the shofar and trumpets and other musical instruments is always very clear.[220]

However, the "Minchat Elazar" (who lived in the 19th-20th centuries in Munkács in Czechoslovakia) did not agree and wrote: However in the case of the shofar in which I was asked if one can fulfil the mitzvah via the telephone, it is obvious in my opinion that one cannot fulfil the mitzvah, since it is not preferable than blowing inside a pit or inside a well for listeners who are standing on the outside.[221]

Rabbi Tzvi Reisman of Los Angeles USA (21st Century) in his book "Ratz Hatzvi" suggested that the reason for the different opinions between the "Minchat Elazar" and the "Neta Shaashuim" is that at the period that the "Minchat Elazar" wrote his responsum, the degree of technology "was not so advanced and one heard on the telephone the sound of an echo together with the sound of the shofar. However, in our days, since the technology has improved tenfold, one is able to hear that the sound of a tekia or a shevarim or a terua via the telephone as a pure sound with no other sounds mixed in with it.[222]

Rabbi Reisman also gives his opinion that the hearing the sound of the shofar via a telephone during the month of Elul would be permitted even according to those who do not allow it on Rosh Hashanah.[223]

A question to be asked is "does a person wearing a hearing aid have to remove it [on Rosh Hashanah] in order not to hear the sound of the Shofar via the hearing aid." Rabbi David Chai HaCohen (21st Century) replied: "There is no obligation to lower the device, because it only increases the sound. However, since the sound of the shofar is loud and one can assume that he will be to hear the sound of the Shofar even without the hearing aid, then it could be removed as an act of piety."[224] On the other hand, Rabbi Shlomo Zalman Auerbach ruled: "One who hears the sound of the shofar via a hearing aid, does not fulfill his obligation since by hearing via a hearing aid, one hears only the vibrations of the aid and not the sound waves of the shofar.[225] It seems that because there are differences of opinion on this subject, in the month of Elul it is not necessary to remove one's hearing aid.

There are other devices for amplifying sounds like microphones, load-speakers etc. Because it is forbidden to use these instruments on Rosh Hashanah, this is probably the reason why the Poskim do not discuss them in connection with the blowing of the shofar. Since the reason for blowing the shofar in the month of Elul is to stir up people to repentance, it seems that it does not matter if you hear the blowing of the shofar with some distortions in the sound. According to this reasoning, it should be possible to hear the shofar via various methods of recordings during the month of Elul.

Blessings over Mitzvot and the Blessing Shehecheyanyu

There are mitzvot from the Torah such as blowing the shofar on Rosh Hashanah, eating matzah on the first night of Pesach, dwelling in a sukkah; a berachah (blessing) is recited before performing such mitzvot. In addition, there are mitzvot instituted by the Rabbis, such as washing the hands before eating bread, lighting of Shabbat candles, and recitation of Hallel, over which one likewise recites the berachah over performing these mitzvot. Furthermore, there are actions which one performs as a custom, and the question is does one recite a berachah over these customs? The answer is that on some of them one does, examples being lighting of Chanukah candles in the Synagogue,[226] and the recitation of hallel on Rosh Chodesh (according to the custom of the Ashkenazim and some of the Sefaradim).[227] However, there are also those customs in which one does not say a berachah, an example being beating the willows on Hoshana Raba.[228]

On blowing the shofar during the month of Elul, the Poskim give several reasons why one does not recite a berachah. The "Imrei Emet" (father of Yechiel Michel Goldschlag) writes that although it is a Rabbinical decree to blow the shofar during the month of Elul, nevertheless one does not recite a berachah, perhaps because the blowing is to remind a person to do repentance, and just as one does not recite a berachah on afflicting oneself or sounding the shofar because of a distress, even though they are commands from the Torah, one does not do so in this case.[229]

A further reason can be found in a book written by Rabbi Moshe Sternbuch (20th-21st centuries) who writes: "I have heard it questioned that since one recites a berachah on a custom such as hallel on Rosh Chodesh, why does not one say a berachah on blowing the shofar in the month of Elul, and in particular on Rosh Chodesh?" He answers that it seems that just as one does not say a berachah when sitting in the sukkah on Shemini Atzeret, the reason being that there should be recognition (that the seven-day festival of sukkot has finished) and thus not to transgress the prohibition of adding a mitzvah. Likewise here, one does not recite a berachah so that there should be a recognition that one is only performing a custom and not a mitzvah.[230]

There is another reason which was given by the Rebbe Avraham Bornsztain of Sochatchov who thought that because the blowing in the month of Elul is only for preparation for the shofar mitzvah on Rosh Hashanah, these blowings "are meant to prepare one's heart to hear the sound of the shofar on Rosh Hashanah; one does not recite a berachah on the preparations for mitzvot, but one relies on the berachah recited on Rosh Hashanah."[231]

A question which can be asked is if it were the custom to recite a berachah, whether it would be recited every day during Elul, or would it be only for the first time one blew in Elul. Since it is the custom to recite a berachah on the lighting of candles in the synagogue every evening of Chanukah, one should logically then recite the berachah on blowing the shofar every day in the month of Elul. Another question is what will be the wording of the berachah in the month of Elul "to hear the sound of the shofar" or "to blow the shofar"? Because there is no mitzvah to hear the sound of a shofar in the month of Elul and it is only a custom, the blessing will be "to blow the shofar." There is no need to add that it is only theoretical because one does not recite a berachah at all!

Another question is does one recite the berachah shehecheyanu on blowing the shofar in the month of Elul, and if so, does one repeat the berachah on Rosh Hashanah? It is written in the Gemara concerning a person building a sukkah: "He who makes a sukkah for himself [meaning before the holiday] he recites the berachah shehecheyanu."[232] But in practice according of the opinion of the Shulchan Aruch: "We rely on the shehecheyanu that is recited over the cup (of wine) during kiddush."[233] There is also the same subject and conclusion with regards to the binding of the lulav before the festival of sukkot. The reason to recite the berachah shehecheyanu before sukkot[234] is that these actions are preparations for the festival of sukkot.

The question is, is it possible to use this reason in order to say the berachah shehecheyanu for the blowing the shofar in the month of Elul? According to the reason that blowing the shofar in Elul is to awakening oneself for repentance, and in accordance to some of the reasons for blowing the shofar given by Rav Saadia Gaon (9th-10th centuries), especially his second, seventh and eighth reasons,[235] whose content is awakening one to repentance, there is room for discussion on whether to recite the berachah shehecheyanu in Elul. In addition, there is the reason that the purpose of blowing in Elul is to learn how to blow on Rosh Hashanah, and therefore the blowing is a preparation for this mitzvah.

In all mitzvot connected with the Festivals, one only recites the berachah shehecheyanu only once in a particular year, and this includes the blowing of the shofar (according to the Sefaradim who only say shehecheyanu on the first day of Rosh Hashanah). Therefore, should one recite the berachah shehecheyanu in the month of Elul, then one would not recite it again on Rosh Hashanah. However, all this is sophistry and in practice one does not recite the berachah shehecheyanu in the month of Elul for the shofar blowing!

Pausing during the Service to Hear the Shofar Blowing

We will study a situation where a person is praying shacharit in the synagogue, and he prays at a slow pace, or he arrives late at the synagogue, and as a result he is in the middle of shacharit when the shofar blowing is reached. The question is, is it permissible, or even obligatory, for this person to stop his praying and listen to the shofar blowing, or, in a situation where he is the person blowing the shofar, is it permitted for him to blow?

In fact, this question does not only relate to blowing the shofar, but is related to all other things, such as answering "amen" over berachot, or answering on borechu, kaddish, kedushah, etc. (these are prayers during a service when the congregation have to respond). There are many details in the halacha that are related to what stage of the service he has reached, or on the nature of his required response.[236] There are siddurim with "Table of a summary of the Laws of Interruptions during Prayers."[237]

In places where all speech is forbidden, it is sometimes permissible to do certain things without saying even a word, for example, between the words "Go'al Yisrael" (which is the conclusion of the berachah before the amidah) and the amidah, it is permissible to put on tefillin without saying the berachot over them.[238] Another example is, if the person leading the service comes (for example) to the kedushah, while a certain worshipper has not yet finished his amidah, the worshipper should stop, be silent and concentrate on to what the person conducting the service is saying.[239]

The Poskim discuss the situation when a person, who in the month of Elul is in the middle of the amidah, whether he is permitted to stop praying and listen to the blowing of the shofar? There is a similar situation in the mussaf service of Rosh Hashanah in synagogues who blow the shofar during the silent amidah, and many of the worshipers are at a different place in their prayers while the shofar is being blown. The answer is that these worshipers may stop praying in order to listen to the shofar blowing.[240]

Some Poskim explicitly discuss this situation in the month of Elul. Rabbi Chaim Kanievsky answered the question of whether it is permissible to stop during the amidah in order to hear the blowing of the shofar in Elul and answered "it is possible."[241] Also, the "Shevet Halevi" (Rabbi Shmuel Halevi Wosner, 20th-21st centuries) wrote on this subject: "If he is unable to hear the shofar later on, it is permissible to stop in order to listen to it, since it is an ancient instituted custom ... and therefore it is an interruption for the sake of a mitzvah. However, if it is possible to finish the berachah that he is then reciting, it is better for him to wait silently after his has finished that berachah."[242]

On the other hand, Rabbi Menashe Klein rules: "He who is in the middle of the amidah whilst the shofar is being blown, even though he will not be able to hear the blowing later on in the day, he does not have to stop in the middle of the amidah in order to hear the shofar."[243]

Rabbi Ephraim Greenblatt was apprehensive that if the blowing of the shofar would interfere with a person's concentration whilst saying the amidah, perhaps he should stop and listen to the shofar.[244] He also brings the different opinion of Rabbi Chanoch Henich Karelenstein who writes that if the blowing "disturbs his concentration on the amidah, it appears obvious that there is no need to stop."[245]

In a similar answer, the "Nitei Gavriel" wrote explicitly: "If one is in the middle of the amidah [during Elul], it is obvious that one should not stop to listen to the shofar."[246] Furthermore, the Maharam from Brisk (Rabbi Mordechai Brisk, 20th century) wrote: "An individual does not stop in the middle of the amidah even just to listen."[247]

It follows from this that during the month of Elul should the service in the synagogue reach the shofar blowing whilst an individual is in the middle of the amidah, there are differences of opinion whether he should stop and listen to the blowing of the shofar. If we hold that the answer is positive, then how much more so it will be permitted to pause during all other prayers. However, if the answer is negative during the month of Elul, then, further study is required to determine the halacha regarding other prayers.

What is the halacha regarding the person who blows the shofar, but is in the middle of shacharit in the month of Elul. Is he allowed to act as the person who blows the shofar?

Rabbi Avraham David Wahrman of Buchach in Ukraine (18th-19th centuries) writes in his book "Eshel Avraham" about this that it appears that it is permitted for him to blow if he is in the middle of the 'pesukei d'zimra" (preliminary Psalms) and perhaps even if he is in the 'bircat kriat shema' (blessing before and after reciting the shema), because this is not regarded as a pause but only his breathing without any sound coming out his mouth, and the sound which is heard is only from the shofar.[248] From this one can see that one allows this, but will some difficulty, and it is obvious that this does not include the amidah. In addition, the "Nitei Gavriel" although he permits this during "bircat kriat shema," this is limited to between the various chapters, and in addition, that there is no other person to blow the shofar except him.[249]

Even though it is forbidden for a person to talk about other things whilst he is praying the amidah, there are certain actions he can do. An example is that if a person is saying the amidah and he not sure about a certain halacha, such as he omitted to say something in the amidah, he is permitted to walk to another location to verify the halacha from a book.[250]

It follows from this, that in the case where the person blowing the shofar is in the middle of the amidah, and the shofar is out of reach, and the worshipers do not know where the shofar is, the blower may go to where the shofar is in order to blow.[251]

Conclusion: Blowing the shofar will also be heard before the coming of the Moshiach, and we pray that we will reach this period speedily

References

Abbreviations
SA = Shulchan Aruch
OC = Orach Chaim
MB = Mishnah Berurah
(1)   Pirkei Rabi Eliezer, Rabbi Eliezer Hagadol the son of Hyrcanus, (Venice, 5368), chap.46'
(2)   Talmudic Encyclopedia, vol.2, (Talmudic Encyclopedia Publishing, Jerusalem, 5716), p.2
(3)   Rabbi Maharam ben Baruch Rothenburg, Sefer Minhagin (Yisrael Elfenbein, published by the "Bet Hamidrash Harabanim b'America", New York USA, 5698), Seder Rinos l'Erev Rosh Hashanah. It should be

noted that during the period of the Geonim, including Rav Saadia Gaon
and Rav Amram Gaon, the blowing of the shofar during the month of Elul
is not mentioned (Rabbi Shemtob Gaguine, Keter Shem Tob, (London,
5716), p.10 par.2. The Avudraham (Rabbenu David ben Yoseph
Avudraham), (Warsaw, 5638), writes about Moshe going up Mount Sinai
in the month of Elul, but does not mention the blowing of the shofar. Also,
the Rambam does not mention the blowing of the shofar in the month of
Elul, and it is surprising that the Rabbinical literature does not indicate
that the Rambam did not mention this fact.

(4)   Pirkei Rabi Eliezer, op. cit., chap.46
(5)   Babylonian Talmud, Rosh Hashanah 19b; Beitzah 6a. However, it seems
       that there were a few cases when there were 30 days in Elul (Mishnah
       masechet Sheviit chap.10, mishnah 2; masechet Eruvin chap.3, mishnah 7)
(6)   e.g. Magen Avraham, SA OC chap.581 par.2
(7)   Rabbi Shlomo ben Yechiel Luria (Maharshal), sefer Chochmat Shlomo,
       (Krakow, 5342), masechet Shabbat chap.9 p.89a; Rabbi Yaakov ben
       Moshe Levi Moelin, sefer Maharil, (Frankfurt on the Main, 5448),
       Hilchot Yamim Noraim; Magen Avraham SA OC chap.581 par.2
(8)   Rabbi Yoel Sirkis (the Bach), commentary named Bayit Chadash (Bach)
       on the Tur OC chap.58, first words: tanya b'pirkei
(9)   Rabbi Moshe ben Nachman, the Ramban, commentary on the Torah,
       Shemot, chap.18 verse 13
(10)   Rashi on Shemot chap 34, verse 2; in addition, there is a translation to the
       word "nachon" in Targum Onkelos and Targum Yonatan ben Uziel
(11)   Bach, op. cit., OC chap.581, first words: uma shecatav Rashi
(12)   Bach, op. cit., OC chap.581, first words: tanya b'pirkei
(13)   Rabbi Eliyahu Mizrahi, Elya Mizrahi al peirush Rashi, (Krakow, 5355),
       parashat Vehoya Eikev, first word: vehitnapal. In addition, Rabbi
       Mordechai ben Avraham Yaffe (haLevush) writes that the going up of
       Moshe on Mount Sinai occurred on 29 Av, (Berdychiv, 5578), Hilchot
       Rosh Hashanah, chap.581 par.1
(14)   Bach, op. cit., chap.581, first words: tanya b'pirkei
(15)   Rabbi Yosef ben Meir Teomim (Pri Megadim), OC Eshel Avraham,
       (Warsaw 6549), Hilchot Rosh Hashanah chap.581 par.2,
(16)   Magen Avraham, SA OC chap.581, par.2
(17)   Tosafot on Masechet Bava Kama 82a, first word: kedai
(18)   Masechet Bava Kama 82a, comments of Bach par.4
(19)   Midrash Tanchuma, (Amsterdam, 5493), Parashat Ki Tisa p.33b
(20)   Ramban on the Torah, Parashat Yitro, Shemot chap.18 verse 13
(21)   Rabbi Yehuda ben Eliezer, Daat Zkenim, (Ofen, 5594), Sefer Deverim
       Parashat Eikev chap.10 verse 10
(22)   Rabbi Binyamin Aharon Slonik (Solnik), Mas'at Binyamin. (Metz, 5536),
       responsum 2
(23)   Siddur Otzar Hatephilot, Sefard, (Ream: Vilna, 5688), Peirush Derech
       Hachaim, dinei Chodesh Elul v'Erev Rosh Hashanah, par.1, p.513; also,
       Rabbi Eliyahu ben Binyamin Wolf Spira (Shapira), Elia Raba, (Sulzbach,

5517), OC chap.581 par.3 Levush; and also, Rabbi Moshe Feinstein, Igrot Moshe, OC vol.4, (Bnei Brak, 5742), chap.21 par.5, are of this opinion

(24) Rabbi Moshe ben Avraham of Przemysl, Mateh Moshe, (Warsaw, 5636), vol.5, Inyan Rosh Chodesh Elul, par.778

(25) Kitzur Shulchan Aruch, chap.128, par.2

(26) Aruch Hashulchan, OC chap.581 par.1

(27) Magen Avraham, SA OC chap.581, par.2

(28) Luach l'Eretz Yisrael for the year 5781 (Harav Tucazinski), Jerusalem, Elul p.83

(29) Luach Dvar B'ito – Each Thing in its Proper Time, (Achiezer: Bnei Brak), 1 Elul, p.1015

(30) Luach Laws and Customs "Dvar Yom Beyomo" 5749 (Kehal Machzikei Hadas in Eretz Yisrael, Hasidei Belz), month of Elul, p.301

(31) Luach Laws and Customs 5750, ("Heichal Shlomo: Ihud Batei Knesset in Israel, Jerusalem), Elul. p.90

(32) Lechu Lachmo b'Lachami, Luach Limud Torah Yomi leHachayal haShem, 5763, (Yeshivat Shavei Hevron), second day of Rosh Chodesh Elul

(33) Luach Hasidei Belz, op. cit., chodesh Elul, p.301

(34) Rabbi Avraham Danzig, Chayei Adam, Rule 138 par.1. MB quotes from the words of the Chayei Adam, chap.591 par.3

(35) Rabbi Ephraim Zalman Margoliot, Mateh Ephraim, (Warsaw, 5680), chap.581 par.7

(36) Rabbi Yehuda Ashkenazi, Ba'er Heteiv, chap.581 par.3 (this is to be found in the editions of the MB)

(37) Rabbi Shneur Zalman of Liadi, Shulchan Aruch Harav, appendix to the customs for the month of Elul, p.68 (p.721)

(38) Rabbenu Yerucham ben Meshullam, Sefer Toldot Adam v'Chava. (Venice), Sefer Adam, section 6 part 1

(39) Rabbenu Simchah from Vitry, one of Rashi's pupils, Machzor Vitry, (Shimon Halevi Ish Horowitz, Nurenberg, 5683), chap.323

(40) Rabbenu Asher ben Yehiel (the Rosh), Masechet Rosh Hashanah, chap.4

(41) Rabbi Eliezer ben Yoel Halevi, Sefer Raviah, vol.2, (Avigdor Aptowitzer, Hevrat Mekitze Nirdamim, Berlin, 5696), Hilchot Rosh Hashanah, chap.542

(42) Rabbi Menachem ben Aharon ibn Zerah, Tzeda laDerech, (Pirara, 5314), rule 5 in Hilchot Rosh Hashanah, chap 2

(43) Tur, OC chap.581

(44) Rabbi Shemtob Gaguine in his book Keter Shem Tob wrote (p.10) that the Tur added the words "and all the month," but in fact there were others who did this at an earlier date.

(45) Rosh, op. cit., Masechet Rosh Hashanah, chap.4

(46) Rabbi Eleazar of Garmiza, Harokeach, Hilchot Rosh Hashanah chap.207

(47) Rabbi Yosef Karo in his commentary called the Bet Yoseph on the Tur, OC chap 502 first words: umashecatav vekorin

(48) Rabbi Chaim Zvi Ehrenreich, commentary Ketze Hamateh on Mateh Ephraim, (Kisvarda Hungary, 5683) chap.581 par.18

(49) Rambam (Maimonides) Mishneh Torah, Hilchot Teshuvah, chap.3 halacha 4

(50) Rabbi Tzvi Hirsch Kaidanover, Kav Hayashar, vol. one, (Lemberg, 5623), chap.48; the same subject is brought in the book Mateh Moshe, op. cit., inyan Rosh Chodesh Elul v'Rosh Hashanah, chap.778

(51) Rambam, Mishneh Torah, Sefer Mada, Hilchot Teshuvah chap.2 halacha 6

(52) SA OC chap.602 par.1 and Rema

(53) SA OC chap.603 par.1 and Rema

(54) Tur OC chap 581

(55) In all books of Ashkenazi selichot

(56) Rabbi Zedekiah ben Avraham Anav Harofe, Shibbolei Haleket, (Venice, 5306), Rule 12 Hilchot Chag haSukkot, chap.121

(57) Zohar, Parashat Tzav p.31b

(58) SA OC chap.664 par.1, Rema; MB chap.664, par.9

(59) Luach Dvar B'ito, op. cit., 5758, Hoshana Raba, Shacharit p.301

(60) Seder Tephillot Hashanah l'Minhag Rumania, (Venice), tephilot shel Sukkot, first word: baboker

(61) Rabbi Aharon Shmuel Kaidanover, Peirush Tiferet Shemuel al Piskei haRosh, masechet Berachot, chap.1 par.40

(62) Machzor l'Sukkot v'Shemini Atzeret v'Simchat Torah k'minhag Bnei Roma, second edition, (Chevrat Yehudei Italia b'Yisrael: Jerusalem, 5775), amidah l'mussaf shel Sukkot, p.138; Rabbi Mordechai ben Avraham Yaffe, Levush Ohr Yehorat (Lvov, 5641), parashat Shelach; Rabbi Menachem Recanti, peirush al haTorah al Derech Haemet, (Venice, 5283). Shelach Lecha; Rabbi Betzalel Landau, Hoshana Raba, (daat, Michlelet Herzog, 5723)

(63) Machzor Lesukkot … Bnei Roma, op. cit., amidah l'mussaf shel Sukkot p.138

(64) Jerusalem Talmud, masechet Rosh Hashanah, chap.4 halacha 8

(65) Korban haEdah on Jerusalem Talmud, first words: veoti yom yom yidroshun

(66) Rabbi Yaakov Chaim Sofer, Kaf haChaim, OC seder yom Hoshana Raba, chap.664 par.1

(67) Luach Dvar B'ito, 5758, op. cit., Hoshana Raba, p.302

(68) Rabbi Moshe ben Machir, Seder Hayom, (Venice, 5365), seder hahakafot... p.94b

(69) Rabbi Meir Horowitz, Imrei Noam, vol.2, (Krakow, 5646), responsum 1

(70) What is the source for blowing the shofar during Hoshannot, October 2015, mi yodea, (Internet)

(71) Luach Dvar B'ito, 5758, op. cit. Hoshana Raba, pp.299-300

(72) Rabbi Yoseph Chaim, Ben Ish Chai, first year, parashat Nitzovim, introduction, first words: uma shetokin

(73) Sefer Tehillim (Psalms), chap.81 verse 4

(74) Rabbi Yitzchak Isaac Tyrnau, Sefer haMinhagim, (Warsaw, 5629), Minhag shel Rosh Chodesh Elul 72 v'Erev Rosh Hashanah; Mateh Moshe, op. cit., Inyan Rosh Chodesh Elul v'Rosh Hashanah, chap.778

(75) Aruch Hashulchan, OC chap.581 par.1

(76) SA OC chap.588 par.5

(77) Babylonian Talmud, Rosh Hashanah 29b; MB chap.588 par.13

(78) Kaf Hachaim, op. cit., chap.583 par.31; MB chap.583 par.8; Ben Ish Chai, op. cit., Parashat Nitzovim par.12; Rabbi Chaim Yoseph David Azulai, Birkei Yoseph, (Jerusalem, 5729), OC chap.583 par.6

(79) Babylonian Talmud, Rosh Hashanah 29b

(80) Rambam, Mishneh Torah, seder Zemanim, Hilchot Shabbat, chap.23 halacha 4

(81) Ibid.

(82) Sheal et Harav – Rabbi David Lau, keli negina b'Yom Tov, (city Modi'in), Moreshet Shut, Elul 5765, (Internet)

(83) SA OC chap.596 par.1 Rema. There are those who dispute this: Rabbi Yoseph Yuzpa Ostreicher (who lived in the fifteenth century) who wrote in his book Leket Yosher, (Berlin, 5663), p.125, the rulings of his Rabbi, the Terumat Hadeshen, and brings the case of blowing the shofar on Rosh Hashanah but not to fulfill the mitzva. He writes: "He who is not proficient in blowing the shofar is permitted to blow on the morning of Rosh Hashanah in order to learn how to blow."

(84) Levush haHod, op. cit., Hilchot Erev Rosh Hashanah, chap.581 par.1; Rabbi Yitzchak Isaac Tyrnau, Sefer haMinhagim, op. cit., Minhag shel Rosh Chodesh Elul

(85) Sefer Tehillim (Psalms), chap.81 verse 4

(86) Minhagim shel haMaharam m'Rothenburg, op. cit., Seder Rinos l'Erev Rosh Hashanah

(87) Levush haHod, op. cit., Hilchot Erev Rosh Hashanah, chap 581 par.1; Hagahot Harav Avraham Klausner (MaHaRIK) al sefer Maharil, op. cit., Hilchot yamim noraim)

(88) MB chap.581 par.24

(89) Ibid.

(90) Sha'ar Hatziyun on MB, chap.581 par.35

(91) The Authorised Daily Prayer Book, translated by Rev. S. Singer, published under the sanction of the Chief Rabbi Dr. Nathan Marcus Adler: for example, editions of 1929 p.85, and 1962 p.90. However, in earlier editions, for example 1904 p.85 the blowing of the shofar during the month of Elul is not mentioned.

(92) The Authorised Daily Prayer Book, with commentary and notes by the Chief Rabbi Dr. J. H. Hertz, edition from year 1976, p.232

(93) Seder Avodat Yisrael, published by Yitzchak ben Aryeh Yoseph Dov, also known as Seligman Baer, (Rodelheim: Frankfurt on the Main, 5661), p.383. The first publication of this Siddur was in 5628.

(94) Minhagim shel haMaharam m'Rothenburg, op. cit., Seder Rinos l'Erev Rosh Hashanah; Rabbi Binyamin Halevi, Mahzor Maglei Tzedek, (Sevonto, 5317-5320), Elul

(95) (Editor) Rabbi Avigdor Berger, collection of Torah articles Zechor l'Avraham, (Bet Hamidrash Yeshivat Eliyahu, branch of Machon Yerushalayim, Holon, 5750), on SA OC chap.581 par.3

(96) Yisrael Ben Gompil, Seder Minhagim of our Community (Piorda – Furth), (Piorda, 5527), par.79

(97) Rabbi Natan Halevi Bamberger, Likutai Halevi colel Minhagei Wurzburg, (Berlin, 5667), p.26 and footnote 84

(98) Elia Raba, op. cit., OC chap.581 par. 4 Levush; it is also brought by MB chap.581 par.24, and there, in place of Sefer Amacral, it is written Sefer Amarcal

(99) Siddur Bet Yaakov (Emden) Minhag Ashkenaz, part one, (Warsaw, 5641), commentary Shaar Shalechet, Shaar Hadelek, par.3, p.76

(100) Babylonian Talmud: Shabbat 117b, Shabbat 131b, Rosh Hashanah 29b

(101) Minhagim shel haMaharam shel Rothenburg, op. cit., Seder Rinos l'Erev Rosh Hashanah

(102) Babylonian Talmud, Pesachim 6a

(103) Rabbi Chaim Eliezer Ashkenazi ben Or Zarua, Teshuvot Maharach Or Zarua, (M. Avitan: Jerusalem, 5762) chap.33, dinei Rosh Hashanah, p.62

(104) Rabbi Avraham Yitzchak Hakohen Kook, Mitzvat Raayah, (published by Mosad Harav Kook: Jerusalem, 5730), OC chap.581

(105) Rabbi Chaim Bleier, Likutei Upiskei Halachot "Chukei Chaim", Parashat Shoftim 5780, no.56, dinei Elul par.3

(106) Rabbi Avigdor Unna, Minhagei Aidot Yisrael, m'Minhagei Yahadut Ashkenaz b'Eretz Yisrael, (daat, Michlelet Herzog), chap.3 chagim umoadim. Chodesh Elul par.1

(107) Rabbi Betzalel Landau, Minhagei Aidot Yisrael, m'Minhagei Yahadut Ashkenaz (Perushim) b'Eretz Yisrael, (daat, Michlelet Herzog), chap.3, chagim umoadim. Chodesh Elul par.1

(108) Luach Dvar B'ito, 5758, op. cit., 1 Elul, p.1015; Tekiyot shel Elul, Pinat hahalacha uminhag, (Chabad Eretz Yisrael)

(109) Rabbi Tuvia Blau, Minhagei Aidot Yisrael, m'Minhagei haChasidim, (daat, Michlelet Herzog), chap.3, Chodesh Elul par.2; Tekiyot shel Elul, Pinat hahalacha uminhag, op. cit., (Internet)

(110) Luach Hasidei Belz, op. cit., 5749, Chodesh Elul, p.301

(111) Al Mehut Chodesh Elul, Breslov, 26 Av 5779, (Internet)

(112) Rabbi Gavriel Zinner, Nitei Gavriel, Hilchot Rosh Hashanah, (Jerusalem, Elul 5761), chap.4 footnote 8

(113) Ibid.

(114) Ibid.

(115) Ibid.

(116) Rabbi Shammai K'hat Hakohen Gross, Shevet Hakahati part 1, (Jerusalem 5747), Hilchot Rosh Hashanah, chap.185; Rabbi Chaim Elazar Shapira,

Darkei Chaim V'Shalom, (Jerusalem, 5734), Chodesh Elul, chap.690. (It was first published in Munkatch in the year 5700.)

(117) Rabbi Zalman ben Aharon Geiger, Divrei Kehilot – Minhagei Tephilot Frankfurt on the Main, (5622), fourth booklet day 66… first day of Rosh Chodesh Elul, par.1

(118) Shulchan Aruch Harav, appendix to the customs for Chodesh Elul, p.68 (p.721)

(119) Luach Dvar B'ito, op. cit., 5758, 1 Elul, p.1015

(120) Ibid.

(121) Likutei Upiskei Halachot "Chukei Chaim", op. cit., Parashat Shoftim 5780, no.56, dinei Elul, par.3

(122) Ibid.

(123) Luach Dvar B'ito, op. cit., 5758, 1 Elul, p.1015

(124) Bach on the Tur, OC chap.592, first words: uma shekatav dechol

(125) Darkei Chaim v'Shalom, op. cit., Chodesh Elul, chap.690

(126) Shevet Hakahati part 1, op. cit., Hilchot Rosh Hashanah, chap.185

(127) Question submitted to Machon Halacha Chabad, (Internet)

(128) SA OC chap 590 par.4 and Rema

(129) SA OC chap.585 par.3

(130 Shulchan Aruch Harav, OC chap.585 par.8

(131) Mishpacha English edition issue 791, 25 December 2019 "Swiss Account," p.145

(132) Information from Rabbi Eliyahu Ben Pinchas., Kislev 5781, Kiryat Arba

(133) Kaddish "tayanu v'tayatru" is a special kaddish recited according to the Sephardi and Taimoni rites after the selichot

(134) Rabbi Mordechai Eliyahu, Hilchot Chodesh Elul, 5763, (Yeshivah website, Bet Hamidrash), (Internet)

(135) Kaf Hachaim, op. cit., OC chap.581 par.13

(136) Keter Shem Tob, op. cit., Minhagim l'Chodesh Elul, p.10

(137) Yalkut Minhagim, Rabbi Yaakov Elazar, Minhagim of the Sefaradim of Jerusalem, (Ministry of Education and Culture, Religious Education Department and Department for the Integration of Jewish Heritage:Jerusalem, 5749), chap.3, chagim umoadim, chodesh Elul, par.5

(138) Keter Shem Tob, op. cit., Minhagim l'Chodesh Elul, p.10

(139) Ibid., p.11, first words: v'ata noda

(140) Rabbi Raphael Moshe Deluya, Zocher Brit Avot, (Har Bracha, Eretz Yisrael, 5770), Chodesh Elul v'Aseret Yemai Teshuvah, par.1 and footnote 295

(141) Rabbi Maor Kayam, Harchavat Peninei Halacha, chap.1: Tekiat Shofar b'Chodesh Elul, par.7: Custom of the Sefaradim, (Internet)

(142) Rabbi Kalfon Moshe Hakohen, Berit Kehuna, (Djerba, 5701), OC part 1, maarechet letter "samech", par.6

(143) Rabbi Yitzchak Amadi, Minhagei Aidot Yisrael, m'Minhagei Yahadut Kurdistan (daat, Michlelet Herzog), chap.3 chagim umoadim, Chodesh Elul par.6; Rabbi Maor Kayam, op. cit.

(144) Shtilei Zeitim, Rabbi David Mashraki (Mizrahi), OC, Hilchot Rosh Hashanah, chap.581
(145) Rabbi Maor Kayam, op. cit.
(146) Rabbi Yoseph Kapach (Qafih), Halichot Teiman, (Machon Ben-Zvi for the study of Jewish Communities in the East: Jerusalem, 5762), Yamim Noraim, p.30
(147) SA OC chap 581 par.1 Rema; this is also brought in Mahzor Maglei Tzedek, op. cit., Elul
(148) MB chap.581 par.2
(149) Rabbi Binyamin Beinish Hakohen, Shem Tov Katan, (Sziget, 5651), Hanhagot v'dinim u'tephilot kedoshot v'noraot m'Rosh Chodesh Elul ..., p.10. (The first edition was published in Sulzbach in the year 5466)
(150) Rosh, op. cit., Rosh Hashanah, chap.4, towards the end of the chapter
(151) Tur OC chap.581. This is also mentioned by Rabbenu Yerucham, op. cit., sefer Adam, section 6 part 1
(152) Chayei Adam, op. cit., rule 138 par.1
(153) Rabbi Moshe Leib Tziltz, Mili D'Avot, (Bardejov Slovakia, 5685), OC part 5, par.20 part 2
(154) Igrot Moshe, op. cit., vol.4 chap.21 part 5
(155) Bintivot Hahalacha, 5774, (vehigito, Irgun Olami l'Chizuk Hahalacha: Jerusalem), Responsa of Rabbi Chaim Kanievsky, p.168
(156) Igrot Moshe, op. cit., vol.4 chap.21 part 5
(157) Bintivot Hahalacha, booklet 24, Elul 5771, op. cit., rulings of Rabbi Menashe Klein, Chief Rabbi of Ungvar Ukraine, par.3 p.38
(158) Igrot Moshe, op. cit., vol.4 chap.21 part 5
(159) Elia Raba, op. cit., OC chap.581, par.3 Levush
(160) Rabbi Yoseph Yuzpa ben Moshe Kashman, Nohag Katzon Yoseph, (Hanau, 5478), Hilchot chaluk haparshiot, parashat Re'ai
(161) Sefer Minhagim Piorda (Furth), op. cit., chap.79
(162) Nitei Gavriel, op. cit., chap.4, Tekiat Shofra b'chodesh Elul, end of footnote 2
(163) Nitei Gavriel, op. cit., chap.4, Tekiat Shofra b'chodesh Elul, par.15
(164) Shibbolei Haleket, op. cit., rule 11, Hilchot Yom Hakipurim, first word: Yerushalmi
(165) Mili D'Avot, op. cit., OC part 5, par.20 part 2
(166) Tur OC chap.624
(167) Bach on the Tur, OH chap.624, first words: uma shekatav v'tokin
(168) MB chap.623 par.12
(169) Luach l'Eretz Yisrael for the year 5781, op. cit., Chodesh Tishrei p.27
(170) Mateh Ephraim, op. cit., chap.623, par.7
(171) Aruch Hashulchan, OC chap.623, par.8
(172) Machzor Vitry, op. cit., chap.356, first words: v'shaliach tzibur, first word: shamati (173) Rabbi Shmuel ben Natan Neta Halevi Loew Kelin, Machazit haShekel, part 2, (Hrubieszow, 5578)   chap 623 at the end
(174) Mili D'Avot, op. cit., OC part 5, par.20 part 2; Elef Lamateh" to the book "Mate Ephraim, op. cit., chap.581, par.8

(175) SA OC chap.689, par.1
(176) SA OC chap.689, par.2
(177) Shevet Hakahati part 1, op. cit., Hilchot Rosh Hashanah, chap.185;
(178) Nitei Gavriel, op. cit., chap.4, Tekiat Shofar b'Chodesh Elul, par.8; Rabbi Yehuda Lev, Why does one not say a berachah on blowing the shofar during Elul? Is a minor allowed to blow? (Lev Hamoadim, halachot ta'amim uminhagim)
(179) MB OC chap.589 par.2
(180) Sha'ar Hatziyun on MB, chap.589, par.2
(181) SA OC chap.589, Dirshu par.5
(182) It is obvious that those who permit a minor, must allow any child who has reached the age of 13 to blow.
(183) SA OC chap.589 par.3
(184) SA OC chap.589 par.4
(185) An Evaluation of Capuchin Monkeys Trained to Help Severely Disabled Individuals, Journal of Rehabilitation Research and Development, vol.28, no.2, pp.91-96
(186) SA OC chap.159 par 12
(187) Rabbi Moshe Sofer, Chidushei Chatam Sofer al Masechet Gitin, Gitin 22b; Rabbi Yitzchak Zilberstein, Chashukei Chemed al Masechet Megillah, (Jerusalem, 5767), Megillah7a
(188) SA OC chap.409 par.8
(189) SA OC chap.586
(190) Rabbi Yom Tov ben Avraham Asevilli (the Ritva), Chidushei haRitva, (Jerusalem, 5727), Rosh Hashanah chap.3, p.26b
(191) Nitei Gavriel, op. cit., Hilchot Rosh Hashanah chap.4, Tekiat Shofar b'Chodesh Elul, par.13
(192) Shevet Hakahati, vol.1, op. cit., Hilchot Rosh Hashanah, chap.185
(193) SA OC chap.586 par.1
(194) MB chap 586 par.10
(195) SA OC chap.586 par.1 Rema
(196) MB chap.586 par.8
(197) SA OC chap 586 par.2
(198) MB chap.586 par.9
(199) Aruch Hashulchan OC chap.586 par.14
(200) SA OC chap 586 par.3-4
(201) Babylonian Talmud, Sukkah 31b
(202) SA OC chap.586 par.7-9
(203) SA OC chap.586 par.10
(204) Aruch Hashulchan OC chap.586 par 24
(205) SA OC chap.586, par.10
(206) SA OC chap.586 par.12
(207) MB chap.586 par.60
(208) Bintivot Hahalacha, 5774, op. cit., Rabbi Chaim Kanievsky, p.168
(209) Bintivot Hahalacha, 5771, op. cit., Rabbi Menashe Klein, p.38
(210) Magen Avraham, SA OC chap.581 par.14

(211) Brought in the book Tzitz Eliezer by Rabbi Eliezer Yehuda Waldenberg, (Jerusalem, 5736), vol.12 chap.48
(212) Tzitz Eliezer, op. cit., vol.12 chap.48
(213) Nitei Gavriel, op. cit., Hilchot Rosh Hashanah, chap.4 par.9 and footnote 14
(214) SA OC chap.581, Dirshu par.6
(215) Rabbi Ephraim Greenblatt, Rivevot Ephraim, vol.1, OC, (Memphis Tennessee, USA, 5735), chap.394
(216) SA OC chap.587 par.1
(217) Rabbi Avraham Yitzchak Hakohen Kook, Shut Orach Mishpat, (Mossad Harav Kook: Jerusalem, 5745), OC chap.140
(218) Rabbi Nate Shlomo Schlissel, Shut Neta Shaashuim, (Munkatch, 5685), chap.4, p.9
(219) Ibid.
(220) Ibid., pp.10, 14
(221) Rabbi Chaim Elazar Shapira, Minchat Elazar, vol.2 (Jerusalem, 5756), chap.72. This is also the ruling of Rabbi Netanel Hakohen Fried in his book Penei Mevin, (Brooklyn USA, 5731). OC chap.103.
(222) Rabbi Tzvi Reisman, Ratz Hatzvi, vol.2, (Los Angeles, USA, 5765), chap.10 par.2
(223) Ratz Hatzvi, op. cit., chap.10 par.11
(224) Rabbi David Chai HaCohen, the mitzvah of hearing the sound of the shofar for those hard of hearing, (Yeshiva, Sh'al et Harav, 21 Adar II 5774), (Internet)
(225) Rabbi Shlomo Zalman Auerbach, Rosh Hashanah – Shofar (Shut Halacha Actuali), (Internet); There is a similar responsum by Rabbi Aharon Boaron, Rosh Hashanah – Shofar (Shut Halacha Actuali), (Internet)
(226) SA OC chap.671 par.7
(227) SA OC chap.422 par.2
(228) SA OC chap.664 par.2
(229) Rabbi Yechiel Michel Goldschlag, Imrei Emet, (Pietrekov, 5681), likutim b'mitzvot takanot uminhagim, par.16
(230) Rabbi Moshe Sternbuch, Teshuvot Vehanhagot, vol.4, (Jerusalem, 5762), yamim noraim, chap.133
(231) Rabbi Yehuda Lev, Why does one not say a berachah on blowing the shofar during Elul? op. cit.
(232) Babylonian Talmud, Menachot 42a
(233) SA OC chap.641 par.1
(234) SA OC chap.651 par.6
(235) Encyclopedia Yehudit, Tekiat Shofar, Ten reasons for blowing the shofar – Rav Saadia Gaon, (daat, Michlelet Herzog), (Internet)
(236) SA OC chap.66; Chayei Adam, Hilchot tephila uberachot, rule 20
(237) e.g. Siddur Hashlem Kol-bo, (Eshkol Publishing: Jerusalem), at end of the siddur
(238) SA OC chap.66 par.8; Chayei Adam, Hilchot tephila uberachot, rule 20 par.7

(239) SA OC chap.104 par.7; Chayei Adam, Hilchot tephila uberachot, rule 25 par.10

(240) Rabbi Moshe Feinstein, Igrot Moshe, vol.1, (New York USA, 5719), OC chap.173

(241) Bintivot Hahalacha 5774, op. cit., Rabbi Chaim Kanievsky, p.168

(242) Shevet Hakahati, vol.3, op. cit., chap.183

(243) Bintivot Hahalacha 5771, op. cit., Rabbi Menashe Klein, p.38

(244) Rabbi Ephraim Greenblatt, Rivevot Ephraim, vol.2, (Memphis Tennessee, USA, 5738), chap.159

(245) Ibid.

(246) Nitei Gavriel, op. cit., Tekiat Shofar b'Chodesh Elul, chap.4 par.6 and footnote 11

(247) Rabbi Mordechai Brisk, Shut Maharam Brisk, vol.3, (Lovinger: Tasnad Rumania, 5699), chap.26

(248) Rabbi Avraham David Wahrman of Buchach in Ukraine, Eshel Avraham, second edition, (Buchach, 5666), Hilchot Birkot Hashachar, chap.51 par.4

(249) Nitei Gavriel, op. cit., Hilchot Rosh Hashanah chap.4 par.7

(250) Chayei Adam, Hilchot tephila uberachot, rule 25 par.9; MB chap.104 par.2

(251) This is according to the opinion that it is permitted for the person blowing to do so whilst he is in the middle of the amidah.

THE READING OF THE TORAH ON SHEMINI ATZERET (SIMCHAT TORAH) IN ERETZ YISRAEL

(Shemini Atzeret (Simchat Torah) = eighth day of Festival of Tabernacles)

<hr/>

Reading the Torah on the Festivals according to the Gemara

The Gemara in Masechet (tractate) Megillah[1] gives the readings of the Torah appertaining to the Festivals during the course of the year. In the Diaspora there are two days of Yom Tov for every Festival, and the Gemara also gives the reading of the Torah for the second day of the Festival. It is clear from the Gemara that in Eretz Yisrael where there is only one day of Yom Tov, one reads the parashiot designated for the *first* day of the Festival, and there is no suggestion by the Rishonim (great Rabbis who lived approximately between the 11th and 15th centuries) to do otherwise.

According to the above-mentioned Gemara, the portion of the Torah beginning "kol habechor" is read on the last (eighth) day of Pesach, on the second day of Shavuot and on Shemini Atzeret. A question which can be asked on this is why on the second day of Yom Tov in the Diaspora of Pesach and Shavuot, one reads "kol habechor," but on Shemini Atzeret one reads it on the first day of Yom Tov?

This question is answered by Rabbi Chaim David Azulai known as the "Chida," (18th century) in his book the "Birkei Yosef":[2] "Today we are knowledgeable on the fixing of the calendar," namely there is no doubt as which day is Yom Tov. We know which day is the real seventh day of Pesach, and therefore we read on that day the splitting of the Red Sea, since this was the day that it was split. Also, on the first day of Shavuot we read the portion dealing with the giving of the Torah, for it was on that day that the Torah was given. We therefore postpone reading "kol habechor" until the second day of Yom Tov. In contrast, the reading of "vezot haberachah" is not at all connected with that Festival, and in the words of the "Birkei Yosef" the section of the Torah "Chag haSukkot" which is contained in the reading of "kol habechor" is connected with the Festival, since it states "ach sameach" which is specifically to include the last day of Sukkot. It is read on Shemini Atzeret, because since today we have a fixed calendar, we in fact know it to be Shemini Atzeret. Furthermore, the reading of "vezot haberachah" is in no way connected to this Festival.

Furthermore, the answer given by Rabbi Mordechai Yoffe,[3] known as the "Levush" (16th-17th centuries), is also related to the reason that there is no connection between "vezot haberachah" and the Festival. The "Levush" explains that since the completion of the reading of the Torah and its restarting from "Bereshit" were postponed from Rosh Hashanah in order to deceive the Satan (the Devil) of the day that Rosh Hashanah occurs, we postpone reading the end of the Torah "until the end of all the Festivals so that one does not have to make a break in the parashiot read during the Festivals."

Rabbi Shlomo ibn Ezra, the "Maharash"[4] with the consent of his Rabbi, the "Knesset haGedolah," brings another explanation, which is from the words of the Tur,[5] namely, that the parashiot that are read on the Festivals, need to be read in the order that they are written in the Torah, and if one makes changes, there will have to be a reason for it. On Pesach, "kol habechor" is read on the eighth day which is after (nearly) all the other readings during the Festival, and likewise on Shavuot. But on Sukkot one reads "kol habechor" on the eighth day and "vezot haberachah" on the ninth day, because "vezot haberachah" occurs in the Torah after "kol habechor."

The language of the Gemara[6] concerning the reading of the Torah on Shemini Atzeret is: "On the last day one reads 'kol habechor' with the commandments and statutes." We can see that unlike the last day of Pesach and the second day of Shavuot, here the Gemara adds the words "commandments and statutes." On this, Rashi[7] explains that we begin reading from the words "aser t'aser" since this section of the Torah contains many commandments and statutes, such as tithes and gifts to the poor which are practiced on Sukkot, because this is the time of the year for the gathering in of the harvest. From this, a number of Poskim (Rabbinical arbiters) ruled that even if Shemini Atzeret occurs on a weekday, one begins to read from "aser t'aser. [In contrast, if the eighth day of Pesach or the second day of Shavuot occurs of Shabbat, the reason that one begins reading from "aser t'aser" is since it is impossible to divide up the verses from "kol habechor" until the end of the reading for that day for seven people who are called up to the Torah on a Shabbat.]

Here it is worth noting that there is a difference between the reading of the Torah for the Festivals and the reading of the Torah for Simchat Torah. The readings on the Festivals are not connected to the readings that are made during the year on the Shabbat cycle. On the other hand, the reading on Simchat Torah is from the parashah "vezot haberachah" which is the last parashah in the Torah. This is the reading whether or not Simchat Torah occurs on Shabbat or a weekday. [In the Diaspora Simchat Torah cannot occur on Shabbat.] From the above, it follows that the reading of "vezot haberachah" is only relevant to places where one concludes reading the Torah in one year, and this is done at the end of the festival of Sukkot.

Halachic ruling of the Shulchan Aruch and comments of the Acharonim

The ruling of the Shulchan Aruch[8] concerning the reading of the Torah on Shemini Atzeret in Eretz Yisrael states: "One takes out three Torah scrolls and in the first one reads from 'vezot haberachah' ..." Rabbi Moshe Rivkash known as the "Be'er Hagolah" [9] (17th century), comments that the author of the Shulchan Aruch does not state the source of his ruling, and in his opinion one must investigate whether it is likely that the source was from the Talmud.

On the above-mentioned words of the Shulchan Aruch, the Vilna Gaon (Gr'a)[10] wrote only: "As with Simchat Torah in the Diaspora." On this Rabbi Gershon Stern[11] in his book "Yalkut Gershoni" wrote: "This is a big innovation of the Vilna Gaon whose methodology is to give the source of the rulings of the Shulchan Aruch."

In order to explain the Vilna Gaon, Rabbi Eliezer ben Shmuel Landau,[12] the "Damesek Eliezer" distinguished between the two words: "v'ha'idna" (today) and "ulemachor" (tomorrow) which appear in the Gemara. The language of the Gemara is: "On a certain Festival one reads … but today when there are two days…" From this the "Damesek Eliezer" concludes that the first passage refers to Eretz Yisrael, and the continuation from the words: "And today when there are two days" refers to the Diaspora. However, regarding the last days of the Festival of Sukkot, the language of the Gemara is: "One reads 'kol habechor' and on the following day ("tomorrow") one reads 'vezot haberachah'." The change in the language of the Gemara (namely the use of the word "tomorrow" instead of the word "today") is brought by the "Damesek Eliezer" who says that the intention is: "The reading of 'kol habechor' refers to the Diaspora which is not the case in Eretz Yisrael where one reads 'vezot haberachah' and from this comes the ruling brought by the Shulchan Aruch."

However, one needs to investigate the words of the Damesek Eliezer. It is stated in the Gemara:[13] "On the last day of Pesach one reads 'vayehi beshalach' and on the following day 'kol habechor'." According to the reasoning of the "Damesek Eliezer" because the Gemara uses the word "tomorrow" and not "today," it will be necessary to interpret that here too the Gemara refers to the Diaspora and according to this, it seems that on the seventh of Pesach in Eretz Yisrael they should read "kol habechor"!

Rabbi Shmuel Strashun[14] known as the "Rashash" (19th century) explained the Gemara in a similar way as to that of the "Damesek Eliezer." He quotes Rashi[15] on the Gemara and according to the words of Rashi the authors of the Gemara interposed into the Baraita (a teaching not incorporated into the Mishnah) the reading of the Torah for the second day of Yom Tov in the Diaspora. Rashi writes that the Baraita was taught for Eretz Yisrael where one observes just one day of Yom Tov, and he concludes that it finishes with the words "and the other days of the Festival ..." namely "the last day (of Sukkot)" is not part of the Baraita but is an addition by the authors of the Gemara. On Rashi's comments, the Rashash argues giving some examples, that "it seems to me that they did not write this exactly." He states that his proof is derived from the language of the Gemara which uses the word "tomorrow" instead of "today."

However, the comments of the Rashash and the "Damesek Eliezer" do not accord with the version of the Mishnah that was in the possession of Rabbi Yitzchak ben Moshe of Vienna[16] (13th century), author of the book "Or Zarua," and perhaps also in the possession of a number of other Rishonim. The "Or Zarua" writes: "The Mishnah states that on the last day of Sukkot one reads commandments and statutes," namely according to the version of the Mishnah which was in the possession of the "Or Zarua" the reading of the Torah for Shemini Atzeret was part of the *Mishnah*, and not an addition by the authors of the Gemara. According to the "text checker" who added in the margin of the "Or Zarua" his sources from the Talmud, wrote there "page 30 side 2," namely the source is in the Mishnah, and he did not suggest that there was a scribal error. In addition The Rishonim, namely the "Ritz Ghiyyat"[17] (Rabbi Yitzchak ibn Ghiyyat, 11th century), the "Manhig"[18] (Rabbi Avraham ben Natan haYarchei, 12th-13th centuries), and the "Orchot Chaim"[19] (Rabbi Aharon ben Yaakov Hakohen m'Lunel, 14th century), all wrote "It is taught ("tanan") that on the last day of Sukkot, one reads 'commandments and statutes'." The expression "tanan" is the language of the Mishnah. It is true that this is not conclusive proof, because the language of the Rishonim is when mentioning a Baraita to also use the expression "tanan." However, in the case before us, in addition to the three Rishonim who used the expression "tanan," the "Or Zarua" used the word "Mishnah," and it is possible that also all the others had this version of the *Mishnah*.

According to Rabbi Moshe Sofer[20] known as the "Chatam Sofer" (18th-19th centuries) it is possible to reconcile the Shulchan

Aruch with what is written in the Tosafot. In the Tosafot[21] it is stated that it is a custom to take out a second Sefer Torah on Festivals and read from it the sacrifices appertaining to that day,[22] and this is an innovation of the Geonim (great Rabbis who lived between about the 6th to the 11th centuries), and is not mentioned in the Talmud. The "Chatam Sofer" explained that in those days, since they did not read the sacrifices of Shemini Atzeret, it was necessary to read something dealing with Yom Tov namely "aser t'aser," but after it was instituted to read something from the second Torah scroll in connection with the Festival, there was no longer an obligation to read "aser t'aser."

The "Aruch Hashulchan"[23] just quotes from the Shulchan Aruch the reading of the Torah on Shemini Atzeret in Eretz Yisrael, but makes no comment on it. In the Shulchan Aruch Harav, all the chapters in the Orach Chaim section from chapter 652 are missing, and therefore one does not know what it would have been written about this subject.

The Triennial Cycle in Eretz Yisrael

One of the reasons given by the "Birkei Yosef," that one does not read "vezot haberachah" on Shemini Atzeret, is that in Eretz Yisrael they completed reading the Torah after three years, and therefore they did not read "vezot haberachah" at the end of the Festival of Sukkot (as part of the regular Shabbat cycle).

The triennial cycle of Eretz Yisrael is mentioned in the Gemara:[24] "The people of Eretz Yisrael who complete the reading of the Torah in three years." In some Chumashim (books whose content is the Five Books of Moses), at the end of each of the five books of the Torah are given the number of sedarim in the triennial cycle.[25] The Chumash published by "Koren" (a Jerusalem publisher of Jewish religious texts) even gives details of the division of these sedarim. In total, the Torah is divided into one hundred and fifty-four sedarim which are read over the course of three years.

However, in Masechet Soferim[26] it states that "one hundred and seventy-five sedarim were established in the Torah to be read on successive Sabbaths." From this number it appears that they completed the reading of the Torah in three and a half years, namely twice in every seven years, and therefore not just in three years. The same number also appears in the Talmud Yerushalmi.[27]

Until which period did this triennial cycle continue? There are almost no sources in the books of the Sages who can answer this question.

At apparently the beginning of the eighth century, a book ("Hachilukim shebein...") was written in Eretz Yisrael, dealing with the differences of customs between the inhabitants of Eretz Yisrael and the inhabitants of Babylon. Paragraph 48 of this book[28] reads: "The people of the East (Babylon) observe Simchat Torah every year and the people of Eretz Yisrael every three and a half years." From this we see that in the eighth century the Torah was completed in Eretz Yisrael only after three years (or three and a half years). According to one version[29] of this book, this paragraph continues with the words: "The date when they finished in one location, was not the same date as in another location." From this we can see that there was no uniform custom for all the inhabitants of Eretz Israel, and it is possible to reconcile from this the different opinions, which were brought about the various numbers of sedarim in the Torah.

Rav Hai Gaon,[30] who lived in the tenth and eleventh centuries, wrote: "There are those who read the portion 'ki hamitzvah hazot' which accords with what is taught in the Mishnah that on the last day of Sukkot one reads the portion of the Torah 'mitzvot v'chukim,' and this is the custom in Eretz Yisrael and Jerusalem." [There are variant versions of this Gemara where in this quoted version the words "kol habechor" do not appear, so it is possible to understand the meaning, according to which the words "mitzvot v'chukim" refer to the parashah "ki hamitzvah hazot."] Since in Eretz Yisrael they read 'ki hamitzvah hazot' on Shemini Atzeret, and they did not read "vezot haberachah," it shows that they had not yet accepted the custom of completing reading the Torah every year on Shemini Atzeret.

In the eighth century, following the Arab conquest of the area, both the Jews of Eretz Yisrael and the Jews of Babylon immigrated to Egypt, and each community built its own synagogue in Fustat. [Fustat was the ancient Cairo.] After the conquest of Eretz Yisrael by the Crusaders in the late eleventh century, the Jews who had been captured by the Crusaders were redeemed, by the Jews of Egypt and brought to Egypt.

Rabbi Binyamin of Tudela[31] wrote in the year 4930 (1170), that in a large city, sitting on the banks of the Nile, there were two synagogues one for the people of Eretz Yisrael and one for the people of Babylon, and he wrote: "And they do not have the same custom regarding the parashiot and sedarim of the Torah, for the people from Babylon are accustomed to read every week a parashah as they did in the whole of Spain, and every year they completed the Reading of the Torah, but the people from Eretz Yisrael do not have this custom but read three sedarim from each parashah and complete reading the Torah at the end of three years."

About ten years later, the Rambam (Maimonides)[32] wrote: "And there are those who complete the Torah in a three-year cycle and it is not a widely accepted custom." A number of years later, Avraham the son of the Rambam[33] wrote in his book "Kifayah Al'abidin" that in his city (Fustat) there were two Synagogues, one according to the customs of Babylon and the other according to the customs of Eretz Yisrael. In connection with the Reading of the Torah he wrote that in the Babylonian Synagogue they read a parashah each week and in the Eretz Yisraeli Synagogue a seder each week.

There is also a manuscript[34] from the year 4971 (1211), whose content was an official recognition of the rite of the Eretz Yisraeli Synagogue in Fustat. From this document we learn that they still used to complete reading of the Torah after every three years. It states that every Shabbat they read the parashah from a Chumash, and after that, the seder from a Sefer Torah.

Two manuscripts detailing the haftarot of the triennial cycle were discovered in the Cairo Genizah. One of them[35] gives the beginnings and ends of about seventy haftarot beginning with "Noach" (which is towards the beginning of the book of "Bereshit") and finishing with the beginning of the book of

"Vayikra." In the second manuscript[36] are to be found a number of haftarot from the beginning of the book of Devarim. These discoveries serve as further evidence, that in the period of the end of the Geonim (exact dates of these manuscripts are not known), there were synagogues in Fustat, which read according to the triennial cycle.

Another important source to identify the haftarot in the triennial cycle is the piyyutim (liturgical poems) of Rabbi Yannai.[37] We do not know the exact period that he lived, and the average of the various opinions is the sixth century. Yannai composed piyyutim that are added in the first three berachot of the Shabbat amidah for shacharit (morning prayer). These piyyutim are built on the triennial cycle. The first verse of each seder is brought at the end of the first stanza to each piyyut; the second verse is brought at the end of the second stanza; the first verse of each haftarah is brought at the end of the third stanza.[38]

From the various manuscripts of the piyyutim of Yannai and from other sources R' Ben-Zion Axelrod (in his introduction to the book of Professor Jacob Mann) prepared a list of almost all the haftarot in the triennial cycle.[39]

Also the "Encyclopedia Judaica"[40] gives a list which contains about eighty-five percent of these haftarot, but there are some differences between the list of R' Ben-Zion Axelrod and the list in the "Encyclopedia Judaica." However, the origin of all the haftarot quoted in this source is not clear. Also, according to this list, there is a *uniform* date for the reading of all the haftarot *throughout* Eretz Yisrael, and this is not in accordance with the version (quoted above) in the difference in customs between Eretz Yisrael and Babylon which says: "The day that they finish reading the parashiot in one location in Eretz Yisrael is not the same as for another location." It should be noted that this list does not include the special haftarot read during the three weeks between Shivah Asar b'Tammuz and Tisha b'Av, and the seven haftarot read after Tisha b'Av.

One of the problems in the division of the parashiot during the year arises in the case where the eighth day of Pesach, or the second day of Shavuot, occurs in the Diaspora on Shabbat.

In Eretz Yisrael these days are "isru chag" (the day after the Festival), and therefore one reads on them the regular parashah for that Shabbat. On the other hand, in the Diaspora they are considered as Yom Tov and one reads on them the special reading for the Festival. It therefore follows that after the Festival it will be necessary to equate the parashiot of the Diaspora with the parashiot of Eretz Yisrael, and for this purpose one has to divide a parashah in Eretz Yisrael into two at the same time that in the Diaspora they are connected.

It is clear that this problem began only with the transition in Eretz Yisrael from the triennial cycle to the annual cycle. According to the book "Kaftor vaFerach" which was written following seven years of research, by Rabbi Ishtori Haparchi who lived during the thirteenth to fourteenth centuries, one can see that already in the year 5082 (1322) this problem was already resolved. According to the author of Rabbi Haparchi's book,[41] when the eighth day of Pesach falls on Shabbat in a non-leap year, it was arranged that "here in Eretz Yisrael that the parashah Tazria was not joined to the parashah Metzora, and in a leap year parashat Matot was not joined to parashat Massey. In connection with Shavuot, he wrote that one would arrange that parashat Matot would not be joined to parashat Massey when the first day of Shavuot occurred on Friday." The words of the "Kaftor vaFerach" prove that by the beginning of the fourteenth century, Eretz Yisrael had already gone over to an annual cycle for reading of the Torah.[42]

However, this was not the case in the Eretz Yisraeli Synagogue in Fustat! Rabbi Yissachar ben Mordechai Susan,[43] author of the book "Ibur Shanim – Tikkun Yissachar," who lived in Egypt in the sixteenth century, wrote that at the time there were only "about ten poor people" in Fustat, and they continued to live there "in honour of the two Synagogues [Babylonian and the Eretz Yisraeli, which were still in existence ...". He also writes that "every Shabbat a minyan (ten) of men went to pray [in the Babylonian synagogue] in honor of the place, and I went and prayed there several times."

It can be seen from this, that on Shabbat there were about twenty Jews in Fustat, namely the ten poor people who lived there, and ten who went there from the new city. It was stated above that the poor remained there in honour of the two synagogues, so it is very possible that a service with a minyan was also held on Shabbat in the Eretz Yisraeli synagogue.

The "Tikkun Yissachar"[44] furthermore writes about the two communities that were in Fustat: "Each of them held according to the custom of their ancestors, and they did not squabble with each other." He also praised the "Bnei Eretz Yisrael, the original inhabitants of the city, since they held, according to the custom of their ancestors."

From all the above, it appears that during the period of the "Tikkun Yissachar," the triennial cycle of the Eretz Yisraeli Synagogue in Fustat was still in practice.

In the year 5432 (1672), the chronicler Yosef Sambari[45] wrote about both the Eretz Yisraeli Synagogue and the Babylonian Synagogue in Fustat: "They do not have a unified custom regarding the parashiot and sedarim regarding reading of the Torah, since the people from Babylon are accustomed to read every week the parashah as is done in Spain, and every year they complete reading the Torah, but the people from Eretz Yisrael do not do so, but divide each parashah into three sedarim and complete the reading of the Torah at the end of three years." Also from here one can see that hundreds of years after the transfer to an annual cycle in Eretz Yisrael, the Eretz Yisraeli Synagogue in Fustat continued with the triennial cycle. However, Sambari uses *exactly* the same words as Benjamin of Tudela,[46] and even went on to say, "These are what used to be the practice in these two synagogues." It is thus possible that Sambari is referring to the custom which was observed *in the past*.

The Transition from a Triennial Cycle to an Annual Cycle in Eretz Yisrael

From the above we see that over the course of generations the custom of Babylon eventually prevailed in Eretz Yisrael. How and when did this change occur?

It was during the seventh century, that the Arabs conquered Eretz Yisrael and Babylon. After the opening of the border between these two countries, a large influx of immigrants from Babylon arrived in Eretz Yisrael. Rabbi Shaul Hanna Kook (the younger brother of Rabbi Avraham Yitzchak Hakohen Kook) who researched the subject brought a ruling from one of the Geonim[47] regarding the Babylonians who immigrated to Eretz Yisrael: "If the immigrant intends to return even though he has spent many years in Eretz Yisrael, he must follow the stringiness of both Eretz Yisrael and Babylon." According to this ruling, the Babylonian immigrants endeavored to establish synagogues in Eretz Yisrael according to the Babylonian custom, even after having lived for many years in Eretz Yisrael. Rabbi Shaul Kook[48] held that even after they had decided to remain in Eretz Yisrael it was difficult for them to change their customs and they continued to pray in their Synagogues in accordance with the customs of Babylon.

One can see from the book giving differences of customs between Jews of Eretz Yisrael and of Babylon ("Hachilukim shebein...")[49] that already in the eighth century there was an influence of customs of the Babylonian Synagogue on synagogues that acted according to the customs of Eretz Yisrael. According to paragraph 47 of this book: "People from the east (Babylon) both the Reader of the Torah and the individuals read the parashah, and the people from Eretz Yisrael read the parashah and the Reader of the Torah (just reads) the seder." This means that every Shabbat every native of Eretz Yisrael read the parashah from a Chumash (namely according to the Babylonian custom), and then the Reader read the seder from the Torah scroll (namely according to the Eretz Yisraeli custom).

In a responsum of the Geonim,[50] which was discovered in the Cairo Genizah, and which was written later than the above quoted book, (possibly about fifty years later), it seems clear that it was more than just an influence! The immigrants that had arrived from Babylon even forced the native Eretz Yisraelis to accept the customs of Babylon. This responsum speaks of the "kedushah" recited in the repetition of the amidah. They did not recite the "kedushah" in Eretz Yisrael except on Shabbat and on Festivals, and even this was only at the shacharit service, with the exception of in Jerusalem and in any country who had Babylonian Jews who had caused arguments and quarrels until the native Eretz Yisraelis agreed to say the "kedushah" every day.

However, from the answer of Rav Hai Gaon, mentioned above, it seems that the Babylonian immigrants did not succeed to cancel the triennial cycle in the synagogues of Eretz Yisrael.

At the end of the eleventh century, the Crusaders conquered Eretz Yisrael. Many Jews were murdered, and the Jewish community in Eretz Yisrael almost disappeared. The Crusader rule in the country ended towards the end of the twelfth century. In the year 4971 (1211), three hundred Rabbis from England and France immigrated to Eretz Yisrael, which included some of the Tosafists. In the year 5017 (1257) Rabbi Yechiel from Paris immigrated to Eretz Yisrael and established a Yeshivah in Acre called the "Midrash Hagadol shel Paris." Ten years later, Rabbi Moshe ben Nachman (Nachmanides), known as the "Ramban" arrived in Jerusalem and found only two Jews there, and established the "Bet Haknesset Haramban" (Ramban Synagogue) there.

The Sages who came from Provence in France during this period, changed the ancient custom in Eretz Yisrael regarding the number of days of Rosh Hashanah,[51] namely instead of observing one day, they instituted in Eretz Yisrael two days of Rosh Hashanah.[52]

We know from the book "Kaftor vaFerach," that already at the beginning of the fourteenth century it was the custom in Eretz Yisrael to complete reading the Torah in the course of one year. Therefore, it is very possible that those who immigrated from Europe during the thirteenth century changed the Eretz Yisraeli custom regarding the Reading the Torah from the triennial to a yearly cycle.

Changing the Reading of the Torah of Shemini Atzeret in Eretz Yisrael

After the immigrants arrived in Eretz Yisrael, they celebrated only one day of Yom Tov. With the change of the Eretz Yisraeli custom, namely in which the Torah cycle is completed in one year and this occurs at the end of Sukkot, a problem arose. This was that the ninth day of Sukkot (Simchat Torah in the Diaspora), is a weekday in Eretz Yisrael and on the twenty-second of Tishrei it is necessary to read the appropriate reading of the Torah for Shemini Atzeret. If so, when would one read the parashah "vezot haberachah" in Eretz Yisrael?

To solve the problem, one would have to "cancel" the reading of the Torah designated for Shemini Atzeret, and move the reading of "vezot haberachah" from the ninth day to the eighth day. Then the question arose, namely, on what did they rely on when they came to change the reading of the Torah from what appears in the Talmud? Indeed, there are some "precedents" for this. If one makes a comparison between the parashiot and the haftarot that appear in the Gemara with those which appear in the Shulchan Aruch, one will find a number of differences: [53]

a) Both according to the Mishnah[54] and according to the Gemara[55] on fast days one reads the blessings and curses which appear in parashat Bechutotai. According to the ruling of the Shulchan Aruch[56] however, one reads the portion of the Torah beginning with the word "veyechal." The source for this change is the Tur in the name of Rav Sar Shalom,[57] one of the Geonim who lived in the twelfth century, and who wrote:[58] "On all public fast days and on all fast days which are declared because of a lack of rain and for any other reason, one reads 'vayechal' in both the shacharit and minchah services."

b) It is stated in the Gemara[59] that the haftarah on Simchat Torah is "vayamod Shelomo" but according to the Shulchan Aruch[60] one reads "vayehi acharei mot Moshe." The source is the Tosafot[61] which states that Rav Hai Gaon ruled that one reads "vayehi acharei mot Moshe".

c) It is stated in the Gemara:[62] "Rosh Chodesh Av which occurs on Shabbat, the haftarah is 'chodsheichem u'moadaichem." The Rema[63] in the Shulchan Aruch rules that the haftarah is "shimu dvar HaShem." This is according to Mordechi[64] who wrote: "And we also act according to the Pesikta." [Some say that the haftarah is "hashamayim kisi," but the Vilna Gaon[65] rejected this custom.]

On these "deviations" from the Gemara, the Tosafot[66] wrote: "And in some things we rely on external books and put aside our Gemara." On this subject there is a responsum from the "Mabit"[67] (Rabbi Moshe ben Yosef Matrani, one of the Sages in the sixteenth century), which speaks of one synagogue in the Diaspora, which has always read on both the eighth and ninth days of Sukkot, parashat "vezot haberachah." The "Mabit" rules that since this Synagogue is accustomed to do so "from ancient times, it is not appropriate to change it, because perhaps since it is an ancient custom it was decreed by some Gaon (great scholar)."

In his reply, the "Maharit"[68] (Rabbi Yosef Matrani, 16th-17th centuries) identifies this synagogue as: "The old synagogue attributed to Eliyahu." It is very possible that he is talking about the Eretz Yisraeli Synagogue in Fustat. We have seen above, according to the "Tikkun Yissachar," that in the sixteenth century they still observed the triennial cycle in the Eretz Yisraeli Synagogue in Fustat, even though in Eretz Yisrael they had gone over to the annual cycle about three hundred years earlier. Therefore it would seem that on the first day of Shemini Atzeret (twenty-second day of Tishrei) they read parashat "vezot haberachah" in accordance with the custom in of Eretz Yisrael, and on the second day of Shemini Atzeret (Simchat Torah) they read it again in accordance with the custom of the Diaspora.[69] Because the Eretz Yisraeli Synagogue in Fustat only completed the Torah after three years, the reading of "vezot haberachah" on two days of Shemini Atzeret was *not* to be regarded as "completing reading the Torah" but for the reading of the special portions designated for Festivals.

It should be noted that the "Knesset haGedolah"[70] suggested that perhaps in the Eretz Yisraeli Synagogue in Fustat, they did not finish reading parashat "vezot haberachah" on the first day of Shemini Atzeret, but instead only on the second day of Shemini Atzeret they read the entire parashah "vezot haberachah" and also a small portion of parashat Berashit.

From the Maharit's answer, we see that there was an old tradition to read parashat "vezot haberachah" on Shemini Atzeret (twenty-second day of Tishrei). It is possible that while synagogues in Eretz Yisrael began to finish reading the Torah as a yearly cycle at the end of Sukkot, and had to find a solution to the problem when to read the parashah "vezot haberachah," they relied on this ancient custom of reading "vezot haberachah" on the twenty-second of Tishrei (Simchat Torah in Eretz Yisrael). From all of the above, it is possible to understand the change in the reading of the Torah, from what should have been read in Shemini Atzeret, to the reading of parashat vezot haberachah, which is the custom in Eretz Yisrael today.

Appendix

During his visit to Egypt, Rabbi Binyamin of Tudela[71] wrote about the two synagogues, namely one of them which followed the customs of Eretz Yisrael, and the other the customs of Babylon. These two Synagogues had a custom and an ordinance to pray together on Simchat Torah and on Matan Torah (Shavuot)."

Why specifically did they pray together on Simchat Torah and Shavuot? To solve this question, I offer the following explanation:

It is clear that the Jews of Eretz Yisrael, who went down to Egypt, observed the second day of Yom Tov in Egypt. Regarding the last day (eighth day) of Pesach and the second day of Shavuot, there was no problem on which portion in the Torah they were obliged to read, and therefore they read in accordance with reading done in the Diaspora, namely "kol habechor." However, because they were accustomed to read the Torah according to the triennial cycle, the reading of "vezot haberachah" on Simchat Torah (the ninth day of Sukkot) was inappropriate, and thus the question arose, regarding what should they read on that day? Therefore they prayed in the Babylonian Synagogue and participated with them in the celebration of the conclusion of reading the Torah.

In a manuscript[72] from the year 4971 (1211), we see that among the other customs of the Eretz Yisraeli Synagogue in Fustat, there was a custom that *every day* they took a Torah scroll out of the Ark, brought it to the reading desk and without opening the Torah, read the Ten Commandments. On Shavuot, the Torah scroll was taken out, and the Ten Commandments were read *from the Torah*. [Shavuot is the only occasion in the year that both the Eretz Yisraeli Synagogue and the Babylonian Synagogue read the Ten Commandments. On the Sabbaths when Parashat Yitro and Parashat Vaetchanan were read in the Babylonian Synagogue, the Eretz Yisraeli Synagogue were reading elsewhere in the Torah.] The custom of the daily reading of the Ten Commandments in the Eretz Yisraeli Synagogue with a closed Sefer Torah on the reading desk, indicates that they gave great importance to the reading of the Ten Commandments. The reading on Shavuot from an *open* Sefer Torah kept it apart from the other occasions when they read from a closed Sefer Torah. Therefore, the Eretz Yisrael Synagogue invited the members of the Babylonian Synagogue to specifically join with them in the Shavuot prayers.

References

Abbreviations

SA = Shulchan Aruch

OC = Orach Chaim

(1)   Talmud Bavli Megillah 31a
(2)   Rabbi Chaim David Azulai (Chida), *Birkei Yosef*, OC chap.668, Shiurei Berachah,
(3)   Rabbi Mordechai Yoffe, *Levush*, OC chap.669 par.1
(4)   Rabbi Chaim Benveniste, *Shu't Knesset haGedolah,* (Kushta (Constantinople), 5493), part 2 responsum 6
(5)   Tur OC chap 490
(6)   Talmud Bavli Megillah 31a
(7)   Rashi on Megillah 31a, first words: korin kol habechor
(8)   SA OC chap.668 par.2
(9)   Rabbi Moshe Rivkash, *Be'er Hagolah*, SA OC chap.668
(10)   Vilna Gaon, *Beur haGra*, SA OC chap.668 par.2
(11)   Rabbi Gershon ben Moshe Stern, *Yalkut ha-Gershuni*, (Munkatch, 5664), OC chap.668 par.5
(12)   Rabbi Eliezer ben Shmuel Landau, *Damesek Eliezer*, (Vilna, 5628), OC chap.668 par.3
(13)   Talmud Bavli Megillah 31a
(14)   Rabbi Shmuel ben Yosef Strashun, *Hagahot v'Chiddushei haRashash*, Megillah 31a
(15)   Rashi on Megillah 31a, first words: hachi garsinon
(16)   Rabbi Yitzchak ben Moshe of Vienna, *Or Zarua,* (Zhitomir, 5622), part 2, chap.393

(17) Rabbi Yitzchak ibn Ghiyyat (Ritz Ghiyyat), *Sha'are Simchah*, (Fürth, 5621), part 1, Mea Shearim, p.117

(18) Rabbi Avraham ben Natan haYarchei, *Sefer Hamanhig*, (Mossad Harav Kook: Jerusalem, 5738), part 2, pp.411-412

(19) Rabbi Aharon ben Yaakov Hakohen m'Lunel, *Orchot Chaim*, (Firenze, 5510), hilchot keriyat Sefer Torah, par.58

(20) Rabbi Moshe Sofer (Schreiber), *Chatam Sofer*, SA OC chap.664

(21) Tosafot on Megillah 30b, first word: ushar

(22) It is written in the Gemara (Yoma 70a.): "One does not roll a Sefer Torah in the presence of a congregation out of respect for the congregants." It therefore follows that on days when the reading is from two separate places in the Torah, two Torah scrolls are taken out of the Ark. The week's parashah on Shabbat Shekalim is in close proximity to the Torah portion for Shekalim. Therefore, Rabbi Menachem Meiri (Megillah, p.106) mentions the opinion that only one Torah scroll should be taken out on such a Shabbat Parashat Shekalim. This was the custom in the Synagogue of Rabbi Shmuel Salant. (Luach l'Eretz Yisrael, arranged by Rabbi Yechiel Michel Tucazinsky 5751, p.37)

In a year when Rosh Chodesh Menachem Av occurs on Shabbat there is a similar situation. The Torah Parashah of that week is "Matot-Massey" or only "Massey," and the maftir for Rosh Chodesh is from parashat "Pinchas" which is in very close proximity to that week's Shabbat Parashah. Therefore, the question arises: Are those who are accustomed to take out just one Sefer Torah on Shabbat Shekalim, also take out just one Sefer Torah when Rosh Chodesh Menachem Av occurs on Shabbat?

(23) *Aruch Hashulchan*, OC chap.668 par.7

(24) Talmud Bavli Megillah 29b

(25) The verses in the Torah which were read each Shabbat in Eretz Yisrael were called Sedra (plural: Sedarim). Those which were read each Shabbat in Babylon were called Parashah (plural: Parshiot)

(26) Masechet Soferim chap.16 halacha 10

(27) Talmud Yerushalmi Shabbat chap.16 halacha 1

(28) *Hachilukim shebein anshei mizrach uvenai Eretz Yisrael* (The differences between the People of the East and the Inhabitants of Eretz Yisrael), (Mordechai Margalioth, Rubin Mass: Jerusalem, 5698), p.88

(29) Ibid., variant readings

(30) *Otzar Hageonim*, ed. B.M. Levin, (Jerusalem, 5693) vol.5 p.62

(31) Rabbi Binyamin miTudela, *Masa'ot shel Rabi Binyamin*, (Ascher: New York), pp.97-98

(32) Maimonides, Rambam Mishneh Torah, Hilchot Tefillah chap.13 halacha 1

(33) Rabbi Avraham ben haRambam, *Kifayah Al'abidin*, MS Bodleian, catalogue Neubauer 1274, folio 56

(34) MS Bodleian, catalogue Neubauer 2834-22 folio 41

(35) MS Bodleian, catalogue Neubauer 2727-3, folio 24

(36) MS Bet Hamidrash l'Rabbonim New York, catalogue Adler 2105

(37) Rabbi Yannai, *Machzor Piyutei Rabi Yannai, l'Torah ul'Moadim*, ed. Tzvi Meir Rabinowitz, (Mossad Bialik and Tel-Aviv University, 1985), vol.1 p.45

(38) Ibid., p.12

(39) Jacob Mann, *The Bible as Read and Preached in the Old Synagogue*, (Ktav Publishing House, New York, 1971), vol.1, pp. LI-LXVII

(40) *Encyclopedia Judaica*, vol. 15, (Jerusalem, 1971), pp.1387-88,

(41) Rabbi Ishtori Haparchi, *Kaftor vaFerach*, (Hirsch Edelman: Berlin, 5611), chap.14, p.55

(42) It is interesting to note that two hundred years later, the Beit Yosef who lived in Safed, did not mention in his commentary on the Tur, the fact that in Eretz Yisrael they read on 22 Tishrei parashat vezot haberachah.

(43) Rabbi Yissachar ben Mordechai Susan, *Sefer Ibur Shanim – Tikkun Yissachar*, (Venice, 5339), p.33b

(44) Ibid.

(45) Yosef Sambari, *Sefer Divrei Yosef*, MS Paris, H130A folio 32

(46) There are a number of other cases in which Sambari copied the words of Rabbi Binyamin mi-Tudela, see note by I. Abrahams, The Jewish Quarterly Review, vol. II, 1890, p. 107

(47) *Teshuvot Geonai Mizrach veMa'arav*, (Yoel Hakohen Müller: Berlin, 5648), responsum 39, (p.12)

(48) Rabbi Shaul Chana Kook, *Iyunim u-Mechkarim*, (Mossad Harav Kook: Jerusalem, 5723), book 2 pp.35-36

(49) *Hachilukim shebein ...*, op. cit., p.88

(50) MS Cambridge, Taylor-Schechter collection, general 35.97

(51) Rabbenu Asher ("Rosh"), *Rosh* on Beitza chap.1 par.4; Rabbi Zerachiah Halevi of Gerondi, *Baal Hamaor* on Rif Beitza 3a

(52) Nowadays, the "Fast of Gedalia" is observed on the third day of Tishrei, namely, the day after Rosh Hashanah. Rabbenu Yerucham (Toldot Adam v'Chava, section 18 part 2) writes: "It is said that it was on Rosh Hashanah that Gedalia was murdered and the fast was accordingly postponed to a weekday." The question that can be asked is that at the period when in Eretz Yisrael they observed just one day for Rosh Hashanah, did they fast in Eretz Yisrael on the second day of Tishrei, or on the third day of Tishrei? Did the Sages make the original decree to fast on the third of Tishrei, since in the Diaspora Rosh Hashanah is always of two days duration, or did they decree that the fast is on Rosh Hashanah, but because that day every year is Rosh Hashanah the fast is postponed until after Rosh Hashanah. According to the first option, even at the period when they observed one day Rosh Hashanah in Eretz Yisrael, those in Eretz Yisrael would fast on the third of Tishrei, but according to the second option they would fast on the second of Tishrei. This question is not theoretical today! In the situation when a Brit Milah takes place on the day to which a fast has been postponed (even Tisha b'Av), the "Ba'alei Brit" (father, mohel and sandak) only fast until the earliest time to recite minchah (just after noon). From this we can learn that if the fast of

Gedalia had been established on the third of Tishrei, it would not be classed as a postponed fast," and the "ba'alei brit" must fast until nightfall. On the other hand, if it was originally established on Rosh Hashanah but every year fasting is postponed until the third day of Tishrei, it would then be classed as a "postponed fast," and the "ba'alei brit" would not then only be obligated to fast until just after the time of noon (Taz on SA OC chap.549)

(53) It is obvious that on every change there are discussions amongst the Rishonim and Acharonim. Here we are only bringing the source of the law in the Shulchan Aruch.

(54) Talmud Bavli Megillah 30b

(55) Talmud Bavli Megillah 31b

(56) SA OC chap.566 par.1

(57) Tur OC chap.566

(58) The order of the reading of the Torah for a public fast is that the Kohen reads from "vayechal" until "asher diber la'asot l'amo." One then omits "parashat ha'egel" (section on the golden calf) and continues reading from "pesol lecha" for the Levi."

One of the tragedies that occurred on the seventeenth of Tammuz is that the Tablets of Stone were shattered by Moshe Rabbenu as a result of the sin of the golden calf (Masechet Ta'anit, chapter 4, Mishnah 6). Therefore the question can be asked why should one miss out "parashat ha'egel" on 17 Tammuz since it is relevant to that day.

On this, Rabbi Zedekiah ben Avraham Anaw the "Shibbolei haLeket" (chap.263) writes that there are places where they are accustomed on 17 Tammuz to read this portion from "vayechal" without any omissions, since the chapter dealing with the golden calf and the shattering of the Tablets of Stone are relevant to that day. The book "Tanya Rabati" [inyan arba'ah tzomot] quotes the words of the "Shibbolei haLeket," adding: "And this is a proper custom." Rabbi Yissachar ben Mordechai Susan the "Tikkun Yissachar" (p. 5) also writes similar words: "And it looks good and great," but he continues that this custom did not spread to our countries, and it seems to me that it is only for the Torah reading of shacharit, but not for minchah, since minchah is the time for prayer and supplication and not for the mentioning of sins that were committed in the past.

In "Lekutei ha-Pardes," a book whose authorship is attributed to Rashi (Inyan Ta'anit, p.28, published in Zhovkva, 5542) it is written that one does not omit the incident of the golden calf on 17 Tammuz. On this, Rabbi Yaakov Emden writes about this in his Siddur [Siddur Beit Yaakov, part two, p.35]. It seems to me that they are referring to the first part of the incident of the golden calf, namely where it begins with "vayiten el Moshe" and continues until "vayishak et Bnei Yisrael." However, from "vayomer Moshe el Aharon" is the second part of the incident of the golden calf, and one does not translate it in order not to besmirch the name of Aharon. Today however when everyone is proficient and one has already read the first part of this incident in which there is forgiveness and shame to the Jews, it is preferable to omit from here onwards until "pesal lecha." It would be good to act this way, but this was already aborted by these generations and thus the omission is the same by everyone.

Today, the Jews of Rome do not omit the affair of the golden calf on the morning of the seventeenth of Tammuz. The Kohen reads from "vayechal Moshe" until "asher diber la'asot." The Levi then reads from "vayifen Moshe" until "el mul hahar hahu *without any omissions*, and the Yisrael begins to read from "vayifsol" (Machzor according to the Italian rite, vol.1 pp.162-63). However, at the minchah service the Jews of Rome read as on other fast days (ibid.)

(59) Talmud Bavli Megillah 31a
(60) SA OC chap.668 par.2
(61) Tosafot on Megillah 31a, first word: lemachar
(62) Talmud Bavli Megillah 31b
(63) SA OC chap.425 par.1, Rema
(64) Mordechi, Megillah chap. Benai ha'ir par.831
(65) Vilna Gaon, *Beur haGra*, SA OC chap.425 par.1
(66) Tosafot on Pesachim 40b first word; aval
(67) Rabbi Moshe Mitrani ("Mabit"), *Shu't Mabit,* (Lemberg, 5621), vol.2 responsum 129
(68) R' Avraham Yaari, *Toldot Chag Simchat Torah*, (Mossad Harav Kook: Jerusalem, 5724), p.34
(69) There was also another synagogue who wanted to read parashat "vezot haberachah" both on Shemini Atzeret and Simchat Torah. However, a scholar who was appointed to oversee various congregations who wanted to institute new customs, would not allow it and ruled that that they should not change the customs practiced by all other congregations, and they should therefore accordingly read "kol habechor." on Shemini Atzeret (Mabit, op. cit.)
(70) *Knesset haGedolah,* op. cit.
(71) *Masa'ot shel Rabi Binyamin*, op. cit. p.98
(72) MS Bodleian 2834-22, op. cit.

EATING CHEESE AND LATKES
ON CHANUKAH
(latkes = levivot = pancakes)

It is written in the Gemara:[1] The mitzvah of lighting Chanukah candles is obligatory on women for they too were included in that miracle." The "Ran" (Rabbenu Nissim of Gerona, 14th century) explains[2] that the Greeks decreed that every newly married woman was to have her first sexual intercourse with the minister, and it was because of a woman that a miracle took place, as it is stated in the Midrash that the daughter of Yochanan fed the head of the enemy cheese. He became drunk and she then cut off his head, and as a result all the enemy ran away. Because of that, it is customary to eat cheese on Chanukah. In a similar language, the "Kol-Bo"[3] (author unknown, 13th or 14th century), and also the "Orchot Chaim"[4] (Rabbi Aharon Hakohen m'Lunel, 13th-14th centuries) wrote: "And because of a woman a great miracle took place and her name was Yehudit. It is explained in an Aggadah that she was the daughter of Yochanan the High Priest that she was very beautiful and the king of the Greeks said that she should lie down with him. She then gave him a cheese dish to eat until he became thirsty, and he then drank a lot until he became drunk, and he lay down and fell asleep. She then took his sword and cut off his head and brought it to Jerusalem, and when the Greeks saw he was dead they fled. It is therefore customary to eat cheese dishes on Chanukah."

If we analyze the words of the "Kol-Bo" and the "Ran," we will arrive at the following:

1) Yehudit was the daughter of the Kohen Gadol - the High Priest.
2) The king ordered that due to her beauty, Yehudit sleep with him (Kol-Bo). The Greeks ruled that the newly married

313

women should first have sexual relations with the minister (Ran).
3) Judith fed the chief of the enemy with a cheese dish.
4) The drunkenness of the chief of the enemy.
5) The head of the enemy was cut off by Yehudit.
6) The fleeing of the enemy.

The questions which need to be asked are: What is the source of the action of Yehudit which is brought by the "Kol-Bo" and the "Ran," and also if all the above points appear in this source?

In answer to these questions, there is among the Apocrypha, namely the "external books," (namely books not included in the Tanach) a book called Yehudit, and its contents, as we shall see below, in general, correspond to the words of the "Ran" and the "Kol-Bo."

A comparison between the book of Yehudit and the words of the "Ran" and the "Ko-Bo"

Because the paragraph stating "Yehudit fed the head of the enemy with a cheese dish" is very relevant to the subject before us, we will begin by making a detailed study of this statement.

The book of Yehudit mentions in two places the foods that Yehudit had with her. The first is in chapter 10 verse 5, and this gives the list of foods that Yehudit took with her when she went to the city of Beit-Eloha (Bethulia) to meet with Holofernes (name in Greek: Olofarna), who was the minister in charge of the king's army, or according to some versions, the king himself. The second source is in chapter 12 verse 19, where it is written that she took these foods when she went to meet with Holofernes. Logically, by looking at the list of foods, we will get an answer to the question: Is "cheese" mentioned in this list? But the matter is not so simple! Today we do not have the original version of the book of Yehudit written in Hebrew! We only have various translations and there are significant differences between them.

The earliest translation from the original version is a translation into Greek. Today, however, we have three versions in

Greek,[5] namely: (i) The translation into Greek known as the Septuagint,[6] (ii) that which is known as "Codex 58," (iii) the version which became known as "Codex 19" and "Codex 108." The wording of the Septuagint for Chapter 10 verse 5 is (translated into English):[7] "And she gave her maid a leathern bottle of wine, and a cruse of oil, and filled a pouch with barley-groats, and a cake of figs, and loaves of fine bread and she so she carefully wrapped up all her dishes and put them upon her." It can be seen that cheese is not included in this list! However, in Codices 19, 58 and 108, "cheese" appears after the words "fine bread."[8] The question arises, why are there three versions of the book of Yehudit in the Greek language? In answer, the researcher Frank Zimmerman[9] suggested that there was no uniform text in the Hebrew version of the book of Yehudit.

From the Greek text there were translations of the book of Yehudit into the Syrian language (the translation is called the "Peshitta"), and into the ancient Latin language, and in these two translations the word "cheese" appears.[10] At a later date, Hieronymus (Jerome), who was one of the fathers of the Christian Church in the fourth century, prepared a Latin translation called the Vulgate. Hieronymus wrote that he translated this book from a text that was in Aramaic, but it can be suggested that he also used translations in Greek and ancient Latin in the course of his work. In the Vulgate the word "cheese" appears.[11]

There are at least five Hebrew translations (not including the modern translations) of the book of Yehudit which were made over the last thousand years, with some of them being free translations. An ancient translation is found in the "Chronicles of Yerachmeel"[12] which was written in the eleventh or twelfth century. This verse is translated: "And she gave to her maid a sack of bread, a skin bottle of wine, and she got up and went."[13] We can see that not only that cheese is not included, but also oil, wheat and figs, which appear in all the other translations, do not appear there!

Another ancient translation into Hebrew can be found in the book "Hemdat Yamim." This is a free translation, and according to the scholar Joshua Grintz[14] (20th century) was composed

315

before the generation of Rashi. This verse reads: "And Yehudit put into the hand of her maid a leather bottle of milk, a skin bottle of wine, a small bottle of oil, and flour, and bread, and cheese, and she then went."[15] We can see that in addition to "cheese," "milk" is also mentioned! The question arises, what is the source for the translator who used the word "milk"? Perhaps the source is the incident of Yael and Sisera in the book of Shoftim: "And she opened a bottle of milk and gave him to drink...[16] He asked for water and she gave him milk."[17] This act is somewhat similar to that of Yehudit, and it is possible that the translator mixed up the two incidents.

In addition, there are also translations into Hebrew in which the translators wrote from which source they did their work. Of course, in the case when "cheese" was originally mentioned, it also appears in the translation! One of these translations was made by Moshe Mildonado in the year 5312(?) (1552) from the Latin.[18] We have seen above that in the Latin translation the word "cheese" appears, and therefore also in this translation it appears.[19] About 130 years later, in the year 5439 (1679), Akiva Levy of Halberstadt prepared a translation from the German.[20] This translation was made from the "Bible" (Tanach, New Testament and the Apocrypha) of Martin Luther. Matin Luther did not mention "cheese"[21] and therefore in the translation of Akiva Levy of Halberstadt, cheese is not mentioned.[22] There is also an Aramaic version of this book, and in the year 5579 (1819) a translation was made from this Aramaic version.[23] Rabbi Moshe ben Nachman (Nachmanides), known as the "Ramban" (13th century,) in his commentary on the Torah quotes from an Aramaic translation of the Book of Yehudit.[24] However, when we compare this quote brought by the Ramban and this translation, we can see that this was not the Aramaic version which was before the Ramban. "Cheese" is not mentioned in this translation.[25]

In addition to the five translations quoted above, there are also a number of short versions in Hebrew ("Midrashim") of the acts of Yehudit.[26] The list of foods that Yehudit took with her does not appear at all in any of these short versions.

Chapter 12 verse 19 speaks of the food Judith took while appearing before Holofernes. In the Septuagint it is written:[27] "Then she took and ate and drank before him what her maid had prepared." This means: Yehudit herself ate from her own food, and it is not written that she fed Holofernes. In addition, this is likewise stated in the other translations, namely those in Greek, Syriac, Latin and in most of the Hebrew translations. However, in a few versions there is a change. One of them is a manuscript from the year 5162 (1402). There it is written[28] that Yehudit instructed her maid to prepare levivot and "charitzei chalav" (cuts of milk – cheese), which Yehudit then took to Holofernes and he then ate them. [The subject of "latkes" will be discussed below.] The phrase "charitzei chalav" appears in the Book of Shmuel[29] and according to the targum (Aramaic translation) it is "gudvin dahlada."[30] The "Metzudat Zion" (Rabbi David Altschuler and his son Yechiel, 18th century) explain the translation:[31] "And perhaps it is called this by virtue of the size the cheese was cut into while it was being made." This is the only source that explicitly states that Judith fed Holofernes with cheese.

In "Hemdat Yamim" it is written:[32] "And she sat down and she ate and opened the leather bottle of milk and she drank and also gave the king to drink," namely Yehudit feeding Holofernes with milk; (it is similar to Yael feeding Sisera with milk). However, as we wrote above, this is the only source that mentions milk, and in addition, it is explicitly written that she made Holofernes drink.

In conclusion, we see that according to a large number of versions, Judith took cheese (along with the other foods) with her while she went to the city of Bethulia (Beit-Eloha), for her meeting with Holofernes. In one version it is explicitly written that she fed Holofernes with "charitzei chalav" (cuts of milk – cheese), and it seems that the writings of the Rishonim (great Rabbis who lived approximately between the 11th and 15th centuries) on this subject, are in accordance with this version.

We will now discuss the question of whether there is a comparison between the other sections quoted by the "Ran" and the "Kol-Bo," and the paragraphs that appear in the book of Yehudit.

The "Ran" and the "Kol-Bo" write that Yehudit was the daughter of the High Priest. However, in the book of Yehudit it is written that she came from the tribe of Shimon![33] However, in two versions it states that she was from the Kohanim. In one of them she says: [34] "I and my brothers are Kohanim" and in the second one she says:[35] "And my brothers and my father's house are kings and High Priests."

The "Kol-Bo" writes that because Yehudit was beautiful, Holofernes wanted to sleep with her. This appears in all versions of the book of Yehudit. However, the version of the "Ran" that the king made her have sexual relations from the outset, does not appear in the book of Yehudit. However, there are several manuscripts of the shortened version that begin with the decrees of the Greeks against the Jews. Amongst them was: "Any new married girl before she had relations with her husband was brought before the king who had sexual relations with her, and only afterwards she returned to her husband." However, the Jews succeeded in finding tricks in order not to observe this decree. When the daughter of the High Priest (the name Yehudit is not mentioned!) got married, the Jews were not able to hide her. However, she displayed self-sacrifice and because of this, her brother Yehudah decapitated the minister.[36] [There is another version that Yehudah, his sister and his friend came to the minister and one of them (it is not known whether it was Yehudah, his sister or friends!) who decapitated the minister.[37]] After this incident, these manuscripts continue discussing the act of Yehudit. Therefore, it is possible that the version of the "Ran" mixed up the names of "Yehudah" and "Yehudit"! However, because the "Ran" states that he found his version in a Midrash, it is possible that this version is no longer extant. Indeed it should be noted that some hold that about half of the books written in the past are no longer extant![38]

However, according to the version before us, Rashi is more precise than the "Ran" on this issue. According to Rashi:[39] "The Greeks decreed that all the married virgins should first have sexual relations with their chief, and because of a woman a miracle occurred." One should note that Rashi does not mention the name Yehudit, and thus one can interpret the words "because of a woman a miracle occurred," that the miracle occurred because of the self-sacrifice of a woman.

The decapitation of the enemy and the fleeing of the enemy appears in all versions.

From all of the above, it can be concluded that the various versions of the book Yehudit, as well as the shortened versions, are basically in accordance with the words of the "Ran" and the "Kol-Bo."

However, Rabbi Azaria Min haAdumim, (16th century), author of the book "Meor Einayim," rejected learning this subject from the book of Yehudit.[40] He claims that the "Ran" "got mixed up between two events which appear in Megillat Ta'anit." The first of these is that on the seventeenth of Elul, it is written[41] that there was a daughter of Matityahu son of Yochanan Kohen Gadol, and when it came to the time for her to marry, Kastrin came to defile her but they did not let him, and Matityahu and his sons defeated the kingdom of Greece and as a result the Greeks were captured and killed. The second occurred on the twenty-fifth of Kislev, which is the miracle of Chanukah.[42] Only with the abolition of Megillat Ta'anit,[43] did the sages of the generation agree to include in the miracle of Chanukah a remembrance of how many miracles were performed by Yehudit who was the daughter of a Hasmonean, and these occurred on the seventeenth of Elul, and the miracle of Chanukah was on the twenty-fifth of Kislev. The common factor in both of them was that the Jews were saved from Antiochus. [44]

The words of the Acharonim

We saw above that due to the miracle performed by Yehudit, the "Kol-Bo" writes that it is customary to eat a "cooked cheese dish" on Chanukah, whilst the "Ran" writes just "cheese," and "Hemdat Yamim" writes that Yehudit gave the enemy "milk" to drink. The question arises, what do the Acharonim (great Rabbis who lived from about the 16th century) write on this subject?

The "Ben Ish Chai" (Rabbi Yosef Chaim of Baghdad, 19th century) brings only the version of the "Hemdat Yamim." According to him:[45] "Eating milk foods on Chanukah is a reminder of a miracle performed when Yehudit fed the enemy milk."

On the other hand, the "Kaf Hachaim" (Rabbi Yaakov Sofer, 19th-20th centuries) brings the "Kol Bo":[46] "Some say that one should eat cheese on Chanukah ... and she fed him a cooked cheese dish." (He then brings the words of the "Ben Ish Chai and he then adds that "there are explanations etc." namely he also accepts other explanations.)

In the sixteenth century, the Rema (Rabbi Moshe Isserles), brought what appears to be a mixture of the various versions. According to him:[47] "Some say that one eats cheese on Chanukah on account of the miracle performed when Yehudit fed the enemy milk." However, in his commentary to these words of the Rema, the Mishnah Berurah is more precise and writes:"[48] "And she fed the head of the oppressors cheese."

The first to bring the eating of cheese as well as milk is the "Levush" (Rabbi Mordechai Yoffe) who lived at the same period as the Rema. According to him:[49] "Some say that one should eat cheese and milk at the same meals since they cause a man to fall asleep, and this is in memory of the miracle which occurred when Yehudit fed the enemy milk." On these words of the "Levush," the "Eliyahu Zuta interprets:[50] "The language of the "Kol-Bo" is a cooked cheese dish." It is not clear whether his intention is to be more precise than the "Levush" in the wording of the "Kol-Bo," or to explain that the intention of the "Kol-Bo" in his expression "a cooked cheese dish" is "cheese and milk."

The incident with Yehudit was also quoted by the "Chayei Adam" (Rabbi Avraham Danzig, (18th-19th centuries), who writes:[51] "and she fed him a cooked cheese dish so that he would be thirsty ... and because the miracle occurred as a result of a milk cooked dish, some are accustomed to eat a milk cooked dish on Chanukah to commemorate the miracle that occurred as the result of milk." In the "Kitzur Shulchan Aruch" (Rabbi Shlomo Ganzfried, 19th century)[52] a similar language appears, but it is not exactly identical. The main difference between them is that

320

when the "Chayei Adam" writes "milk cooked dish," the "Kitzur Shulchan Aruch" writes "milk foods."

A question that could be asked is whether in the language of the Poskim (Rabbinical arbiters), cheese is to be regarded as a "milk food" and "milk cooked dish"?

The phrase "milk foods" is mentioned in the laws of Shavuot. The Rema writes:[53] "Eat milk foods on the first day of Shavuot," and from the commentary of the "Ba'er Heteiv"[54] one can see that it includes "hard cheese." Defining cheese as a "milk cooked dish" is more difficult, because in the making cheese from milk there is no cooking! However, it is possible to bring a proof from the final meal eaten before the fast of Tisha b'Av. At this meal, it is permissible to eat only one cooked dish. In the Rabbinical literature[55] it states that latkes (pancakes) filled with cheese are regarded as two cooked dishes. The "Machazit haShekel"[56] (Rabbi Shmuel Loew (Kalin), 18th century) explains that since one normally eats cheese raw it is not regarded as a cooked dish. However, when one cooks it, it is then considered a cooked dish. The "Ben Ish Chai"[57] also explains this in the same way. Therefore if one eats cooked cheese, one could define it as a "cooked milk dish."[58]

Even though all the Acharonim bring the custom of eating dairy foods on Chanukah, from the language of "Chayei Adam" and the "Kitzur Shulchan Aruch,"[59] it states that "a few are accustomed." From this one can see that this custom has not spread in over the entire Jewish people. We also see from a poem written by Rabbi Naftali haKohen from 1757 that the main thing is to remember the actions of Yehudit act, and eating cheese is only secondary. The language of his poem is:[60]

Eating cheese is not necessary,
It is customary to remember and not forget,
The actions of Yehudit act she did on purpose,
To feed him milk and make him sleep.

Among those who actually observed the custom of eating milk products on Chanukah were the Gaon from Munkacz (Rabbi

Yechiel Michael Gold, 20th century)[61] who was particular to eat "on his afternoon meal on Chanukah milk foods," and the Jews of Algeria[62] who ate "bread with cheese" on Rosh Chodesh occurring during Chanukah.

Levivot

Earlier in this paper we referred to a manuscript from the year 5162 (1402) which describes an act by Yehudit in which she fed levivot (pancakes) and "charitzei chalav" (cuts of milk – cheese) to Holofernes.

"Levivot" are mentioned in the "Account of Amnon and Tamar" in the book of Shmuel:[63] "Amnon said to the king let my sister come and make me a couple of levivot that I may eat from her hand ... and she took the dough and she kneaded it and made levivot in her presence and she cooked the levivot. And she took the pan and poured them out before him."

On the word "and she kneaded it," Rashi[64] comments: "And she first immersed semolina into boiling water, and then into oil." On the words "and he cooked," the "Metzudat David" comments:[65] "again on the pan of oil." On the words "and she took the pan," the Gemara in Masechet Sanhedrin writes:[66] "She made for him some sort of pancakes." From all this we see that levivot are what are today called "donuts." It should be noted that among the foods that Yehudit had, there were the ingredients for making levivot, namely wheat and oil.

Because of the feeding of "charitzei chalav" to Holofernes, the Rabbinical authorities write that it is customary to eat cheese (or milk products) on Chanukah. The question arises whether, and if so, where is eating levivot mentioned with regards to Chanukah?

In books of Halacha and of Customs (not including books printed in our generation!), the eating of levivot on Chanukah is not mentioned at all, but in songs and zemirot (Jewish hymns) it is mentioned.

The first to mention eating levivot on Chanukah was Rabbi Kalonymus ben Kalonymus (13th-14th centuries). In one of his poems he wrote[67] in a rhyming format, that in the ninth month which is Kislev in honour of Matityahu and the Hasmoneans, important women would gather together to make nice looking large round levivot, whose size was that of the frying pan, and whose colour would resemble that of a rainbow. They would then bake the dough and make all sorts of tasty foods from this mixture. They would especially take fine wheat flour and make donuts. They would also drink joyfully the same way as is done on festivals in order that they may transfer evil and suffering to their enemies who included Antiochus.

A few hundred years later, at the beginning of the seventeenth century, the book "Shetei Yadot" by the poet Menachem di Lonzano was published. This book is divided into two, and in the first part "Yad Ani," there are five "fingers." It is a moral poem and the fifth finger is called "tova tochachat." In the fourth part of this poem he mentions "Latkes with cheese on Chanukah."[68] One can thus see that according to Lonzano there was the custom to eat levivot with cheese on Chanukah; perhaps his intention is levivot filled with cheese, a dish which was known at that period (as mentioned above).

There is also a song for Shabbat Hanukkah written by "Avraham."[69] Some say that the author was Avraham ibn Ezra whilst others reject that he was the author.[70] This song first appeared in a siddur printed in Thessaloniki[71] in the year 5315 (1555), and at a later date in other books,[72] and also in Shabbat zemirot (Jewish hymns) booklets.[73] The song begins "Eat fatty foods and semolina" and brings all kinds of foods for the Shabbat Chanukah meal. From the Shabbat zemirot that are sung throughout the year, one can see that some of the foods that appear in this song, for example: "fatty foods"[74] and "barburim" (swans?)[75] are those eaten especially on Shabbat. It seems that "semolina" belongs especially to Chanukah, but what is meant by the word "semolina"? According to Rashi's interpretation of the word "and she kneaded it" (quoted above), it is possible that the meaning of "semolina" as mentioned in this song is levivot. There is another opinion that it is "kuskus" (rolled wheat semolina) which is eaten on Chanukah by the Jews of Algeria.

In the book "Ze haShulchan" about the customs of the Jews of Algeria, Rabbi Eliyahu Gig (19th-20th centuries) writes:[76] "It is customary amongst certain Jews on Rosh Chodesh occurring during Chanukah to include a cooked dish which is called kuskus with meat."

It should be noted that in the responsa of Rabbi Yisrael of Bruna (15th century) it is written[77] not to sing this song because "the meal of Chanukah is not regarded as a mitzvah meal, and from the content of this zemer it would make it appear that this meal was indeed a mitzvah meal." However, there is a difficulty since the Rema writes: "It is customary to sing a lot of zemirot and praises at this meal, and then it becomes a mitzvah meal." Also, according to the customs of Chernobyl and Skver, this zemer is sung on Shabbat Chanukah.[79] This shows that not everyone accepted the opinion of Rabbi Yisrael of Bruna on this issue.

The question remains, why is only eating of cheese on Chanukah mentioned by the "Kol-Bo" and the "Ran," but they do not mention the eating of levivot? In answer, there is room to suggest, that the reason the "Kol-Bo" writes "a cooked cheese dish" and not just "cheese" is that it refers to the dish of "levivot" filled with cheese."

Eating levivot (and also donuts) is mentioned in the various books of customs (as opposed to books of songs and zemirot). However, one needs to stress that this is limited to books published in today's era![80] The reason that is given for this custom, but without the quoting of any source, is that of the miracle of the oil. This is difficult because the mitzvah is to light oils on Chanukah but not eat them!

The Date that Yehudit Lived

From the language of the Gemara: "Women are obligated to light Chanukah candles since they were included in the miracle," it seems that the Gemara is referring to a miracle that happened during the time of the Maccabees. However, there are disagreements amongst the Rabbis on this issue.

There are, of course, those who set the date of the action of Yehudit during the time of the Maccabees. On the other hand, Rabbi Yaakov Emden known as Yaavetz, (18th century, Altona) in his book "Mor Uktzia" writes[81] that those who fixed the action of Yehudit during the time of the Maccabees, mixed up two different miracles, namely the original miracle of Chanukah and the miracle of Yehudit. In fact they occurred at completely different time periods which were not even near to each other. Because they did not want to commemorate the possibility that the miracle of Yehudit occurred at the period of the destruction of the First Temple, and since they did not find a specific time for Yehudit's action, they combined it together with the miracle of Chanukah. In addition, the "Ben Ish Chai" writes[82] that this miracle of Yehudit took place many years before the miracle of Chanukah, but since in both cases the enemy was the kingdom of Greece, who also wanted to prevent the Jews from observing the Torah, they connected this miracle to that of Chanukah.

What do historians say about the date of the composition of the book of Yehudit, and the period of Yehudit's action?

Regarding the date of the composition of this book, there are three different opinions.[83] The first is that the book was written in the middle of the fourth century BCE, the second is that it was composed during the Hasmonean period, and the third explanation (a very forced one!) is during the time of Bar-Kochba.

The second opinion (namely, in the time of the Hasmoneans) is the accepted opinion.[84] From this it is possible to understand why the book of Yehudit was not included among the books of the Tanach. In the Tosefta to Masechet Yadayim it is written:[85] "The Book of Ben Sira and other books which were written from then on, do not make one's hands ritually unclean." (This principle applies to books not included in the Tanach.) The book of Ben-Sira was written in the second century BCE and therefore the book of Yehudit which was written later (or at the same period), could not be included in the books of the Tanach.

Now that we have determined the possible date of the composition of the book of Yehudit, the question arises as to whether the event regarding Yehudit and Holofernes took place during the period of the book's writing, or instead, up to several hundred years before it. There is a discussion on the subject by the historian Robert Piper,[86] and he came to the conclusion that the action of Yehudit could have been from the seventh century BCE, namely, the end of the period of First Temple, until the second century BCE, namely, the Hasmonean period. All this accords with the various opinions proposed by the Rabbis for the act of Yehudit.

However, it should be noted that only one manuscript[87] mentions a connection between the book of Yehudit and Chanukah. This manuscript begins with a quote from the "Sheiltot d'Rav Achai Gaon" (about the 8th century) in the Derashot for Chanukah, that the Jew must give praise to G-d for every miracle such as Chanukah and Purim. It then continues with the action of Yehudit, and concludes by saying: "And they returned to Jerusalem with praise and song and great joy, and they searched and found one cruse of oil which was sealed with the seal of the High Priest and there was only sufficient oil for one day and a miracle occurred and it burned for eight days."[89]

In addition, the piyyut recited on (the first) Shabbat of Chanukah, "Odecha Ki Anafta,"[90] written by the poet Rabbi Yosef ben Shlomo of Carcassone in the eleventh century, relates the action of Yehudit, and concludes how the sages realised the great miracle that happened and they therefore established reading the entire Hallel all eight days and lighting the candles each year.

However, the question still remains, would the Rabbis have used the principle found in a book that is not included in the Tanach, in order to establish a custom in Israel?!

We see that in the Gemara,[91] the Zohar,[92] the Midrash,[93] the Mefarshim (commentators),[94] etc., great use is made of the "Apocrypha." In addition, paintings of Yehudit holding Holofernes' head also appear in two ancient Passover Haggadot.[95]

Since the subject before us is related to the book of Yehudit, we will give an example of the use of this book by the Rabbis. In his commentary on the Book of Devarim, the Ramban quotes from the Aramaic translation of this book:[96] "And the king of Assyria sent to all the servants of Ninevah." [In his commentary,

the Ramban calls the book of Yehudit "Megillat Shushan."
However, "Megillat Shushan" is the book Shoshana in the
Apocrypha. Grintz[97] explains that "in the translation which was
in the possession of the Ramban, the story Shoshana was joined to
the story of Yehudit, and since it appeared first, the whole book
was thus named."]

There is also a Rabbinical principal that Aggadot and
Midrashim cannot contradict what is written in the Talmud, but
they can add, learn from them and rely on them.[98] Moreover,
it is written in the Tosafot:[99] "In a number of cases, we rely on
external books and not the Gemara." The Tosafot gives a number
of examples in which the halacha follows the external books
and is thus contrary to the Gemara. [It should be noted that
"external books" which are mentioned by the Tosafot are the
"Minor Tractates" and the Pesikta.]

From all of the above, it is conceivable that the Rabbis would
not have refused to use the book of Yehudit to establish eating
customs on Chanukah, since it does not contradict anything
written in the Gemara.

References

Abbreviatons
SA = Shulchan Aruch
OC = Orach Chaim
(1)   Masechet Shabbat 23a
(2)   Ran on Shabbat 10a first words: amar Raba
(3)   *Kol-Bo* chap.44, hilchot Chanukah and tephillah
(4)   Rabbi Aharon Hakohen m'Lunel, *Orchot Chaim*, (Jerusalem, 5716),
       hilchot Chanukah, par.12
(5)   R. H. Charles, The Apocrypha and Pseudepigrapha of the Old Testament
       in English, (Oxford, 1913), p.243,
(6)   Biblia Sacra Polyglotta, Complectenia, Textus Originales, Tomus Quartus,
       Judith, Brian Walton, (London, 1657)
(7)   *Hasefarim hahitzonim*, edited by R' Avraham Kahana, (Jerusalem, 5730),
       vol.2, Book of Yehudit, p.368
(8)   Charles, op. cit., p.259, Morton S. Enslin, *The Book of Judith*, (Dropsie
       University, Philadelphia, 1972)
(9)   Frank Zimmermann, "Aids for the Recovery of the Hebrew Original of
       Judith", Journal of Biblical Literature, (Philadelphia), vol.lvii, 1938, p.68
(10)   Charles, op. cit ; Enslin, op.cit.

(11) Biblia Sacra Polyglotta, op. cit.

(12) M. Gaster, The Chronicles of Yerachmeel, (Ktav Publishing House: New York, 1971), pp.3-4

(13) MS Bodleian, catalogue Neubauer 2240.5, folio 170a; MS Bodleian, catalogue Neubauer, folio 262a

(14) R' Yehoshua Grintz, Sefer Yehudit, (Mosad Bialik: Jerusalem, 5717), p.198

(15) Sefer Hemdat Yamim l'Rosh Chodesh, (Venice? 5523), vol.2, chap.2, Ma'aseh haYehudit, p.57a

(16) Biblical Book Shoftim, chap.4 verse 19

(17) Biblical Book Shoftim, chap.5 verse 25

(18) Ma'aseh Yehudit, (Kostantiniye (Constantinople), 5312?), p.1b

(19) Ibid., p.11

(20) Ma'aseh Yehudit v'Nes Chanukah, (Berlin, 5526), introduction

(21) Dr. Martin Luther's Bibelubersetzung nach der letzten Original-Ausgabe, kritisch bearbeitet von Dr. Heinrich Ernst Bindseil und Dr. Hermann Agathon Niemeyer, (Halle, 1848), Das Buch Judith, Cap.10 v.6 (p.13)

(22) Ma'aseh Yehudit v'Nes Chanukah, op. cit., p.10a

(23) Megillat Yehudit, (Vienna 5579-1819), introduction

(24) Ramban, Devarim chap.21 verse 14 first words: lo titamaer bo

(25) Megillat Yehudit, op. cit., pp.19b-20a

(26) e.g. Rabbenu Nissim ben Yaacov, Chibur Yafe Mehayeshua l'Chacham, (5611), pp.22b-23b; Oseh Pele, vol.1, (Livorno, 5624), pp.14b-15b; MS London-Montefiore, 279/11, folios 47b-49a; MS Manchester-Gaster 82, folios 172a-173a; MS Budapest- Kauffman 259.5, folios 88-92

(27) Hasefarim hahitzonim, op. cit., p.372

(28) MS Bodleian, catalogue Neubauer 2746/6, folios 70b-71a

(29) Biblical book Shmuel I, chap.17 verse 18

(30) Targum Yonatan, Shmuel I, chap.17 verse 18

(31) Metzudat David on Shmuel I chap.17 verse 18

(32) Hemdat Yamim, op. cit., p.57b

(33) Yehudit: chap.8, verse 1, chap.9 verse 2. It is interesting to note that a mistake was made in the Vulgate (Latin translation of Bible) where it states "Ben Shimon ben Reuven", and in the Hebrew translations from the Latin, the translators did not think to correct this mistake! The error that Yehudit comes from the tribe of Reuben also appears in the section "zulat" of the liturgical poems for the second Shabbat of Chanukah (R. Yitzchak ben Aryeh Yosef Dov (Ber), Seder Avodat Yisrael, (Shukan: Berlin, 5697), p.644, and MS Bodleian catalogue Neuberger, 2669, folio 39a.

(34) MS Parma, catalogue deRossi 2557 (1094)

(35) MS Manchester-Gaster 82, folio 172b

(36) MS Parma, calalogue deRossi 2557 (1094); Halachot v'Agadot, edited by Dr. Michael Higger, ("Devai-Rabbanon": New York, 5693), Ma'aseh Yehudit. pp.95-98

(37) MS Budapest-Kauffman 188.2, folio 91

(38) see: *Books & People*, Bulletin of the Jewish National & University Library, Jerusalem, no.2, January 1992, p.9

(39) Rashi on Shabbat 23a, first words: hayu b'oto nes

(40) Rabbi Azaria de Rossi min haAdumim, *Meor Einayim*, (Vilna, 5626), pp.430-31

(41) *Megillat Ta'anit*, (Warsaw, 5634), chap.6, p.12b

(42) Ibid., chap.9, p.14a

(43) Masechet Rosh Hashanah 19b

(44) *Meor Einayim*, op. cit., p.431

(45) Rabbi Yosef Chaim, *Ben Ish Chai*, first year, parashat Vayeshev, par.24

(46) Rabbi Yaakov Chaim Sofer, *Kaf Hachaim*, SA OC chap.670, par.17

(47) SA OC, chap.670 par.2, Rama

(48) Mishnah Berurah, SA OC, chap.670 par.10

(49) Rabbi Mordechai Yoffe, *Levush* OC chap.670 par.2

(50) Eliahu Zuta, *Levush*, OC chap.670, par.12

(51) Rabbi Avraham Danzig, *Chayei Adam*, section 154 par.3

(52) *Kitzur Shulchan Aruch*, chap.139, par.3

(53) SA OC, chap 494 par.3, Rama

(54) *Ba'er Heteiv*, SA OC chap.494 par.8

(55) *Tanya Rabati*, (Warsaw, 5639), chap.59; Rabbi Zedekiah ben Avraham Anaw, *Shibbolei Haleket*, (Vilna, 5647), chap.265; Rabbi Avraham Abele Gombiner, *Magen Avraham*, SA OC chap.552, par.4

(56) Rabbi Shmuel Loew, *Machazit haShekel*, SA OC chap.552 par.4

(57) Ben Ish Chai, first year, parashat Devarim, par.19

(58) A difficulty in the words of the "Chayei Adam" is, why does this paragraph end with the words: "The miracle was done as a result of milk." He has already written "that the miracle was done as a result of a milk dish." So why does he repeat these words, and why for the second time he changed the words a "milk dish" to "milk"? The matter requires further investigation!

(59) *Chayei Adam*, op. cit.; *Kitzur Shulchan Aruch*, op. cit.

(60) Rabbi Naftali Katz Hakohen, *Sha'ar Naftali*, (Bruenn, 5517), Zemer shel Chanukah, p.36b

(61) Rabbi Yechiel Michael Gold, *Darkei Chaim Veshalom*, (Munkacz, 5700), Yemai Chanukah, par.817

(62) Rabbi Eliyahu ben Yosef Gig, *Ze haShulchan*, (Algiers, 5649), vol.2, dinei Chanukah, chap.32 par.4

(63) Biblical Shmuel II, chap13 verses 6-10

(64) Rashi on Shmuel II chap 13 verse 8 first word: vatelabev

(65) Rabbi David Altschuler, *Metzudat David*, Shmuel II chap.13 verse 8, first word: vatevashel

(66) Masechet Sanhedrin 21a

(67) Rabbi Kalonymus ben Kolonymus, *Even Bohan Vaderech Hayashar*, (Sulzbach, 5465), chap38 (p.15b-16a)

(68) Rabbi Menachem di Lonzano, *Shetei Yadot*, (Venice, 5378), Tovah Tokhahat, vol.4 p.134b (Acknowledgments to the Academy of the Hebrew Language for their help in locating this source for me.)

(69)   *Likutei Tzvi*, (Zhitomir, 5630), p.184 (367)

(70)   Rabbi Chaim Elazar Spira, *Sha'ar Yisaschar,* (Munkatch, 5700), vol.3, chap.75 p.29

(71)   *Machzor haGadol Mikol Hashanah according to the Ashkenazi rite,* (Salonika, 5315), p.29

(72)   *Likutei Tzvi*, op. cit.

(73)   Die hauslichen Gesange fur Freitag Abend, (Seder Zemirot l'leil Shabbat), (Frankfurt am Main, 1922), Zemer Naeh l'Shabbat Chanukah, p.40

(74)   Zemirah "Yom ze Mechubad" and Zemirah "Shabbat Hayom Lashem"

(75)   Zemirah "Mah Yedidut Menuchatach"

(76)   *Ze haShulchan,* op. cit.

(77)   Rabbi Yisrael of Bruna, *She'elot u-Teshuvot,* (Jerusalem, 5720), chap.137, (p.93)

(78)   SA OC, chap 670 par.2 Rama

(79)   *Luach Davar b'Ito,* (Every Thing in its Proper Time), 5753, (Achiezer: Bnei Brak), p.375

(80)   e.g. Rabbi Eliyahu Kitov, *Sefer haToda'ah,* (Jerusalem ,5736) p.171

(81)   Rabbi Yaakov Emden, *Mor Uktzia,* (Altona, 5521), vol.1, OC chap 670 p.76a

(82)   *Ben Ish Chai,* first year, parashat Vayeshev, par.24

(83)   e.g. Rabbi Tzvi Perez Chajes (Hayot), *Sefer Yehudit,* (Florence, 5669); Enslin, op. cit., pp. 26-31

(84)   Chajes, op. cit., p.2; Enslin, op.cit., p.26; Robert H. Pfeiffer, History of New Testament Times, (Greenwood Press: Connecticut, 1949), p.295 fn.13

(85)   Tosefta Masechet Yadayim, chap.2. par.5

(86)   Pfeiffer, op.cit., pp. 292-296

(87)   MS Bodleian, catalogue Neuberger 2669, folios 38a-39b

(88)   *Sheiltot d'Rav Achai Gaon*, parashat Vayishlach, sheilta 26, derashah l'Chanukah. The quote is from MS Bodelian, catalogue Neuberger 2669, folio 39a

(89)   During Chnukkah, the daily readings from the Torah are the section of the Torah concerning offerings brought by the 'nesiim' on the occasion of the consecration of the alter in the Tabernacle (SA OC chap.684 par.1). In the Synagogue, the reading is divided up as follows: for example, on the second day of Chanukah, the Kohen reads the first three verses of "bayom hasheni" the Levi then reads the following three verses of "bayom hasheni." Since the total verses in this section of the Torah is just six verses, the Yisrael then goes back and reads all these six verses again. (op. cit.). On the eighth day, however, the reading continues until the beginning of parashat Baha'alocha and it is customary for the Yisrael to read from 'bayom hateshii" onwards (Mishnah Berurah, op. cit., par.4). However there is a discussion in the Sdei Chemed (vol 9, p.200, maarechet Chanukah, par.12) whether the Yisrael should go back and read from "bayom hashemini" in order that all the three readings should include "bayom hashemini," since this is indeed the eighth day of Chanukah.

According to this reasoning, on the first day of Chanukah the Kohen should read until at least until the middle of "bayom harishon" (instead of stopping before "bayom harishon"). There is room for investigation why the Sdei Chemed does not discuss the reading of the Torah for the first day of Chanukah. However, 'Luach Dvar b'Ito' for the year 5755 writes that many place the Kohen reads until the end of the first three verses of "bayom harishon" but it gives no sources for this.

(90)   Seder Avodat Yisrael, op. cit., pp.629-633; Peirush nechmad... on the Yotzer for Shabbat Chanukah, (Venice, 5366)
(91)   e.g. Masechtot: Chagigah 13a, Yevamot 63b, Bava Kamma 92b, Nida 16b; Talmud Yerushalmi Berachot chap.7 halacha 2
(92)   Zohar, Parashat Noach, p.72b (par.295)
(93)   e.g. Bereshit Rabba parashah 8, first words: letter bet Rabbi Chama (Vilna); Bereshit Rabba parashah 91 first words: letter gimmel vayovou benai (Vilna); Midrash Tanchuma parashat Vayishlach chap.8 (Warsaw); Midrash Tanchuma Parashat Chukat chap.1 (Warsaw)
(94)   Ramban, introduction to commentary on the Torah; Tosafot Yom Tov, Megillah chap.3 mishnah 6 first word: banesiim
(95)   Ramban, Devarim op. cit.
(96)   Haggadah shel Pesach, around the words "Sh'foch Chamatcha" (Prague, 5287); Haggadah shel Pesach, around the words "Keha Lachma Anya" (Lublin, 5370)
(97)   Grintz, op. cit., p.207
(98)   Clalai haTalmud l'Knesset Hagedolah chap.70; Talmudic Encyclopedia (Talmudic Encyclopedia Publishing: Jerusalem, 5719), vol.9 p.253
(99)   Tosafot Pesachim 40b, first word: aval

Appendix

Follow on to publication of my paper.

The material below is taken from my unpublished autobiography

I wrote in my paper that there was no source for the eating of oil dishes on Chanukah. However, many years later I came across a book which said that there is a manuscript which states that the father of the Rambam says that is customary to eat oil dishes on Chanukah. Although I then made searches for this manuscript, I did not find it.

It was at the beginning of January 2006 – namely during Chanukah - that I received an e-mail from Mordechai Honig of Airmont, New York. In this e-mail he gave me the source for this custom. Apparently the Rambam's father had written it in his commentary on the Prayers which was in Arabic. Honig also quoted from a poem by Emmanuel Haromi which mentioned "levivot" on Chanukah and that this poem was in fact similar to that written by Kalonymus, which I brought down in my paper.

I then searched in the Internet for Honig's telephone number and duly thanked him for this information.

WHEN DOES ONE OBSERVE THE MITZVOT OF PURIM IN A DOUBTFULLY WALLED CITY?
(mitzvot = commandments)

It is written in the Mishnah,[1] that in "large cities" the megillah is read on the fourteenth of Adar, and in "cities walled from the days of Yehoshua bin Nun" on the fifteenth of Adar. Regarding cities which it is doubtful if they were surrounded by a wall from the days of Yehoshua, the Gemara says:[2] "Hezekiah read [the megillah] in Tiberias on the fourteenth and on the fifteenth ... Rav Asi read the megillah in Huzal on the fourteenth and on the fifteenth."

The following question then arises: Is there a difference in a "city doubtfully surrounded by a wall" between the reading of the megillah and the other mitzvot of Purim? The Gemara does not mention this question at all, and therefore one does not know whether Hezekiah and Rav Asi, who read the megillah on both the fourteenth and on the fifteenth, observed the rest of the mitzvot of the day – namely "mishloach manot" (sending of food presents) and "matanot laevyonim" (gifts to the needy) and the "seudat Purim" (the special Purim meal), on the fifteenth of Adar.

Are the Mitzvot of Purim from the Torah?

There is a principle "Safaika d'Oraita l'chumrah, Safaika d'Rabbanon l'kula." (If there is a doubt regarding a Torah commandment, one acts strictly; with a Rabbinical commandment one acts leniently). Possibly there a connection between this principle and the observance of the Purim mitzvot in a doubtfully walled city.

There are several questions that can be asked on this:

1) Are the mitzvot of Purim considered mitzvot from the Torah or mitzvot from the Rabbis?
2) Do all Purim mitzvot fall into the same category or not? For example: Is the mitzvah of reading Megillat Esther on Purim considered a mitzvah from the Torah, whilst the other mitzvot of Purim, mitzvot from the Rabbis?

It is written in the Talmud Bavli:[3] "Our Rabbis taught, Forty-eight prophets and seven prophetesses prophesied to Israel, and they neither took away from nor added anything to what is written in the Torah, except for the reading of the megillah. How did they derive this from the Torah? Rav Hiyya ben Abin said in the name of Rav Yehoshua ben Korcha. If for being delivered from slavery to freedom, we chant a hymn of praise, should we not do so all the more for being delivered from death to life." Namely only the mitzvah of reading the megillah is regarded as the mitzvah from the Torah.

However, according to the Talmud Yerushalmi:[4] "Rav and Rav Hanina and Rav Yonatan and Bar Kapra and Rav Yehoshua Ben-Levi said that this megillah was told to Moshe at Sinai, since there is no chronological order in the Torah." It is also written in Megillat Ta'anit[5] (a book mentioned in the Mishnah and several times in the Gemara): "Rav Yehoshua ben Karcha said that from the time of the death of Moshe, no prophet has arisen and added a mitzvah for the Jewish people with the exception of the mitzvot of Purim." It does not state the mitzvah of reading the megillah, but the mitzvot of Purim, namely *all the mitzvot* of Purim are regarded as mitzvot from the Torah.[6]

The "Binyan Shlomo" (Rabbi Shlomo ben Yisrael Moshe Hakohen, one of the leading Rabbis in Vilna in the 19th century), discusses this issue at length and writes:[7] "And it seems to me that the practical difference in halacha between the words of the Talmud Yerushalmi and that of the Talmud Bavli is that according to the Talmud Yerushalmi all the laws of Purim, feasting, rejoicing, mishloach manot, and matanot laevyonim are regarded as mitzvot from the Torah, and from this we can learn that in doubtfully walled cities the inhabitants are obligated to observe for two days all the mitzvot of Purim. However according to the Talmud Bavli there is no proof using the principle of 'kal v'chomer' (inference from a minor to a major) except for reading the megillah, but for the mitzvot of feasting and joy there is no 'kal v'chomer' and obviously it is therefore only a decree of Mordechai and Esther, and is not preferable to other decrees of the prophets, (for example, the four fasts during the year commemorating the destruction of the First Temple, and the mitzvah of the arava (willow) on Hoshana Raba), and one is thus not obligated to observe them on the fifteenth of Adar."

Is it Possible to Observe the Mitzvot of Purim before Purim?

There is room to investigate whether it is possible to observe the mitzvot of Purim before Purim itself! If the answer is positive, it will be possible to observe the mitzvot of Purim in a doubtfully walled city on the fourteenth of Adar alone, even if it turns out that the city was surrounded by a wall and the correct date of Purim is in fact the fifteenth!

(In this chapter we shall study performing the various mitzvot of Purim on the fourteenth of Adar for a person in an undoubtedly walled city.)

Reading of the Megillah: It is written in the Talmud Yerushalmi:[8] "Everyone can fulfil the mitzvah on the fourteenth which is the time of its reading." According to this, a person living in a doubtfully walled city, even if he were to read the megillah only on the fourteenth, there would be no reason for him to read it also on the fifteenth! However, Rabbi Avraham Yeshaya Karelitz known as the "Chazon Ish"[9] (19th-20th centuries) writes about this Yerushalmi that this is only if he erred reading it, but if he read it deliberately on the fourteenth, the Rabbis obligate him to read it again on the fifteenth. Furthermore, both the Rambam (Maimonides) and the Shulchan Aruch omit this Yerushalmi.

Matanot laevyonim: It is written in the Gemara[10] that in villages where the reading of the megillah is advanced to the eleventh, or twelfth or thirteenth of Adar, one collects and distributes the matanot laevyonim on the day of the reading, "because the poor are waiting anxiously for the reading of the megillah." The Rambam[11] (Maimonides) learns from this Gemara that one is not obligated to advance it but it is permitted. Rabbi Zerachiah Halevi of Girona,[12] known as the "Baal haMaor" (12th century) writes that now that the Jews are suffering from poverty, there is an apprehension that the poor will eat the food before Purim, and will have nothing to eat at their seudat Purim on the fourteenth day. From this, Rabbi Avraham Gombiner known as the "Magen Avraham"[13] (17th century) deduces that "one does not give them [matanot laevyonim] before Purim." However, Rabbi Aharon Alfandari the "Yad Aharon"[14] (one of the sages of Turkey in the 18th century) learns from the words of "Baal haMaor": "This means that even if one does not certainly know that they will eat it on Purim, one will fulfil the mitzvah even if one gives it not at the obligated time, and the reason is that one needs to be joyful on that day." However, his reasoning is difficult, since the advancing of giving matanot laevyonim applies only for the day of reading the megillah in the villages, because of the reason that "the poor are waiting anxiously for the reading of the megillah," but this *is limited to the day of the reading of the megillah*.[15] Therefore one needs to investigate whether one can learn from the words of the Baal haMaor whether the advancing of matanot laevyonim applies today, since the law regarding the villages is not applicable today.

Mishloach Manot: On the advancing of the giving mishloach manot before Purim, there is a conclusion which stems from the opinion of the "Yad Aharon,"[16] which could be relevant to our subject, namely, if one were to send mishloach manot before Purim and they would arrive on Purim (namely on the day of Purim for both the sender and the recipient[17]), according to the "Yad Aharon," the sender fulfils his obligation. It follows from this that a person living in a doubtfully walled city who sends mishloach manot to a person of his city *only* on the fourteenth, but *twice*, once for it to be delivered on the fourteenth, and once to be delivered on the fifteenth, fulfils the mitzvah of mishloach manot! Thus either way he observes the mitzvah, namely if the city was not surrounded by a wall, the date of Purim is fourteenth Adar, and he observes the mitzvah according to all the opinions, and if the city was surrounded by a wall and he sends the mishloach manot on the fourteenth, namely before Purim in his city, and they arrive on the fifteenth which is Purim for both the sender and the recipient, according to the opinion of the "Yad Aharon" he fulfils the mitzvah!

Feast and Joy – the Seudat Purim: According to the Gemara,[18] even in the villages, who advance matanot laevyonim to the day of reading the megillah, "being joyful is not accomplished except in its proper time," namely the day of Purim itself.

In the novella of Rabbenu Meir Simcha (19th-20th centuries) on the Talmud it is written:[19] "One certainly does not fulfil the mitzvah of feasting and being joyful if one does them on the fourteenth, since the primary day is the fifteenth. Only the reading of the megillah on the miracle which occurred on the fourteenth (one fulfills one's obligation)." From his words "*only the* reading of the megillah," namely mishloach manot, matanot laevyonim and the seudat Purim one is obligated *only* on the fifteenth. Likewise, Rabbi Yoel Elitzur[20] writes that there is a principle that according to the halacha a person living in a walled city only fulfils the mitzvah of reading the megillah on the fourteenth, *but on the other mitzvot of the day he is only obligated on the fifteenth.*

Mishloach Manot, Matanot Laevyonim and the Seudat Purim

What were the rulings of the Poskim (Rabbinical arbiters) during the course of the generations about the date of the observance of these mitzvot in a doubtfully walled city? It was stated above that the Gemara did not discuss this subject at all.

The first to discuss the subject was the Midrash "Lekach Tov." On the verse in the Book of Esther:[21] "And the other Jews

337

who were in the king's provinces ..." the Midrash writes:[22] "And in a place where one it is not known whether or not it was surrounded by a wall in the days of Achashverosh, it is customary for two days of feasting and rejoicing, as in a walled city and also in an unwalled city." According to Rabbenu Hananel (11th century):[23] "Joy means a Purim meal," and Rashi writes:[24] "Joy: food and drink." [One needs to investigate why it is written "from the days of Achashverosh," and not from "the days of Yehoshua bin Nun"? In one of the manuscripts[25] which is still extant, the words "from the days of Achashverosh" do not appear at all.]

The author of the Midrash "Lekach Tov" was Rav Tuvia son of Rav Eliezer who was one of the sages in Greece and lived at the end of the 11th century. He took material for his composition from the Gemara, Halachic Midrashim, and Aggadic Midrashim; however some of them are not extant today.

In the thirteenth and fourteenth centuries a number of Rishonim (great Rabbis who lived approximately between the 11th and 15th centuries) ruled on the subject. The first was Rabbi Elazer ben Yehuda of Garmiza. He was an Ashkenazi sage who was born around 1165 and died around 1238. In his book "Harokeach" he writes:[26] "In a place where there is doubt as to whether or not their city was surrounded by a wall from the days of Yehoshua bin Nun, one reads the megillah in the fourteenth, but on the fifteenth one reads it without a berachah and one rejoices on both days.

Some years later, Rabbi Zedekiah ben Rabbi Avraham Anaw, who was one of the sages of Italy in the thirteenth century, in his book "Shibbolei haLeket"[27] writes in an almost identical language as did the author of "Harokeach," and he states that his source is Rabbenu Shlomo (Rashi).

The question arises, why the Midrash "Lekach Tov," the "Rokeach" and "Shibbolei haLeket" do not mention mishloach manot and matanot laevyonim? One could answer this as follows: We can learn from the Poskim that there is a strong connection between the seudat Purim and mishloach manot. For example, Rabbi Levi ibn Habib known as the "HaRaLBaCh" (15th-16th centuries) writes:[28] "Because these food dishes are for the seudah," and the "Terumat haDeshen" (Rabbi Yisrael Isserlin, (15th century) writes:[29] "The reason for mishloach manot is that everyone should have sufficient food to enable them to have a proper meal." From the words of the "Terumat haDeshen" Rabbi Ovadia Yosef in his responsa "Yehave Daat" writes:[30] "And according to this, it seems that the time for mishloach manot is dependent on the time of the mitzvah of the seudat Purim."

Above was brought that the "Baal haMaor" links matanot laevyonim with the Purim meal. In addition, the Gemara[31] states that on the day the megillah is read, matanot laevyonim are collected and distributed to the needy. From this, it can be deduced that in a doubtfully walled city, it is possible that matanot laevyonim are given also on the fifteenth, and this may explain why the above-mentioned Rishonim did not explicitly mention matanot laevyonim.

During the same period of the "Shibbolei haLeket" lived Rabbi Yeshaya ben R. Eliyahu Matrani known as the "Riaz," who was one of the greatest sages of Italy. He wrote the "Piskei Halachot," which includes rulings in accordance with the order of the Masechtot (Tractates) in the Talmud. He also wrote "Kuntres haRaayot" which included proofs to all the rulings that he wrote, but this book seems not to be extant today. In his rulings for Masechet Megillah he writes on the subject before us:[32] "A city that was doubtful to the early generations, if it was surrounded by a wall from the days of Yehoshua, and if not, one reads the megillah on the fourteenth and on the fifteenth ... and it seems to me that one is to be joyful and give matanot laevyonim on both days, as explained in Kuntres haRaayot. However, since today Kuntres haRaayot is not extant, we do not know from where he learned that one has to be joyful and give matanot laevyonim on the two days of Purim. However, there are extant a large number of manuscripts on his rulings, the oldest being from the fourteenth century.[33] Also, in the middle of the sixteenth century, Rabbi Yehoshua ben Shimon Baruch, one of the sages of Italy, brings in his glosses "Shiltei Giborim"[34] on the Rif, a large number of the rulings of the Riaz, including the ruling quoted above.[35]

We see that the Riaz *explicitly* mentions matanot laevyonim. "Midrash Lekach Tov," "Shibbolei haLeket," and the "Rokeach" do not mention matanot laevyonim and therefore it is possible to suggest that according to them one does not observe them also on the fifteenth. However, from the language of the Riaz, without any shadow of a doubt, and according to his opinion, matanot laevyonim should be given even on the fifteenth. One should note that at the period of the Acharonim (great Rabbis who lived from about the 16th century), when this subject is brought in the halacha, they bring the opinion of the Riaz, but not that of "Midrash Lekah Tov" nor "Shibbolei haLeket" nor the "Rokeach."

In contrast to the above Rishonim, Rabbi Hezekiah da Silva, known as the "Pri Chadash" who lived in the second half of the seventeenth century writes:[36] "A city which is doubtfully walled, even though one reads the megillah on two days because of a doubt, nevertheless one does not recite the berachah except on the fourteenth, since this is the day of reading for the majority of the world. Using a similar reasoning, this is the law for the seudat Purim which one does not observe except on the fourteenth, as is done in the majority of the world and this is indeed sufficient."

There are various opinions that seek to explain why the "Pri Chadash" does not rule on this issue in accordance with the other Poskim. The "Binyan Shlomo"[37] suggests that perhaps the "Pri Chadash" thinks that the reading of the megillah is like a mitzvah from the Torah, and the other mitzvot of Purim are only Rabbinical, and "because berachot are Rabbinical, therefore it is sufficient for just one day [namely the fourteenth to recite the berachah on reading the megillah], and likewise since feasting and joy are also Rabbinical, it is sufficient for just one day. He also suggests that matanot laevyonim is a matter of money and in "a doubt of money for the poor one goes leniently" and in "mishloach manot there is a danger of theft by one's fellow man that if he knows that the city was unwalled he would not want on the fifteenth to send food packages to his friend, and therefore one does not spend money if there is a doubt." According to this reasoning Rabbi Ovadia Yosef in his responsa "Yabia Omer" [38] comments on this and he brings the words of Rabbi Shlomo ben Avraham Aderet[39] who was known as the "Rashba" (13th-14h centuries), that every time that a poor person requests charity, one is obligated to give him, and because of this one should give him matanot laevyonim also because of a doubt, and it is as if a poor person has stood next to him and requested charity. On the question of mishloach manot, the "Yabia Omer" writes:[40] "I saw that someone had written that in a doubtfully walled city, one is obligated to give mishloach manot also on the fifteenth, and this is not relevant to the rule of being lenient in the case of a doubt concerning money, since the obligation to give mishloach manot is a certainty, and there is a doubt if has exempted himself since he gave on the fourteenth, and is like one says loan me money and there is a doubt if he repaid it, he is obligated to pay."

The "Pri Megadim" wondered about the words of the "Pri Chadash" and wrote:[41] But "perhaps it detracts from reading the megillah which has to be read on the two days?" but he then continues, that it is possible to give matanot laevyonim to the needy on the fourteenth day only, because it is no worse than giving before Purim, and we only prevent it because of the fear that the poor will eat the matanot laevyonim before Purim. He concludes: "What happens if the poor person eats it on the fourteenth and he has nothing on which to rejoice on the fifteenth. He concludes that this is left as requiring further study." The "Binyan Shlomo" reinforces this comment of the "Pri Megadim" and writes:[42] "There is a greater apprehension, since the fourteenth is also his Purim and it is more common to eat the food on that day [rather than on the days before Purim]."

Another suggestion to explain the words of the "Pri Chadash" arises from the explanation of the "Shevet Benjamin" Rabbi Binyamin Yitzchak Trachtman (20th century)." He writes:[43] "Mishloach Manot is a mitzvah of joy. Certainly it is not relevant to say that it was established specifically because of the miracle of Shushan Purim; however, for the reading of the megillah it is relevant to say this. Indeed, mishloach manot was instituted for the joy of Purim. Therefore, when one gives mishloach manot on the fourteenth, he fulfils the mitzvah also for the fifteenth. But one needs to investigate this reasoning, since it, apparently refutes the words of the Midrash and the Rishonim (brought above) who write that one rejoices on the fourteenth and on the fifteenth.

Rabbi Yosef Molcho, who was one of the Rabbis in Salonika, in the middle of the eighteenth century is the author of the book "Shulchan Gavoha." There he writes:[44] "A city that is doubtfully surrounded by a wall, in the same way that one is obligated to read the megillah on two days because of a doubt, he is likewise obligated in mishloach manot and matanot laevyonim on the two days. Just as reading the megillah on the fourteenth one does not fulfil one's obligation, were it to be surrounded by a wall at the time of Yehoshua bin Nun; hence his reading on the fourteenth is as if he read it on the thirteenth or twelfth which is not the correct time, and therefore he would not fulfil his obligation. Likewise, by sending mishloach manot on the fourteenth one does not fulfil one's obligation since it is not definitely the day of Purim. However, after that, he concludes that "this is not the custom." About a century earlier, Rabbi Chaim ben Yisrael Benvenist (17th century) author of the book "Knesset haGedolah" had written the exact opposite, when he wrote:[45] "And so is the custom [to have joy and give matanot laevyonim on both days]."

The "Magen Avraham" ruled:[46] "One reads on the fourteenth and fifteenth and it is customary to be joyful and give matanot laevyonim on both days." On these words, the "Pri Megadim" finds a difficulty:[47] "Why did the Rabbi omit mishloach manot on both of the days because of a doubt?" However, the "Magen Avraham" just quoted the "Shiltei Giborim" who brought down the "Riaz." (Above was given an explanation of the reason why the Midrash and the Rishonim did not explicitly mention mishloach manot). This ruling of the Riaz was also quoted by the Mishnah Berurah,[48] Rabbi Yaakov Chaim Sofer[49] in his book "Kaf haChaim"and Rabbi Yichya Tzalach,[50] (18th century) in his book "Shtilei Zeitim."

The question arises: What do the yearly calendars which are published nowadays by various organisations or individuals state on this subject? The calendars that mention that one needs to observe *all the mitzvot* of Purim in cities which were doubtfully walled also on the fifteenth. These calendars include "Luach l'Eretz Yisrael" brought out by Rabbi Yechiel Tucazinsky,[51] "Luach Hechal Shlomo,"[52] "Luach Yad l'Achim,"[53] and "Luach Belz."[54]

The Procedure in Certain Cities

Above we quoted Poskim who discuss this subject in general. However, in the Rabbinical literature one can find this subject in the context of certain cities situated in Eretz Yisrael and in the Diaspora, and here are some examples:

Safed: Rabbi Yitzchak Luria known as the Ari lived for some years in the city of Safed, and his principal disciple Rabbi Chaim Vital (16th-17th centuries) writes in "Shaar Hakavanot":[55] "With the other mitzvot of Purim such as mishloach manot and matanot laevyonim, because of a doubt it was customary to also observe them on the second day." However, today, even though the majority of the Synagogues read the megillah on the fifteenth, only a few Jews in Safed observe the other mitzvot of Purim on the fifteenth.[56]

Damascus: Rabbi Chaim Vital[57] wrote that at the period of the Ari they acted in the same way as in Safed. In answer to a question that had been submitted about the year 5684 (1924) to Rabbi Ezra Hakohen Trab[58] (author of the book "Milei d'Ezra") by Jews originating from Damascus and had gone to live in Buenos Aires in Argentina, it is obvious that at that period "asu" (did) two days Purim. One should note that they used the expression "asu" and not "koru hamegillah" (read the megillah). Therefore, the question arises that maybe it is possible to learn from this that in Damascus on the fifteenth they not only did they observe the reading of the megillah but also observed the other mitzvot of Purim?

Baghdad: Regarding the custom in the city of Baghdad [in Iraq], the "Ben Ish Chai"[59] (Rabbi Yosef Chaim, 19th century) writes: "It is necessary to observe on the fifteenth the seudat Purim, mishloach manot and matanot laevyonim, both by men and women, but on a lower scale than as on the fourteenth."

Aram Tzova: It is written in the book "Derech Eretz" written by Rabbi Avraham Adas:[60] "There was a custom of Aram Tzova [the city of Aleppo in northern Syria] from ancient times, when all the laws of Purim were observed on both days .... This custom continued among the Musta'arabi Jews (veteran residents), until after the arrival of the Jews who had been expelled from Spain. However, recently these two communities merged (the Sephardic community and the Musta'arabi community), and they repealed all the Purim laws concerning the second day, with the exception of the reading of the megillah ... ".

Prague: Although there are no walled cities in Europe from the days of Yehoshua,[61] there are discussions about the city of Prague regarding the observance of two days of Purim. After a long deliberation Rabbi Shmuel Yuda Leib Koider, the "Olat Shmuel" concludes saying:[62] "One must obviously not feel bad about anyone who does not read the megillah on the fifteenth, and one who does read it at least receives the reward like someone who reads from the Torah." Rabbi Elazar Fleckeles,[63] author of "Teshuva me'Ahava," who was a Dayan and later Rabbi of Prague in the late eighteenth and early nineteenth centuries, writes that even though most of the inhabitants who included the "benai Torah" (very religious Jews) only read on the fourteenth, "the G-d fearing read the megillah also on the fifteenth, and also observe mishloach manot, matanot laevyonim and the seudat Purim, and their only difference between the fourteenth and the fifteenth is the recital of the berachah on reading the megillah." The reading of the megillah on the fourteenth and fifteenth in Prague is also mentioned by Rabbi Avraham Danzig,[64] known as the "Chayei Adam." Interesting, he does not mention any other city in the world that observes two days Purim! It would seem that the reason is that the "Chayei Adam" studied in yeshivot (Rabbinical colleges) in Prague, so he personally saw the Purim customs in Prague.

Balkh: Jews settled in Afghanistan in ancient times. Among the Jewish communities in Afghanistan was a community in the city of Balkh. On Purim in that city it is written: "In the city of Balkh the megillah is read on two days, on the fourteenth of Adar and on the morrow, the fifteenth, according to the custom of Jerusalem, because they have a tradition that Balkh was surrounded by a wall from the days of Yehoshua bin Nun."[65] In the language of R' Giora Pozilov: "... Purim is observed on the fourteenth and fifteenth ..."[66] Today there is no Jewish community in Balkh. The former residents of Balkh who live in Eretz Yisrael do not remember if they only read the megillah, or also observed all the Purim mitzvot on the fifteenth.[67] Possibly the reason that they do not remember if they observed all the mitzvot of Purim for two days, is that the *entire* community of the Jews of Afghanistan had the custom that the families extend the joy of the mitzvot of Purim to the second day, namely Shushan Purim,[68] and some write for even more than two days![69]

However, most of the practical questions that have actually arisen are in the renewal of the Jewish settlement in Eretz Yisrael in recent times:

Tiberias: The date of Purim in the city of Tiberias was already in doubt during the period of the Gemara.[70] The city was surrounded by a wall from the days of Yehoshua, but only on three sides of the city; the fourth side is the Sea of Galilee. Regarding Purim in Tiberias, Rabbi Zvi Pesach Frank rules:[71] "One reads the megillah on the fourteenth and fifteenth ... and also joyfulness, the seudat Purim, mishloach manot and matanot laevyonim on both days." However, in practice, in the city of Tiberias today, the observance of Purim on the fifteenth is limited to just the reading of the megillah.[72]

Hebron and Kiryat Arba: There are discussions by the Poskim on the question of whether to observe two days of Purim in Hebron, because the city was a city of refuge. However, there is an ancient custom in Hebron to read the megillah on the fourteenth and the fifteenth.[73] Kiryat Arba, which was established near Hebron in

5731 (1971), is considered "close and visible" to Hebron. Rabbi Dov Lior the Rabbi of the city published a summary of the laws for Purim for the residents of Kiryat Arba and Hebron. According to these halachot:[74] "On the fifteenth one only reads the megillah on the night and the day ... it is customary not to have a seudat Purim and distribute mishloach manot and matanot laevyonim, except for those who want to be meticulous in this matter."

Lod: The city of Lod is mentioned in the Gemara[75] as a walled city from the days of Yehoshua, but one needs to be sure that Lod today is in the same location as the ancient city of Lod. In a question which was asked to Rabbi Yitzchak Yaakov Weiss, the "Minchat Yitzchak" (20th century), it states:[76] "Today the entire area of Lod is large, and numerous Jews live there; however, it is in no way possible to say that there are no Jews living in the area of the ancient city, and therefore one should read the megillah there on two days." In his answer, regarding the observance of the other mitzvot of Purim in Lod, Rabbi Weiss writes:[77] "It is necessary ... to be joyful and matanot laevyonim on both days." In addition, in a letter from Rabbi Israel Mintzberg and the Rabbis of Lod in 5722 (1962), it states:[78] "The joy of Purim, the seudat Purim, mishloach manot, matanot laevyonim on both days." On a chart of the laws of Purim in Lod, the Rabbi of the city Rabbi Natan Ortner writes:[79] "Mishloach manot, matanot laevyonim – send and give; on fifteenth – send and give. Seudat Purim: on fourteenth – required, fifteenth – required."

Beit El: In 5738 (1978), the Beit El settlement was established adjacent to a place called Beitin. Some identify this Beitin as the ancient Beit El. At the time of the foundation of the new settlement, there were discussions and rulings on the date of Purim in the Beit El settlement, including a ruling by Rabbi Ovadia Yosef. He ruled:[80] "It is proper and correct that they should read in Beit El even on the second day but without a berachah .... and it is customary to be joyful and have a seudat Purim and mishloach manot and matanot laevyonim on both days."

Haifa: In 5755, the Chief Rabbi of Haifa, Rabbi She'ar Yashuv Cohen, published a pamphlet entitled "kviyat yemai Purim b'Haifa" (Determining the Days of Purim in Haifa).[81] In it he wrote that there are a few congregations who read the megillah also on the fifteenth. Also, followers of Chabad in the city began reading the megillah on the fifteenth, after receiving a letter from the Lubavitcher Rebbe, which he sent to a conference of shlichim on Mount Carmel in the summer of 5744 (1984). In his letter he wrote: [82] "Haifa: one reads the megillah also on the fifteenth since there is a doubt if it was surrounded by a wall." Furthermore, the followers of Chabad in Haifa observe all the mitzvot of Purim also on the fifteenth.[83] Among the other Synagogues that read the megillah on the fifteenth is the synagogue of Rabbi Shneyer Klopt. The worshipers in this synagogue also give mishloach manot and matanot laevyonim and have a seudat Purim on the fifteenth.[84]

Bnei Brak: According to Rabbi Avraham Yeshaya Karelitz (19th-20th centuries) known as the "Chazon Ish," the city of Bnei Brak is considered "near and visible" to the city of Jaffa, a city that was doubtfully surrounded by a wall, and the custom of the "Chazon Ish" was to observe *all the mitzvot* of Purim also on the fifteenth.[85] About twenty years after the death of the Chazon Ish, Rabbi Binyamin Zilber received a letter from a resident of Bnei Brak asking,[86] among other things:" If one has to observe mishloach manot and matanot laevyonim on both days." In his reply, Rabbi Zilber wrote:[87] "And the law in a doubtfully walled city is that one reads on the fifteenth without a berachah and one also is joyful, and observes matanot laevyonim and mishloach manot in the same way as on the fourteenth." In the "Luach Davav b'Ito" it is written:[88] "Many people today read the megillah in Bnei Brak (which is near to Jaffa) without a blessing, and also observe the rest of the mitzvot of the day, in accordance with the Chazon Ish.

The new neighborhoods of Jerusalem: There are many questions and answers about the new areas of Jerusalem, such as Har Nof and Ramot. The Rishon LeZion, Rabbi Ovadia Yosef discusses the subject and in the conclusion he writes:[89] "The residents of the Ramot neighborhood and the residents of the Har Nof neighborhood in Jerusalem should read the megillah on the fourteenth of Adar with the berachot. Also, they need to have a seudat Purim and observe mishloach manot and matanot laevyonim on the fourteenth of Adar. And if they want as an act of special piety to read the megillah also on the fifteenth without a berachah, and if they want to have a seudat Purim, and if they want to observe the mitzvot of mishloach manot and matanot laevyonim, they will receive a Heavenly blessing." It should be noted that in Rabbi Ovadia Yosef's halachic rulings on Purim in the doubtfully walled cities, his language is clearer and more unequivocal:[90] "One needs to be strict for the mitzvot of mishloach manot, matanot laevyonim and have a seudat Purim on the fifteenth because of a doubt."

There are a number of Rabbinical writings, which mention certain cities where the megillah is read on both days, for example the cities: Gaza,[91] Izmir,[92] the villages near Izmir,[93] Salonika,[94] Batzra,[95] Bombay.[96] However, it is not mentioned whether one only reads the megillah or also observes the other mitzvot of Purim.

The Rav from Munkatch had an interesting custom, about which it is related:[97] "On the day of Shushan Purim he would observe the mitzvot of mishloach manot, matanot laevyonim and make a feast and have rejoicing as on the first day of Purim." Needless to say, the city of Munkatch is not considered a city doubtfully surrounded by a wall from the days of Yehoshua! However, regarding the reading of the megillah on the fifteenth, he wrote:[98] "It is forbidden to read a kosher megillah (especially amongst ten people) and also one may not recite the berachah [in a place where there is no doubt that it is a walled city] because this is like reading for the mitzvah on the fifteenth since it is written "v'lo yaavor" (may not read the megillah in the days after Purim), and it is also written[99] that he "made a noise about the same practice."

The Prayer Al Hanissim and Reading the Torah on Purim

In addition to the specific mitzvot for Purim, there is an addition to the order of the tefillot (prayer service) on Purim. This is

"Al Hanissim" which is added in the amidah and in bircat hamazon. There is also a special Reading from the Torah beginning "Vayavo Amalek" (since Haman was descended from Amalek). It is necessary to investigate if in a doubtfully walled city these additions and Torah readings are made on the fifteenth.

On this, the following questions can be asked: 1) Can reading the Torah at times not instituted by the Sages result in a berachah to be recited in vain? 2) Could the addition of al hanissim in the amidah on the fifteenth of Adar (in an unquestionable non-walled city) be considered an interruption to the amidah?

(1) *Reading the Torah*: Here there are different opinions on this subject. On the one hand, Rabbi Avraham ben Mordechai Halevi, author of the book "Ginat Vradim," who was the head of the Rabbis of Egypt in Cairo at the end of the seventeenth century, writes:[100] "On the subject of Reading from the Torah one may be lenient, since every time that ten men agree to read from the Torah is good. They take out the Torah and recite the berachot out of respect for the congregation, because it is our life and the length of our days, and I wish the Jews would read the Torah every day. "

We also see that a number of Ashkenazi communities read from the Torah on the *night* of Simchat Torah, namely at a time not fixed by the Sages for the reading of the Torah. This custom is mentioned by Rabbi Isaak Tyrnau[101] (at the end of the fourteenth century or the beginning of the fifteenth century), and was brought by the Rema in his "Darchei Moshe"[102] and also in his glosses on the Shulchan Aruch.[103] The "Mishnah Berurah"[104] and the "Kitzur Shulchan Aruch"[105] write that it is now customary to call up three people to the Torah on the night of Simchat Torah and read from the parashah "Vezot Haberachah" (which is the Torah reading on the morning of Simchat Torah).

On the other hand, there are Poskim who are concerned that it might be a berachah recited in vain. Rabbi Naftali Tzvi Yehudah Berlin, the "Netziv" from Volozhin[106] (19th century), writes that according to the Talmud Yerushalmi, whenever one reads from the Torah in public, one must say the berachah, and therefore reading the Torah on a day not prescribed by Sages results in a berachah being recited in vain. In addition, it is also forbidden to read from the Torah on a day that the Sages set for the reading of the Torah, but not at the hour they set, for example at night. However, a question can be asked, that according to the opinion of the Netziv, is it permitted when the fifteenth of Adar occurs on a Monday (since in any case there is a reading of the Torah on every Monday) to read the verses beginning "Vayovo Amalek" in place of the designated reading for that day?

2) *Al Hanissim*: There are discussions about whether reciting al hanissim on the fifteenth in a doubtfully walled city would be an interruption in the saying of the amidah. The earliest mention of this subject was, apparently, in the "Sheiltot d'Rav Achai Gaon":[107] "And to say al hanissim in the berachah 'modim' (of the amidah), but only on the fourteenth." The "Beit Yosef" quotes Rabbi Aharon Hakohen m'Lunel,[108] known as the "Orchot Chaim" (14th century): "To read from the Torah and to recite al hanissim in the amidah is forbidden, except on the fourteenth day." However, the Beit Yosef does not agree with the "Orchot Chaim," and writes:[109] "I do not know what is the prohibition to recite al hanissim [in a non-walled city on the fifteenth.]" Lechatchila (from the outset) in a city which is not walled, one should not say al hanissim on the fifteenth, and there are Poskim who even in rule that b'diavad (in retrospect) one has to repeat the amidah,[110] although there are others who rule that b'diavad one does not have to repeat the amidah[111] since there is some relevance (to the contents of al hanissim) to these days."

Until now, we have discussed saying al hanissim on the fifteenth in a city not surrounded by a wall. It is obvious that the Poskim that rule that one does not have to repeat the amidah in unwalled cities, one would not have to repeat it in a *doubtfully* walled city. However, in addition to this, the "Pri Megadim"[112] and also the Mishnah Berurah[113] write that "doubtfully walled cities say al hanissim in the amidah on the two days, and it is not regarded as an interruption since it was said because of a doubt." Thus from the outset, one says al hanissim in a doubtfully walled city on the fifteenth.

On the other hand, the "Rosh Yosef"[114] (Rabbi Yosef Eskapa who was one of the sages of Turkey and the Av Bet Din in Izmir in the first half of the seventeenth century), holds that the intention of the "Orchot Chaim" (given above) is that in a doubtfully walled city saying al hanissim on the fifteenth is an "interruption in one's prayers," and to read from the Sefer Torah is "not in accordance with the halacha." In addition, the "Sefer Eretz Yisrael" by Rabbi Yechiel Michel Tucazinsky's,[115] as well as most of the various calendars published today,[116] write that today in a doubtfully walled city one does not say al hanissim and also does not read from the Torah on the fifteenth. In contrast however, according to Luach Belz[117] (a Hassidic dynasty founded in the town of Belz in Western Ukraine), one does say al hanissim and read from the Torah.

One solution to the question of whether or not to say al hanissim on the fifteenth in a doubtfully walled city, was proposed by Rabbi Yoel Schwartz, and he based it on the halacha, that whoever forgets to say al hanissim on Chanukah[118] in bircat hamazon should say it in the "harachamons" (the latter part of bircat hamazon), and he who forgets it in the amidah should say it in "Elokai n'tsur" [119] (the supplications at the end of the amidah); there is also a similar halacha for Purim.[120] Rabbi Schwartz writes:[121] "And it is appropriate on the fifteenth to say al hanissim in the prayer "Elokai n'tsur," and in the bircat hamazon in the "harachamons."

The first to discuss this issue in connection with a particular city was Rabbi Yissachar ben Mordechai Susan, who lived in Safed in the sixteenth century, in his book "Ibur Shanim-Tikkun Yissachar," and he wrote about this subject at length. According to his book, all the Musta'arabi Jews who lived in Safed and nearby villages read from the Torah also on the fifteenth, "and we heard that it was also the custom in Gaza and Damascus."[122] In addition, this was also the custom of the Westerners in Safed.[123] After the expulsion from Spain, a large number of Sefaradim arrived in Safed, but they were not prepared to read from the Torah on the fifteenth in accordance with the reasoning that "if one goes up to the Torah one is obliged to recite the berachah for reading the Torah, and one would therefore be saying a berachah on a doubtful reading." He justified both customs and added that

351

although the Sefaradim did not read the Torah on the fifteenth, their Sages did not say anything against the Musta'arabi Jews who did read the Torah.[124]

The "Tikkun Yissachar" rejected the possibility of comparing the omission of the berachah on reading the megillah with the reciting of al hanissim and the reading of the Torah.[125] According to him, in the reading of the megillah the actual reading is the mitzvah itself and the berachot are secondary, and thus even if one omits saying the berachah, one still fulfills the mitzvah of reading the megillah. In contrast, the saying of al hanissim and reading the Torah is the mitzvah itself and by omitting them, needless to say, one is not fulfilling the mitzvah.

From the notes of Rabbi Yissachar Susan on the revision of his book "Tikkun Yissachar,"[126] one can see that the Sefaradic custom of reading the Torah influenced members of the other communities in Safed. He recounts an event that took place in the year 5319 (1559) when the man leading the service began to refuse on the second day to go up to take out the Sefer Torah, because he saw that here the Sefaradim did not take out the Sefer Torah, and he began to say that the reason was because of a doubt regarding the berachot to be recited, and I said briefly to him, that because the congregation were about to take out the Sefer Torah according to their custom, we can follow our custom to read from it and say the berachot, and thus not change our custom. This is in the same way as we say today al hanissim in the amidah, even though there is an apprehension not to say it in the amidah, and they then took out the Sefer Torah as was their custom. One student related this event to Rabbi Yosef Karo, the author of the Shulchan Aruch, and he replied: "They acted correctly and it is a good custom today to say al hanissim in the amidah and to take out the Sefer Torah, but our Sefaradi congregation are not accustomed to take out the Sefer Torah." According to the language of Rabbi Yosef Karo, it is very possible that the Sefaradim in Safed say "al hanissim on the fifteenth.

The Tikkun Yissachar argues that even if one does not read the Torah on the fifteenth, one can say al hanissim.[127]

The "Knesset haGedolah"[128] learns from this that this was not just theoretical, but this was the practice in Safed where "it was customary to say [al hanissim] by even those who were not accustomed to take out the Sefer Torah" but he does not specify which community or communities acted in this way.

During the period of the "Tikkun Yissachar," the Ari lived in Safed, and his custom on the fifteenth of Adar is written about by Rabbi Chaim Vital in his book "Shaar Hakavanot":[129] "And I heard from Mahar'i Hakohen who saw my teacher [the Ari] in the Ashkenazi Synagogue on the fifteenth of Adar, which was the second day of Purim when he was in Safed that they did not say al hanissim. I also saw my teacher (the Ari) on the second day of Purim in Damascus and they did not say al hanissim, and in my humble opinion that the reason was not to make an interruption in the amidah because of a doubt." From the language of Rabbi Vital, it is not clear whether it was customary also in the Ashkenazi synagogue not to say al hanissim.

From all of the above, one needs to investigate if at the beginning of the sixteenth century, members of the Sefaradic community and / or members of the Ashkenazic community in Safed said al hanissim on the fifteenth. However, we know from Rabbi Yisrael m'Sklow (who was one of the disciples of the Vilna Gaon), in his book "Pe'at haShulchan,"[130] which was first printed in 1836, that at that time (and even today!) they did not say al hanissim or read from the Torah on the fifteenth of Adar in the city of Safed.

Rabbi Chaim Satun (19th-20th centuries), author of the book "Eretz Chaim"[131] claims that the opinion of the Ari led to the cancellation of saying al hanissim in Safed. However, we know that those who immigrated to Safed as a result of the expulsion from Spain (and also those from other places) caused changes to be made in various customs,[132] and therefore one needs to study further the language of the "Eretz Chaim."

Rabbi Yosef Molcho,[133] author of the "Shulchan Gavoha" wrote in the middle of the eighteenth century that in Salonica (Thessaloniki), a city where they observed two days of Purim, the people of the city did not say al hanissim nor did they read from the Torah on the fifteenth. The expellees from Spain also arrived in this city at the end of the fifteenth and the beginning of the sixteenth centuries, but it is not known if before the Sefaradim arrived whether the people of that city said al hanissim and read from the Torah on the fifteenth.

The difference in customs on this subject between the Musta'arabi Jews and the Sefaradim was also found in the city of Aram Tzova (Aleppo). The "Machzor Aram Tzova"[134] printed in the year 5287 (1527) writes: "And on the second night like the first, except that no berachah is said on the megillah, and on the second day as on the first, except no berachah is said on the megillah." From this language we can learn that they said al hanissim and read from the Torah on the fifteenth. Rabbi Avraham Entebbe,[135] who was the Rabbi of Aram Tzova in the nineteenth century, pointed out the difference in customs between the Musta'arabi Jews and the Sefaradim in saying al hanissim on the fifteenth: "The Musta'arabi Jews say it on the fifteenth whilst the Sefaradim do not say it." Also, Rabbi Shmuel Laniado,[136] author of the book "Shulchan haMelech" writes that the Sefaradim did not read from the Torah on the fifteenth, but that the Musta'arabi Jews did so. According to Rabbi Yaakov Atiyah,[137] today the Musta'arabi Jews do not say al hanissim, nor do they read from the Torah on the fifteenth.

According to Rabbi Chaim Benvenist, the author of the book "Knesset haGedolah," who served as Rabbi in various cities in Turkey in the seventeenth century, and also in the city of Constantinople (Kushta) stated that there were differences of customs between the various communities on this subject. According to him:[138] "In Constantinople I saw a strange custom, namely the Sefaradim were accustomed not to say it, but those from Romania were accustomed to say it." Regarding the reading of the Torah, he writes: "And I think that the Romanian community in Constantinople were accustomed to take out the Sefer Torah on the second day of Purim," and apparently, the Sefaradim did not do so. The Romaniots (the original Jewish population of the territories of the Byzantine Empire, Constantinople, the Balkans and Asia Minor; however, according to the "Knesset Yisrael" they were from Romania) were the first residents of the Jewish community in Constantinople, and the Sefaradim arrived only after their expulsion from Spain. It is interesting to note that the "Knesset haGedolah" does not write that in Constantinople they observed two days of Purim! However, Rabbi Eliyahu Cohen[139] who moved to Eretz Yisrael from Constantinople stated that they read there the megillah on the two days, but they did not observe the other mitzvot of Purim on the fifteenth. (One could note that there is a comparison between Tiberias and Constantinople in that the city wall together with the sea complete the "circumference" of the city.)

We have seen above that the "Tikkun Yissachar" writes about saying al hanissim even by those who do not read the Torah. However, there are places where the opposite is practiced! One of the places is Baghdad. The Ben Ish Chai writes:[140] "And their custom [in Baghdad] is to read from the Torah "vayovo Amalek" on the fifteenth. He explains that the ancient custom was to say al hanissim on the fifteenth, but "Rabbenu Moshe Chaim abolished the custom of saying al hanissim because of the apprehension of an interruption (whilst saying the amidah); however, regarding the reading of the Torah he did not want to cancel since there was no apprehension of a berachah being recited in vain." The book "Aim Haderech"[141] written by Rabbi Michael Ashkenazi quotes Rabbi Shmuel Chaim ben Estroga, author of the book "Shemen Hamishchah" who was one of the sages of Turkey at the beginning of the nineteenth century. One can see from his writings that in the community of Yanbul [probably Istanbul] and in the community of Varia [probably Tiriya] there was the same custom as in Baghdad. However, he adds: "It is possible that in ancient times it was the custom to say al hanissim," but this stopped after "the number of the worshipers from Spain increased."

Above we saw that the Rav from Munkatch observed mishloach manot, matanot laevyonim and the seudat Purim, even though Munkatch is not a doubtfully walled city! He also recited quietly al hanissim on the fifteenth of Adar in the amidah and in bircat hamazon.[142]

The Three-Fold Purim

The fourteenth of Adar cannot occur on Shabbat. However, the fifteenth namely Shushan Purim can, and in such a case one advances the reading of the megillah and matanot laevyonim to Friday the fourteenth of Adar. On the Shabbat, one says al hanissim and reads from the Torah "vayovo Amalek." The seudat Purim and mishloach manot are observed, according to the Talmud Yerushalmi, on Sunday the sixteenth of Adar. Some are strict, and in addition have a seudat Purim and give mishloach manot also on Shabbat. The question arises, what does one do in a doubtfully walled city in the year that the fifteenth occurs on Shabbat?

On this there are two opinions:

1) Rabbi Moshe Sternbuch writes:[143] "On doubtfully walled cities, in modern times in the new settlements, one should be strict in all the laws of Purim in this threefold Purim, and it is customary to observe this in private when there is no custom to be strict in that location, since one should be apprehensive to the principle of not making divisions amongst the people and showing off.

2) Rabbi Natan Ortner[144] argues that because of a "double doubt," an inhabitant in a doubtfully walled city does not need to send mishloach manot on Sunday, the sixteenth of Adar. He explains that there are three doubts: (i) maybe his city was not in fact walled at the time of Yehoshua, (ii) there are opinions that when the fifteenth occurs on Shabbat a walled city inhabitant is liable to send mishloach manot on the fourteenth, (iii) there are opinions that if an inhabitant of a walled city sends mishloach manot on the fourteenth he has fulfilled the mitzvah. With regards to the seudat Purim, Rabbi Ortner[145] brings Poskim who hold that when the fifteenth occurs on Shabbat, one has the seudat Purim on the fourteenth, and there are other Poskim who hold that one has in on Shabbat. From this and in accordance with the principle of a double doubt, if he has the seudat Purim on the fourteenth,

then on the Shabbat he should add something to his meal and drink a cup of wine or liquor, with the intention that it for the Purim mitzvah. He then does not have to have a seudat Purim on the sixteenth.

With regards to saying al hanissim in the amidah and in bircat hamazon on the fifteenth, there is no reason to make a difference whether the fifteenth occurs on a weekday or if it occurs on Shabbat. However, in the reading of the Torah there is room to discuss the matter, because in a walled city there is a difference in the maftir and also in the haftarah, namely, one takes out two Sifrei Torah and reads for the maftir "vayovo Amalek" and the haftarah is the same haftarah as for Shabbat Zachor,[146] this being in place of the regular haftarah for that Shabbat. The custom of the Musta'arabi Jews in Aram Tzova,[147] when the fifteenth occurred on Shabbat, was the same as in the walled cities. This was also the custom of the Romaniot in (it seems) Constantinople.[148] On the other hand, the "Tikkun Yissachar"[149] writes that in Safed and the surrounding villages, in Gaza and in Damascus, they did not read the maftir and the haftarah for Purim. In addition, the "Knesset haGedolah" writes [150] that according to everybody, cities who would normally read the megillah both on the fourteenth and fifteenth, when the fifteenth occurred on Shabbat do not take out two Sifrei Torah to read "vayovo Amalek."

Epilogue

According to the majority of the Poskim, in a doubtfully walled city, where they also read the megillah on the fifteenth, they must also observe on the fifteenth the mitzvot of mishloach manot, matanot laevyonim and the seudat Purim. Nowadays, however, most inhabitants in a doubtfully walled city only read the megillah on the fifteenth; perhaps the reason is that they rely on the opinion of the "Pri Chadash."

Also, in the past they read the Torah and said al hanissim on the fifteenth in a doubtfully walled city. However, nowadays, perhaps because of the influence of the Jews expelled from Spain

357

who then arrived in the Middle East, they do not read from the Torah on the fifteenth, and in most of the cities they also do not say al hanissim.

In his paper, Rabbi Yoel Elitzur[151] opposes the restricting the fifteenth of Adar to just reading the megillah, and he considers that one must thoroughly investigate the subject, and one will then have a programme for future generations. Rabbi Ovadia Yosef ruled that if one is accustomed to read the megillah on the fifteenth because of the doubt, one then needs to be particular to keep all the mitzvot of that day including reading from the Torah "vayovo Amalek."[152]

References

Abbreviations
SA = Shulchan Aruch
OC = Orach Chaim
(1)   Mishnah Megillah chap.1 mishnah 1
(2)   Talmud Bavli Megillah 5b
(3)   Talmud Bavli Megillah 14a
(4)   Talmud Yerushalmi Megillah chap.1 halacha 5
(5)   Megillat Ta'anit (Warsaw, 5634), chap.12, (p.17a)
(6)   Rabbi Shlomo ben Yisrael Moshe Hakohen, *Binyan Shlomo*, (Vilna, 5649), part 1, chap.58, (p.65)
(7)   Ibid., p.128
(8)   Talmud Yerushalmi, e.g. Megillah chap.1 halacha 3
(9)   Rabbi Avraham Yeshaya Karelitz, *Chazon Ish*, OC chap.153 par.2, first word: badin
(10)   Talmud Bavli Megillah 4b
(11)   Maimonides, Rambam Mishneh Torah Hilchot Megillah chap.2 halacha 14
(12)   Rabbi Zerachiah Halevi of Girona (Baal haMaor), *haMaor haKatan*, commentary on Rav Alfas on Megillah 1b
(13)   Rabbi Avraham Gombiner, *Magen Avraham* on SA OC chap.694 par.1
(14)   Rabbi Aharon Alfandari, *Yad Aharon*, (Izmir, 5494), OC chap.694
(15)   Rabbi Avraham Aharon Yudelovitz, *Shu't Bet Av*, (Warsaw, 5656), book 1, chelek Avraham al Orach Chaim, chap.103
(16)   *Yad Aharon*, op. cit,
(17)   Rabbi Binyamin Yehoshua Zilber, *Az Nedabru*, (Bnei Brak, 5735), part 6, chap.80
(18)   Talmud Bavli Megillah 5a
(19)   *Chidushei Rabbenu Simcha al haShas*, (Yehudah David Grinvald: Brooklyn New York, 5744), vol.1, Megillah 4, (p.246)

(20)   Rabbi Yoel Elitsur, "Zeman Purim b'Ir Lod," *Techumin*, vol.9, (Zomet: Alon Shvut, Gush Etzion, 5748), p.380
(21)   Biblical Esther chap.9 verse 16
(22)   MS Munich 77 folio 39b; MS Parma de Rossi 206 (2879) folio 37a; MS Bodleian catalogue Neubauer 240/8 folio 263b; MS New York Bet Hamedrash l'Rabbonim L825 folio 25a; *Sifrei Agadata al Megillat Esther*, (Shlomo Buber:Vilna, 5747), Midrash Lekach Tov on chap.9 verse 16, (p.110)
(23)   Rabbenu Chananel on Megillah 5a
(24)   Rashi on Megillah 5a, first word: simcha
(25)   MS Bodleian, op. cit., 240/8
(26)   Rabbi Elazar miGarmiza, *Sefer haRokeach*, (Warsaw, 5640), hilchot Purim, chap.239
(27)   Rabbi Zedekiah ben Avraham Anaw, *Shibbolei haLeket*, (Dubno, 5554), hilchot megillah, third principle, chap.54
(28)   Rabbi Levi ibn Habib, *Shu't HaRaLBaCh*, (Lemberg (Lvov), 5625), chap.32
(29)   Rabbi Yisrael Isserlin, *Terumat haDeshen*, (Sadilkav, 5595), chap.111
(30)   Rabbi Ovadia Yosef, *Yechave Daat*, (Jerusalem, 5741), part 4 chap.40
(31)   Talmud Bavli Megillah 4b
(32)   MS Bodleian catalogue Neubauer 648 folio 131a; MS Bodleian, catalogue Neubauer 643 folio 75b; MS London British Museum catalogue Margoliouth 523 folios 79b-80a; MS Paris heb 395 folio 221a; MS Sassoon (732) 707 folio 228; MS Cambridge Add 169/1 folio 84a; MS London Montefiore 92.1 folio 109a; MS Moscow Ginsberg 491 folio 116a
(33)   MS Bodleian, op. cit., 643; MS Sassoon, op. cit., (732) 707
(34)   Rabbi Yehoshua ben Shimon Baruch, *Shiltei Giborim* on the Rif, Megillah 2a
(35)   It is of interest to note that in a number of MSS of Rabbi Yeshaya ben Eliyahu di Tarani known as the Riaz (MS Paris heb 395, op. cit.; MS Sassoon (732) 707, op. cit.; MS London Montefiore 92.1, op. cit.; MS Moscow 491, op. cit.) he adds "They recite the berachah on the Megillah on both days." One needs to investigate if there is a connection between this version of the Riaz and the custom which was observed in the city of Tyre at the period of the Rambam to recite the berachah on the megillah on both days. On this the Rambam wrote in a responsum that they were pronouncing the Divine name in vain, and that if they continued to do this and recite the berachah on the fifteenth, it would be proper for every scholar and G-d fearing person to turn away and refrain from answering Amen on such a berachah, and anyone who were to answer Amen on an in vain berachah, or on a berachah for which there is a doubt, would have to give an accounting for this in the future." (Resposum 84 of the Rambam, published by Hevrat Mekize Nirdamim: Jerusalem, 5694)
(36)   Rabbi Hezekiah da Silva, *Pri Chadash*, (Amsterdam, 5466), OC chap.695 par.4

(37) *Binyan Shlomo*, op. cit., p.65

(38) Rabbi Ovadia Yosef, *Yabia Omer*, part 7, OC chap.60, first word: ul'inyan

(39) Rabbi Shlomo ben Avraham Aderet, S*efer haRaSHBA al Chidushei Masechet Shavuot*, (Prague, 5548), chap.3, p.16

(40) *Yabia Omer*, op. cit.

(41) Rabbi Yosef ben Meir Teomim, *Pri Megadim*, Mishbetzot Zahav, SA OC chap.695 par.5

(42) *Binyan Shlomo*, op. cit., p.65

(43) Rabbi Binyamin Yitzchak Trachtman, *Shevet Binyamin*, (Mishwaka Indiana, 5690), p.31

(44) Rabbi Yosef ben Avraham Molcho, *Shulchan Gavoha*, (Salonika, 5516), OC chap.695, par.8

(45) Rabbi Chaim ben Yisrael Benvenist, *Knesset haGedolah*, (Livorno, 5552), OC Beit Yosef, chap.688

(46) *Magen Avraham*, SA OC chap.688 par.5

(47) *Pri Megadim*, Eshel Avraham, SA OC chap.688 par.5

(48) Mishnah Berurah, SA OC chap.688 par.10

(49) Rabbi Yaakov Chaim Sofer, *Kaf haChaim*, SA OC chap.688 par.22

(50) Rabbi Yichya Tzalach, *Shtilei Zeitim*, SA OC chap.688 par.10. He also brings the opinion of the Pri Chadash (chap.695 par.3)

(51) *Luach l'Eretz Yisrael*, arranged by Rabbi Yechiel Michel Tucazinsky, for example year 5752, p.42

(52) Luach Laws and Customs, for example the year 5739, (Ihud Batei Knesset in Israel, Jerusalem, 5738), p.60

(53) Yad L'Achim Wall Calendar, Jerusalem, for example Adar 5756

(54) Luach Laws and Customs "Dvar Yom Beyomo" for the year 5749, (Kehal Machzikei Hadas in Eretz Yisrael, Jerusalem, Hasidei Belz), p.177

(55) Rabbi Chaim Vital, *Shaar Hakavanot*, (Salonika, 5612), the sixth gate, Derushai haPurim, p.158

(56) Information from Rabbi Shmuel Ben-Eliyahu, Rabbi of Safed, Menachem Av 5756

(57) *Shaar Hakavanot*, op. cit.

(58) Rabbi Ezra Hakohen Trab, *Milei d'Ezra*, (Jerusalem, 5684), OC chap.10, (pp.12a, 14b)

(59) Rabbi Yosef Chaim, *Ben Ish Chai*, first year, parashat Tezave, hilchot Purim, par.17

(60) Rabbi Avraham Adas, *Derech Eretz Adas*, (Bnei Brak, 5750), Minhagei Aram Tzova (Aleppo), section Orach Chaim, Shushan Purim, p.152

(61) Rabbi Mordechai Yoffe, *Levush*, OC chap.688 par.4

(62) Rabbi Shmuel Yuda (Leib) Koider, *Olat Shmuel*, (Prague 5583, OC chap.111

(63) Rabbi Elazar Fleckeles, *Shu't Teshuva meAhava*, (Prague, 5569), part 1 chap.210

(64) Rabbi Avraham Danzig, *Chayei Adam*, principle 155 par.8

(65) Michael Gal, "Purim Sheni b'Afghanistan" *Yeda-AM*, (Tel-Aviv), 2 Nissan 5708, p.42

(66) *Yalkut Minhagim Miminhagei Shivte Yisrael*, arranged by Asher Wassarteil, (Misrad haChinuch vehaTarbut: Jerusalem, 5756), [by mistake it is written there the city "Balkh"]

(67) Information from Rabbi Yonah Yanun, Givat Shaul Jerusalem, Tishrei 5757

(68) *Yeda-AM*, op. cit.

(69) "Purim aitzel Yehudei Afghanistan," newspaper *Omer*, (Jerusalem), 15 Adar 5721, p.6

(70) Talmud Bavli Megillah 5b

(71) Rabbi Tzvi Pesach Frank, *Shu't Har Tzvi*, (Jerusalem, 5733), OC part 2 chap.128 par.27

(72) Information from Rabbi David Peretz, Rabbi of Tiberias, Menachem Av 5756

(73) see for example: *Kaf haChaim*, OC chap.688 par.17

(74) *Leket Hilchot l'Purim 5748*, written by Rabbi Dov Lior, Rabbi of Kiryat Arba, distributed duplicated to the inhabitants of Kiryat Arba and Hebron, (Acknowledgements to R' Itama Shneiweiss of Kiryat Arba who supplied me with this source.)

(75) Talmud Bavli Megillah 4a

(76) Rabbi Yitzchak Yaakov Weiss, *Shu't Minchat Yitzchak*, (Jerusalem, 5743), part 8, par.61

(77) Ibid.

(78) Letter from Rabbi Yisrael Mintzberg and the Rabbis of Lod in the year 5722, *Techumin*, op. cit., vol.9, p.366

(79) Rabbi Natan Ortner, Rabbi of Lod, "Purim b'Ir Lod," summary of halachot, *Techumin*, op. cit., vol.9, p.364

(80) Ruling of Rabbi Ovadia Yosef, "Aimatai Zeman haPurim b'Hitnachalut Bet-El uva'Machanot haTzava shom," *Techumin*, op. cit., vol.1, p.120

(81) Rabbi She'ar Yashuv Cohen, Chief Rabbi of Haifa, "Keviyat Yemai haPurim b'Haifa," Haifa 5755

(82) Kuntres haShlichut, letter from Rabbi Menachem Mendel Scheerson (Lubavitch Rebbe), 5753, p.50. (Acknowledgements to Rabbi Leibel Schildkraut of Haifa who sent me this source.)

(83) Information from Rabbi Gedalya Axelrod, Rabbi of Chabad of Haifa, Nissan 5757

(84) Information from Rabbi Shneyer Klopt, Synagogue Rabbi in Haifa, Menachem Av 5756

(85) *Likut Dinim v'Hanhagot m'Maron ha-Chazon Ish*, (Rabbi Meir Graineman: Bnei Brak, 5748), part 1, OC p.117

(86) *Az Nedabru*, op. cit., (5734), part 5, chap.41, p.88

(87) Ibid., p.90

(88) *Luach Davar b'Ito* (Every Thing in its Proper Time), year 5755, (Achiezer: Bnei Brak), p.643

(89) Rabbi Ovadia Yosef, "Aimotai Chayavin Toshvei haShechunot Ramot v'Har Nof Likro haMegillah, " *Shorashim* – journal of the Union of Torah observant Sefaradim – Shas movement, issue no.8, Adar 5748, p.6

(90) Rabbi Ovadia Yosef, "m'Hilchot uMinhagei Purim", *Kol Sinai*, journal of the ideological non-political group "Ne'emanei haTorah", issue 6, vol.2, Adar 5723, p.15

(91) Rabbi Yaakov Shaul Elyasher, *Shu't Simcha Laish*, (Jerusalem, 5653), OC chap.5

(92) Rabbi Rahamim Nissim Yitzchak Palagi, *Yafe laLev*, (Izmir, 5636), part 2, OC chap.688 par.2

(93) Rabbi Chaim Palagi, *Ruach Chaim*, (Izmir, 5641), OC chap.688, first word: miktze

(94) *Shulchan Gavoha*, op. cit., par.5

(95) *Yalkut Minhagim*, op. cit., p.113

(96) Rabbi Abdallah Avraham Yosef Somech, *Zivchei Tzedek*, (Jerusalem, 5741), part 3, par.10

(97) Rabbi Yechiel Michal Gold, *Darchei Chaim v'Shalom*, (Munkatch, 5700), chodesh Adar, chap.854

(98) Rabbi Chaim Elazar Shapira, *Nimukei Orach Chaim*, (Slovakia, 5690), OC chap.688 par.3; *Darchei Chaim v'Shalom*, op. cit., chap.853

(99) *Darchei Chaim v'Shalom*, op. cit., chap.853

(100) Rabbi Avraham ben Mordechai Halevi, *Shu't Ginat Vradim*, (Kushta (Constantinople), 5477), OC chap.49 principle 1, first words: vechol ze

(101) Rabbi Isaak Tyrnau, *Sefer Minhagim*, (Prague, 5443), minhag shel Shemini Atzeret

(102) Rabbi Moshe Isserles (Rema), *Darchei Moshe* on the Tur, OC chap.669, par.1

(103) SA OC chap.669 par.1 Rema

(104) Mishnah Berurah SA OC chap.669 par.15

(105) Kitzur Shuchan Aruch chap.138 par.7

(106) Rabbi Naftali Tzvi Yehudah Berlin (Natziv), *Shu't Meshiv Davar*, (Jerusalem, 5728), part 1, chap.16, first words: nachzor l'inyan

(107) *Sheiltot d'Rav Achai Gaon*, (Mosad Harav Kook: Jerusalem, 5727), Parashat Vayakhel, chap.67, p.438

(108) Rabbi Aharon Hakohen m'Lunel, *Orchot Chaim*, (Florence, 5511), hilchot megillah u'Purim, seder hatefillah

(109) Beit Yosef on Tur OC chap.693 first words: katav reish-ayin

(110) Rabbi Yoel Sirkis, *Bach* on the Tur, OC chap.693, first words katav resh-ayin; *Pri Chadash*, SA OC chap.693 par.2

(111) Rabbi Eliyahu Spira, *Elia Raba*, (Sulzbach, 5517), OC chap.693, par.2; *Magen Avraham*, op. cit., SA OC chap.693, par.1; Rabbi David Halevi Segal, *Taz*, SA OC chap.693 par.3

(112) *Pri Megadim*, op. cit., Mishbetzot Zahav, SA OC chap.693 par.3

(113) Mishnah Berurah SA OC chap.693 par.6

(114) Rabbi Yosef Eskapa, *Rosh Yosef*, (Izmir, 5418), hilchot megillah, chap.693, first words: katav haRav

(115) Rabbi Yechiel Michel Tucazinsky, *Sefer Eretz Yisrael,* (Jerusalem, 5726), chap.5, (p.470)
(116) Luach Rabbi Tucazinsky, op. cit.; Luach Heichal Shlomo, op. cit.: Yad L'Achim Wall Calendar, op. cit.
(117) Luach Belz, op. cit.
(118) SA OC chap.682 par.1 Rema
(119) Mishnah Berurah SA OC chap.682 par.4
(120) *Ben Ish Chai,* op. cit., par.14
(121) Rabbi Yoel Schwartz, *Adar uPurim,* (Jerusalem, 5743), chap.5, din arim hamesupakot, par.1
(122) Rabbi Yissachar ben Mordechai Susan, *Sefer Ibur Shanim – Tikkun Yissachar,* (Venice, 5339), p.59b
(123) Ibid., p.61
(124) Ibid., p.5
(125) Ibid., p.60
(126) Ibid., p.61
(127) Ibid., pp.60a-60b
(128) *Knesset haGedolah,* op. cit., (Jerusalem, 5726), chap.693
(129) *Shaar Hakavanot,* op. cit.
(130) Rabbi Yisrael m'Sklow, *Pe'at haShulchan,* (Jerusalem, 5728), chap.3 par.15
(131) Rabbi Chaim Satun, *Eretz Chaim,* (Jerusalem, 5668), OC chap.693 par.2
(132) see: Chaim Simons, "hevdelim b'kriyat haparshiot bein benai Eretz Yisrael u'vein benai chutz la'aretz," journal *Sinai,* (Mosad Harav Kook: Jerusalem), vol.106, p.35
(133) *Shulchan Gavoha,* op. cit., chap.693 par.5
(134) *Machzor Aram Tzova* (Aleppo), (Venice, 5287), p.164b, (Venice, 5320), p.115a
(135) Rabbi Avraham Entebbe, *Hochma uMussar,* (Jerusalem, 5741), p.258
(136) Rabbi Shmuel Laniado, *Shulchan haMelech,* (Aram Tzova, 5683), chap.201 par.3
(137) *Derech Eretz Adas,* op. cit., p.153
(138) *Knesset haGedolah,* op. cit., Beit Yosef chap.693
(139) Information from Rabbi Eliyahu Cohen, formerly from Constantinople, Iyar 5756
(140) *Ben Ish Chai,* op. cit., par.14
(141) Rabbi Michael Ashkenazi, *Aim Haderch,* (Salonika, 5661), OC chap.2 par.2
(142) *Darchei Chaim v'Shalom,* op. cit., chap.856, p.317
(143) Rabbi Moshe Sternbuch, *Bircat Hachamah, Erev Pesach shechal b'Shabbat, Purim shechal b'Shabbat,* (Bnei Brak, 5741), par.14 note, first words:v'raoiy ledakdaik
(144) Rabbi Ortner, op. cit., p.362
(145) Ibid., p.363
(146) According to the majority of the Poskim, the reading of Parashat Zachor is a Torah command. There are those who say that the reason for reading

Parashat Zachor once in a year is that one forgets things after twelve months. However, in a leap year there are 13 months and therefore the custom of the Chatam Sofer was in a year prior to a leap year, one should have the intention for fulfil this mitzvah when reading Parashat Ki Tetze (Minhagei Chatam Sofer, appendix to customs of the Chatam Sofer paragraph 12 footnote 12, Jerusalem 5731). The time for reading Parashat Zachor is on the Shabbat before Purim, and it is possible even in a non-leap year that there could be more than 12 months between Shabbat Zachor of that year and that of the previous year. For example, if in a particular year Shabbat Zachor occurs on 9 Adar and in the following year (which is not a leap year) it could occur on 13 Adar, thus 12 months and 4 days will have passed between successive readings. Thus one needs to investigate why in such a situation the Chatam Sofer does not write that one should have the intention to fulfil this mitzvah on Parashat Ki Tetze? One could mention that "Luach Davar b'Ito" of the year 5755 (published by Achiezer Bnei Brak, page 1021) writes that if in a non-leap year, one reads Parashat Zachor on a date which is later than the previous year, one should then have had the intention when reading Parashat Ki Tetze to fulfil the mitzvah of reading Parashat Zachor.

(147) *Shulchan haMelech*, op. cit.
(148) *Aim Haderch*, op. cit.
(149) *Tikkun Yissachar*, op. cit., p.61
(150) *Knesset haGedolah*, op. cit., Beit Yosef, chap.693
(151) Rabbi Yoel Elitsur, "Zeman Purim b'Shaalvim," *Ki Sarita*, (Shaalvim, 5748), (edited by Rabbi Moshe Ganz and Rabbi Uri Desberg), p.124
(152) Ruling of Rabbi Ovadia Yosef, *Techumin*, op. cit., vol.1, p.120

Appendix

The material below is taken from my unpublished autobiography.

Whilst researching this paper I came across a microfilm of a manuscript which had recently arrived from Moscow. It was catalogued as "Piskei Hariaz" but seemed to be much longer than the other manuscripts on this subject. It crossed my mind that possibly it could be "Kuntrus Haraiyot" (from which Piskei Hariaz is a summary) a book regarded as not being extant. This certainly merits investigation!

CPSIA information can be obtained
at www.ICGtesting.com
Printed in the USA
BVHW041033061222
653555BV00004B/20

9 781803 813103